D0151589

AVIATION SAFETY: A BALANCED INDUSTRY APPROACH

AVIATION SAFETY: A BALANCED INDUSTRY APPROACH

Michael D. Ferguson, MS
Sean Nelson, CSP, CSTM, CDT

Australia • Brazil • Japan • Korea • Mexico • Singapore • Spain • United Kingdom • United States

Aviation Safety: A Balanced Industry Approach

Michael D. Ferguson and Sean Nelson

Vice President, Careers & Computing: Dave Garza

Director of Learning Solutions: Sandy Clark

Associate Acquisitions Editor: Nicole Sgueglia

Director, Development-Careers and Computing: Marah Bellegarde

Managing Editor: Larry Main

Senior Product Manager: Sharon Chambliss

Editorial Assistant: Courtney Troeger

Brand Manager: Kristin McNary

Market Development Manager: Erin Brennan

Senior Production Director: Wendy A. Troeger

Production Manager: Mark Bernard

Content Project Manager: Christopher Chien

Art Director: Jackie Bates/GEX, Inc.

Technology Project Manager: Joe Pliss

Media Editor: Deborah Bordeaux

Cover Image: © AlbanyPictures, iStockphoto

Library of Congress Control Number: 2012936305

ISBN-13: 978-1-4354-8823-6

ISBN-10: 1-4354-8823-7

Delmar
5 Maxwell Drive
Clifton Park, NY 12065-2919
USA

Cengage Learning is a leading provider of customized learning solutions with office locations around the globe, including Singapore, the United Kingdom, Australia, Mexico, Brazil, and Japan. Locate your local office at: **international.cengage.com/region**

Cengage Learning products are represented in Canada by Nelson Education, Ltd.

To learn more about Delmar, visit **www.cengage.com/delmar**

Purchase any of our products at your local college store or at our preferred online store **www.cengagebrain.com**

Printed in the United States of America
1 2 3 4 5 6 7 16 15 14 13

CONTENTS

PREFACE

As aviation continues to expand globally, our individual and collective understanding of safety must progress in tandem with the degree of change occurring in the industry. In other words, practical safety in aviation must improve at approximately the rate of growth of the industry. Ideally, the goal should be putting safety in a position where it is *leading*—that is, getting safety initiatives out in the front of the growth that is occurring in order to identify hazards, mitigate and manage risk, protect employees and members of the public, safeguard assets, enhance profitability, and create effective, economically feasible solutions to problems before accidents and incidents can occur. *Prevention* of undesired events is therefore one of the direct and implicit key concepts present throughout this book.

Instead of lagging behind and being *laissez-faire* with respect to safety leadership, current and future leaders in the aviation industry need to ensure that their organizations and personnel are out in front of safety issues. In order to be effective, safety programs and initiatives must have the complete and unwavering support of those in decision-making positions. Therefore, it is imperative that students of aviation (as future leaders) as well as industry professionals obtain a firm understanding of contemporary safety programs and principles in order to set them on the right path, regardless of where they are in their careers. Whether aspiring to or currently positioned as a career pilot, air traffic controller, manager, dispatcher, executive, technician, or any of the myriad other positions in the industry, aviation students and professionals alike must have a strong understanding of the applicability and importance of safety and their role in preventing accidents and incidents.

Although many other books on aviation safety generally contain excellent, accurate information, one significant element missing from most texts is a balanced approach to safety in aviation. Many books cover a number of flight-related safety issues, regulatory organizations, accidents, and other common themes. However, other important aspects such as ground operations safety, ethics, and active accident/incident prevention strategies are not always considered. Many books on safety focus on providing a theoretical background to safety issues, while not always providing a significant practical connection to how the information can be directly applied to the aviation industry. These statements are not intended to be criticisms, but rather an acknowledgment that a truly systemic approach to aviation safety must include the consideration of numerous elements besides the theoretical.

With this in mind, we have worked to structure this book in a manner that actively connects the theoretical to the practical by guiding the reader through a number of pertinent topics designed to provide a solid knowledge base in terms of basic applied safety principles in aviation. In order to accomplish this objective, the chapters of this book have been written in a way that treats each individual topic as significant, relevant, and as a part of a larger, interconnected safety framework. It is our hope that this book will serve educators and students alike by providing them with solid, relevant, industry-based concepts.

We have also worked to enhance the pedagogy of aviation safety by presenting topics and information derived from and directly applicable to various aspects of the aviation industry. The book is not perfect; in fact, there are a number of aviation safety topics and concepts that we did not include due to space and time constraints. However, we have worked to present a number of key elements in an attempt to balance flight safety with ground safety and a number of other topics. It is our hope that you will find the information contained in this book to be useful in furthering your knowledge of applied safety in the aviation industry.

Cheers!

Michael and Sean

ACKNOWLEDGMENTS

The list of people I wish to thank for their contributions to this book, their support, or both is rather extensive, but I will try to be concise.

For their direct contributions to this book, my most sincere thanks go out to the following people: Dan Krueger and Brandon Wild for co-authoring Chapter 8, "Flight Safety Programs," and William "Bill" Towle for co-authoring Chapter 9, "Airport Safety." Guys, your expertise respectively on ASAP, FOQA, and airports was invaluable. Brandon Wild for his *In Focus* section in Chapter 8; Toshia Marshall for her *In Focus* interview in Chapter 10, "Emergency Response"; John Ostrum for his *In Focus* interview in Chapter 9; Dr. Don Arendt for his *In Focus* interview in Chapter 2, "Regulatory Oversight"; Bunty Ramakrishna for her *In Focus* interview in Chapter 4, "Introduction to Safety Management Systems"; and Bryan Ray (B-Ray) for serving as the "model" in the job hazard analysis photos in Chapter 5, "Elements of Effective Aviation Safety Programs." Thanks all—your expertise and contributions to these chapters are greatly appreciated.

Mark Vorzimmer for his support, leadership, and for being one of the most important professional mentors I have ever had. Thanks, Mark!

The faculty and staff of the Department of Aviation at St. Cloud State University (my former colleagues): Dr. Jeff Johnson, Dr. Tara Harl, Angie Olson, Dr. Steve Anderson, Dr. Pat Mattson (Emeritus), and, of course, Sandy Osterholt. Thank you, my friends. The students (past and present) of the Department of Aviation at St. Cloud State University; many of you were openly supportive of this project. You were and are my inspiration for writing this book, and it was a sincere pleasure being one of your professors during my six wonderful years at SCSU. Blue skies and keep fighting the good fight!

Kayla Caufield, Jake Oswald, and Ben Deutsch for your assistance in researching and gathering some of the information used in this book. Thanks all and great job!

My dear friend Dennis Shillingburg for his friendship, strength, support, prayers, and wisdom. Thanks for being a great friend and mentor.

My great friend Kevin Toth for his friendship and support of me and this project. I got your back, and I know you got mine!

Sean Nelson, one of my best friends and co-author, for agreeing to take on this project and for his sacrifice, patience, and gentle prodding as we worked together through this long process and across the miles.

My sons, Jacob (Jake) and Josiah (Si), for their support and understanding for all the times I had to work on the book. Thanks guys, I am immensely proud of you both!

Finally, to the love of my life, my wife Doreen: Words that are adequate to thank you sufficiently for everything and for the person you are fail me. For your unwavering love, strength, support, prayers, assistance, and encouragement, thank you. Without you I would not be where I am and this book would not exist. I thank the Lord for you and for this book.

Michael D. Ferguson, MS

This has been a long journey, from which I have learned a great deal. My wife (Traci) has been there for me through the long nights, with few complaints. A new son (Kai) was added to the family during this process. He and his big brother (Gavin) were very supportive and understanding that Daddy had a lot of work. Truth be told, the boys came first more often than not. These two are truly my greatest accomplishments.

I want to give a special thank-you to James "Skipper" Kendrick for his contribution of the case study found in Chapter 5, "Elements of Effective Aviation Safety Programs." Skipper is a Certified Safety Professional (CSP) and a Past President and Fellow of the American Society of Safety Engineers (ASSE). He has provided training for many safety professionals across the country. I was one of those students early in my career. As my career progressed, I have owed much of my success to leaders like Skipper, and I'm proud to have known him for so long.

Sean Nelson, CSP, CSTM, CDT

The authors and publisher wish to thank and acknowledge the following individuals who reviewed the manuscript:

Jose Ruiz, Southern Illinois University, Carbondale
Fred Hansen, Oklahoma State University
Chien-tsung Lu, Purdue University
Geoff Whitehurst, Western Michigan University
Al Mittelstaedt, Arizona State University
Lisa Whittaker, Western Michigan University

1 Beyond Compliance: Ethics and Aviation Safety

Chapter Learning Objectives

After completing this chapter, the reader should be able to:

- Explain the basics of how the aviation industry is regulated.
- Form and articulate a position about the relationship of ethics and safety in an organization.
- Explain the definition of ethics and understand its application and importance in aviation safety.
- Describe the importance of establishing an ethical organizational culture to promote and enhance safety.
- Explain the basics of some ethical concepts and theories such as rights, respect, egoism, utilitarianism, and duty, and their applicability to safety in an organization.
- Explain the basics of the OSHA General Duty Clause and its purpose.
- Discuss the primary responsibility of employers in providing a safe working environment under the requirements of OSHA standards and the OSH Act of 1970.
- Identify and describe the primary rights of employees under OSHA and the OSH Act.
- Discuss how some basic economic concepts such as opportunity cost and diminishing returns may impact to safety decision making in an organization.
- Discuss the importance of leadership in establishing and maintaining a solid ethical safety culture in an organization.

Key Concepts and Terms

Business Ethics
Cost/Benefit Analysis (CBA)
Diminishing Returns/Diseconomy of Scale
Direct Costs
Duty
Economics
Egoism
Ethical Climate
Ethical Complacency
Ethics
Indirect Costs
Moral Standards
Occupational Safety and Health Act of 1970
Opportunity Cost
Organizational Culture
OSHA General Duty Clause
Respect
Rights
Safety Culture
Safety Management Systems (SMS)
Scarce Resources
Utilitarianism

Figure 1-1.

Introduction to the Chapter

There is much to consider on the complex topics of ethics and safety. As research was conducted for this chapter, it became apparent that different people hold varying views of whether safety is a moral issue and/or an economic issue. As a result, an interesting quandary was presented with respect to how the topic of ethics should be approached in this book. Should safety be presented as a largely ethical and moral issue, or from the straightforward, practical (and real) standpoint that safety is largely an issue of economics? In the latter context (economics), the application of limited financial and other resources rules the day, based upon the potential of receiving tangible benefits from a decision to spend money and time in one area over another. In simple terms, should a dollar be spent on safety or on something else?

After careful consideration, we came to the conclusion that safety should be approached as *both* an ethical and an economic issue, meaning that each of these factors should ideally be taken into consideration during safety policy development, program management, execution of related responsibilities, regulatory compliance, and the development of the overall organizational culture (of which the safety culture is a part). The contention here is that a strong, balanced organizational culture will possess an ethical culture in which a solid safety culture is present and active in the organization—a culture in which safety is not given just "lip service," but is made a priority, not only because regulatory compliance is important, but because it is the right thing to do to protect people and assets. This

Figure 1–2.

assertion may seem idealistic, but a company does have a duty and a legal responsibility to provide a safe and healthy work environment for its employees. For example, the OSHA General Duty Clause helps to establish a minimal baseline for an organization's responsibilities regarding the protection of employees. This clause will be discussed in more detail later in this chapter.

This chapter contains information regarding the perspective of safety as an economic and ethical issue. While some may disagree with our perspective, it is our conviction (gained through education, research, and years of real-world experience in the industry, as well as similar experiences of others) that organizations should not only do their best with respect to safety to ensure compliance, but also that organizations should foster the development of an organizational ethical culture that values safety in order to protect employees, assets, and the public while working to ensure operational efficiency and fiscal responsibility—in other words, a balanced approach. An organization should not approach safety solely for the purpose of regulatory compliance, but should embrace safety because it is the right thing to do for the organization, the employees, the public, and the environment. Herein lies the core of the ethics of safety.

To clarify, the purpose of this chapter is not to examine ethics and safety from a deep philosophical or theoretical perspective that considers all manner of ethical complexity, nor is it intended to be all-inclusive in considering ethics. Rather, this chapter will discuss some of the basics of ethics and their application to safety in aviation organizations. Ethics and morality are very deep, complex issues which simply cannot be covered adequately in one brief book chapter. For this reason, the focus will be on applied ethics as it relates to safety and organizational culture.

The Regulated Nature of Aviation

Beyond question, aviation is one of the most tightly regulated industries in the world. Few other industries are subjected to such intense levels of scrutiny and oversight by such a variety of regulatory agencies. In the United States, the Federal Aviation Administration (FAA), Department of Transportation (DOT), Transportation Security Administration (TSA), Environmental Protection Agency (EPA), Occupational Safety and Health Administration (OSHA), state and local authorities, and internal organizational policies provide a high degree of oversight of aviation activities. It is important to understand this factor: The regulatory oversight imposed upon the airline industry, aviation and aerospace manufacturers, and the wide spectrum of general aviation interests is virtually all-encompassing. Regulation in aviation is top-to-bottom, covering to some degree almost every aspect of aviation, with few exceptions.

Also, the aviation industry is routinely subjected to the scrutiny of the broadcast, online, and print media, especially when a significant (and often highly visible) safety event such as an aircraft accident occurs. At such times, the degree of attention given to the event by the media is often intense and speculative in nature, rather than strictly factual. This has sometimes caused aviation to be cast in a negative light in the eyes of the public in a number of ways.

Thus, aviation operations and interests take place under a tremendous amount of regulatory and media scrutiny, with the very real potential of fines and litigation if an organization is found to be out of compliance. Fines levied against aviation organizations by regulatory agencies can be extremely expensive and may also result in negative publicity if the issues that produced the fine are made public. A recent example is the case involving the maintenance practices of Southwest Airlines. Southwest was fined $7.5 million by the FAA for operating aircraft in revenue service without having conducted fuselage inspections. It follows that it is in the best interest of aviation companies to do their best to make sure they are as compliant as possible in the eyes of the various regulatory bodies, if for no other reason than to reduce the risk of accidents, loss of life, fines, litigation, negative press, or other undesirable consequences.

However, as important as it is for aviation organizations to remain in regulatory compliance, is it simply enough for them to stop at that point with respect to safety? Is it enough to be just "in compliance," or should more be done to achieve safety and ethical practices that go beyond meeting minimum acceptable standards? It is evident that the existence of regulations, standards, and company policy and procedures, and the specter of potentially undesired media attention fail to automatically translate into people following all the rules, procedures, and policies within aviation organizations, nor does it translate into people behaving ethically in the aviation workplace. There are those who continue to make unethical decisions and even coerce others into doing the same. Therefore, ethics, as applied to individuals and companies, is not just a question of obeying laws and following rules. It is more a question of individual and corporate character, code, and culture. In general, people tend to make decisions and choices that are aligned with their personal values, perceptions, and beliefs. Conversely, organizations tend to make decisions based on the internal culture created by the leadership of the organization. If the personal ethics of an individual or the prevailing organizational ethics are misaligned with what is considered to be culturally or socially acceptable, unethical decisions or situations may arise and create problems internally and externally.

However, the perception of what is considered unethical behavior or practice is subject to debate, based on differences in individual values, beliefs, and perspectives, as well as differences between social and geographic cultures and other factors. For the purposes of this chapter, unethical practices, beliefs, and decisions will be defined as those that fall outside of what are considered to be acceptable and appropriate moral standards within the confines of the broader culture and social situation in which the organization and its people are located. This may be an overly simplified manner of contextualizing such a complex issue as ethics, but it least it gives us a basic starting point for determining ethical and unethical practices.

Defining Ethics

Although it is a common word, defining **ethics** is not necessarily easy. Some researchers have indicated that ethics is hard to define because it is complex and includes consideration of corporate management, the environment, and sustainability. While researching this chapter, one definition found for ethics stated that ethics are rules for behavior, based on beliefs about how things should be, and involves assumptions about humans and their capacities, logical rules extending from these assumptions, and notions of what is good and desirable. This same definition went on to assert that ethics are also sets of rules for acceptable behavior and concern the "shoulds" and "should nots" of life, the principles and values on which human relations and interactions are based. Thus, in its most basic form, ethics may be defined as how things should be, and how life should be lived.

Another definition stated that ethics are an individual's collection of personal morals and that an ethical person makes decisions based on what they feel is acceptable or right. Yet another definition asserted that ethics is concerned with rendering moral judgments to determine right and wrong. The Merriam-Webster online dictionary further defined ethics as the discipline dealing with what is good and bad and with moral duty and obligation, and the principles of conduct governing an individual or a group. Still another definition stated that ethics is concerned with the values and morals an individual person or society embraces as desirable and acceptable, and is also concerned with the virtuousness of individuals and their motives. In summary, ethics deals with determining what is right and wrong, making decisions based on moral judgments, and establishing standards governing human conduct.

Since most aviation organizations are businesses, it will be helpful to define business ethics. While research was being conducted on ethics, a good working definition of business ethics was found and has been adopted for the purposes of this chapter. Thus, **business ethics** is the process of examining decisions (before and after they are made) to ensure that the decisions are aligned with the moral standards of a given society and its culture, using the moral standards of a society in making decisions. In short, business ethics is applied ethics. **Moral standards** are considered to be a societal measuring stick for determining moral norms and principles (e.g., those concerned with cheating, theft, lying, bribery, coercion, consequences, honesty, integrity, telling the truth, etc.); they help to establish a baseline for determining whether an action or decision is right or wrong in a society—or, in the context of this chapter, in an organization. Within this context and by application of the working definition of business ethics applied here, regulatory standards and internal company practices and rules could be considered as synonymous with moral standards, as they determine expectations and establish baselines to ensure at least the acceptable minimums for regulatory and company compliance. Internal organizational policies and external regulations form the basis for determining if a given organization or person is inside or outside of the acceptable standards of behavior. Negative consequences may result if the organization or individual is not in compliance with standards.

Corporate Ethics Scandals

In recent years, much has been written about highly visible corporate ethics scandals that have been widely publicized; some of the most notable ones include Enron, Tyco, and WorldCom. The Boeing Company has been embroiled in some rather significant ethics scandals over the years resulting from the direct actions of individuals who either worked directly for Boeing or were employed by a Boeing contractor. More recently, the cases of Bernard Madoff and Tom Petters (former chair of Petters Group Worldwide, whose holdings included Sun Country Airlines) received a significant amount of attention in the media, and for good reason. Both Madoff and Petters were found guilty of committing significant crimes involving enormous amounts of other people's money. Petters was also found guilty of 20 counts of money laundering, wire and mail fraud, and conspiracy to commit further fraud; he was sentenced to 50 years in prison.

The aforementioned scandals and other prominent ethics cases have contributed in part to a surge of renewed interest in business ethics in the last several years—so much so that many organizations now feature ethics programs and training for their employees. This trend is also evident in many higher education degree programs. Increasingly, postsecondary academic programs feature mandatory courses in ethics for their students, particularly in fields such as business, medicine, law, and computer science. Even some academic programs in aviation now feature an ethics course. Although results are ambiguous and debatable with respect to measureable effectiveness, it has been generally recognized that providing ethics education and/or training to the members of an organization and to students is a good idea and may produce some positive benefits. At the very least, an ethics program in an organization demonstrates a good faith effort in working to establish a positive business practice to help build trust among employees, other organizations, and the public.

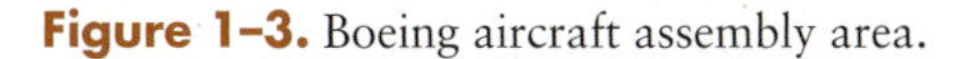

Figure 1–3. Boeing aircraft assembly area.

The Importance of Ethical Behaviors and Decisions

A simple question to ask is this: Why do some individuals and groups engage in practices and make decisions that are unethical? The answer: There are many reasons—far too many to cover in a single book chapter. Many of these reasons are arguably for an individual's personal gain, often at the expense of others. However, the premise asserted here is this: Ethical behavior and decision making are extremely important and are a matter of willful choice. Each person has the ability to choose how they will conduct themselves and may choose to make or not make sound ethical decisions. By the same measure, each company and organization also has the ability to choose to operate as ethically as possible, to make ethical decisions, or to not act ethically. Regardless of the wide variety of opinions and perspectives that people possess with respect to ethics, one thing is certain: Ethical decision making and behavior by organizations are absolutely essential to long-term success in business. Trust is a key component for ensuring long-term success in business, and trust can only be established over a period of time by consistent action and being true to one's word (or contract). Firms or persons who behave in an unethical manner risk the erosion or outright destruction of trust among other organizations, the public, stakeholders, and their employees. And once trust is violated, it is difficult to reestablish.

An extremely important point to understand is that the actions and decisions of individuals and organizations do not affect just the person or group committing the action, but will ultimately impact other people and organizations. Judging by their actions, many people do not seem to take this factor into account, but it is true nonetheless: Our actions and decisions will almost always impact others. If behaviors and decisions are not ethical, the impacts can be devastating. For example, when the deeply rooted financial fraud present at Enron finally took its toll, the entire company went under and tens of thousands of jobs vanished. Many people who worked for companies owned by Enron lost their pensions and 401(k) plans when the once-powerful company ceased to exist. In addition, Enron's accounting firm, Arthur Andersen LLP, virtually went out of business due to their complicity in the Enron fraud. Much more could be said on this, but the basic point has been made: Cases of corporate malfeasance and impropriety can have far-reaching effects both internally and externally.

By the same measure, the unethical practices and decisions of an individual person will impact other people. People do not exist in isolation; one's behaviors and actions impact other people, almost without exception. The chief pilot of a charter company who pushes his or her flight crew members to fly into known hazardous weather conditions in the name of being on time or to turn a profit is acting in a manner that is not only unethical, but also unsafe. The director who encourages a subordinate manager to cook the books in order to give the appearance of budgetary compliance is attempting to coerce the person and manipulate the situation to their advantage. This type of fraudulent action becomes harder to cover the longer it is permitted to occur, as more cooking of the books will most likely occur in order to cover the original fraud. What happens when the whole thing comes to light? The maintenance supervisor who turns a blind eye to unsafe aircraft maintenance practices, for whatever reason, is also acting unethically and most likely illegally as well. Is it worth the risk of an undesired event such as an accident to fulfill the demands of the decision maker who is ignoring the problems? The actions of a single person may even affect entire organizations in some instances. The actions of a few individuals at Southwest Airlines resulted in one of the largest monetary fines ever levied by the FAA against an airline ($7.5 million, as mentioned previously). Unethical practices by individuals and organizations can and sometimes do have significant consequences which impact others.

By the same measure, it is in the best interest of any organization to adopt and abide by specific ethical standards as well. As many have written, the crimes committed by some of the leadership at Enron, Tyco, WorldCom, and other organizations have served (at least in part) as a catalyst for having ethics in the workplace as guiding principles for many other companies. While this is certainly a good thing, if companies are going to have ethics as a core value, it is very important that they back up their words by ensuring that they actively support and expect their leaders and employees to conduct themselves ethically. Unfortunately, this is not easy to do because of the varying beliefs and values possessed by the people who make up the company. Nonetheless, a company espousing ethical values must support them through actions and accountability.

In-Focus: The Importance of Ethics

Michael Ferguson is the Manager of Corporate Safety at an airline headquartered in the western United States. He has many years of experience in various positions in the airlines and other areas of the aviation industry, including several leadership positions. He also taught ethics in a collegiate aviation program for several years. In his own words, Michael briefly discusses the importance of maintaining strong personal and corporate ethical standards.

Being a consistently ethically sound person is not easy to do, especially if you are working in an environment where some members of the leadership and perhaps even your peers are behaving in ways that are ethically questionable or just flat out wrong. Granted, determining what is right and wrong often varies from person to person, but I am not talking about viewing things through a relativistic perspective here. To clarify, I am talking about moral, legal and ethical standards that are in line with company policy, regulatory compliance and what are considered to be acceptable behaviors in a given society. Irrespective of a person's individual beliefs, morals and values, conducting oneself in a manner that puts them at odds with accepted legal, regulatory, organizational and social standards may mean that a person has chosen to act unethically.

Back to the point: It is often during times when others are "behaving badly" that a person's true character—who they really are—comes to the surface. Drawing the figurative line in the sand and standing on one's ethical principles is not always going to make a person popular with their peers and leaders. For this reason (and many others which we do not have the time to go into here) it is extremely important that a person continually develop strong personal ethical standards within themselves. By doing so, one will stand a better chance of being prepared to handle difficult ethical situations when confronted with such things and hopefully be less likely to make poor personal conduct decisions.

Courtesy of Michael Ferguson

Organizational Culture and Ethical Climate

Much of what constitutes the functioning of an organization can be found in the company's actual internal practices. What is the company culture like? Are the stated values, goals, and mission of the organization supported by the actions of leaders, managers, decision makers, employees? Are internal decisions in line with standards? How much priority is given to safety? To ethics? Is the safety message regularly supported within the company by leadership and the employee group? How is the safety message supported?

What we are driving at here is determining the state of the organizational climate or culture. **Organizational culture** is how a company or other organization actually *is*, functionally and environmentally. Culture consists of the values, decisions, actions, beliefs, goals, guiding principles, and other criteria that describe the internal climate of the organization. Organizational culture develops over a period of time and tends to follow practices and systems established by long-standing employees and leadership. These established practices are typically passed along to newer employees. A significant component of an organization's culture is its **ethical climate**, which describes the organization's overall climate with respect to ethical practices and beliefs. The establishment of a solid ethical climate is intentional and willful; it requires vision, effort, and dedication in order to be effective. A strong, desirable ethical culture will not develop by default. Rather, creating and fostering internal ethics must be made a priority by those in a decision-making capacity in an organization and demonstrated through consistency in actions and development of corresponding policies, practices, and decision-making processes. The ethical climate will often follow the ethics of the leadership; thus it is important that leaders in an organization demonstrate a solid, consistent ethical base through their words and actions so that employees receive the proper ethical message through their leaders.

Ethics and the Safety Culture

So what is the connection of the aforementioned ethical information to safety? The answer: It is our contention that a strong organizational culture should foster the development of a strong ethical culture, of which a strong **safety culture** is a key element. If regulations and internal company policies represent a form of moral standards (as stated previously), then a strong safety culture should be the foundation of an organization's business culture. A good safety culture will help to ensure regulatory compliance *and* the protection of employees and assets, including protecting employees who report safety issues from reprisal. Thus, a proper safety culture is at once ethical and economically responsible. It is ethical because it works intentionally to establish a safer working environment for employees and to ensure regulatory compliance. It is economical in nature by virtue of its recognition that accidents are very expensive and can cost the organization significant sums of money each year. Preventing accidents simultaneously protects people and limited economic resources, both of which can add to the bottom line of the organization.

A strong safety culture also demonstrates forward thinking on behalf of the leadership of the organization, by working to get out in front of problems before they occur. In other words, an intentional safety culture is by nature proactive, seeking to be preventive rather than reactive after an undesired event (an accident, incident, etc.) has occurred. Properly implemented and managed, the organizational safety culture establishes the internal practices of hazard and risk identification, risk assessment and management, investigation of events, and reporting of safety-related information and events. This assertion is supported through the recent widespread establishment of **Safety Management Systems (SMS)** in certain segments of the aviation industry, such as the airline industries of some nations, some airports, and in certain segments of corporate aviation. An SMS is a system of policies, procedures, and practices implemented within and by an organization to enhance safety. To be effective, an SMS should be an integrated element of an overall organization. That is, the SMS becomes intertwined as a vital and necessary part of the overall business plan of the organization. When an SMS becomes a part of the business structure, it can help to improve safety by providing a mechanism for identifying and managing the risks present in the company while simultaneously providing the means to monitor and improve the safety culture of an organization. This topic is discussed in more detail in Chapter 4, "Introduction to Safety Management Systems."

Continuing this line of reasoning, a solid ethical culture should include the creation and fostering of a strong internal safety culture. As with ethics, a strong, effective safety culture is the result of willful, intentional decisions, actions, training, and policies specifically designed to create an internal culture where safety is among the foremost of a company's priorities. Employees need to consistently see and hear members of leadership consistently promoting safety in the workplace through their words and behaviors in order to establish a good safety culture.

Unfortunately, many people and organizations suffer from ethical complacency. **Ethical complacency** is the tendency to ignore ethics until something happens and it becomes necessary to

Figure 1–4. Proper aircraft maintenance and inspection are key components of ensuring safety in the aviation industry.

pay attention and take reactive action to the issue that has occurred. In this respect, safety and ethics are quite similar. In the past (and continuing still in some organizations), safety has not always been proactive; in some instances a company may be motivated to respond only after an accident or incident has occurred. However, the modern safety movement is tending to move away from being reactive, and toward being intentionally proactive (under SMS) and even predictive in terms of safety. With respect to the similarity of ethics and safety and ethical complacency, there has been a widespread tendency for ethics to remain in the background, often unseen and largely ignored until an internal ethical issue comes to light. At that point, ethics becomes paramount. Unfortunately, in some instances, by that time it is too late and the damage may have already been done to the organization, its employees, and others. Therefore, to reiterate, it is important that an organization be proactive in establishing a balanced organizational culture in which safety and ethics are integrated throughout the business plan of the company and are intentional and proactive.

Applicable Ethical Concepts

With respect to ethical concepts and theories that are applicable to the information in this chapter, there are a number of topics that could be discussed here. However, we will deal with those concepts that we feel are the most applicable to what we are trying to communicate in this chapter. Thus, the primary ethical concepts that will be considered here are duty, respect, utilitarianism, and rights.

The Ethics of Duty

Duty, or deontological ethics, can be summarized as the belief or perception of what a person or organization should do or is obligated to do. For example, a military combat bomber or fighter pilot will do their best fulfill the requirements of the mission that has been assigned to them. Each will attack a specific target in order to accomplish their objective. In other words, the pilots will work to fulfill their duty by attempting to complete their missions. Another general example: an airline manager who works to determine the most cost- and operationally effective ways to ensure budgetary compliance in order to fulfill what has been required of them by their vice president. The manager's duty is to complete the task presented by their VP within the required time constraints and create the best possible solutions. Duty can be a very powerful motivator in a person's life.

An interesting (and very real) point to consider is this: Duty is often not separate from self-interest; rather, they are often linked together. A person or organization will often perform their duty from the perspective of self-interest or **egoism**, not just from a sense of loyalty or obligation. Egoism is defined as what is good for one's self, or acting in accordance with self-interest. This point has not been made to sound cynical or negative, but realistic. If we are honest with ourselves, we will often work at fulfilling our duty at least in part from the perspective of also fulfilling our self-interest. The airline manager mentioned in the previous paragraph may work out of a sense of loyalty and duty to their VP and company, but may also be motivated by self-interest as well, for a number of reasons. The manager may be concerned about remaining in good standing with the VP and does not want to risk tarnishing her reputation by not delivering, or she may be concerned about job security, or perhaps she realizes that completing specific tasks successfully will position her for possible job promotions. Self-interest may also be displayed by the VP and by the company (doing what is best for the company). Much more could be said about this, but the basic point has been made.

Duty plays a significant role in the realm of safety for a variety of reasons. One of the most fundamental reasons for this is found in the need to actively protect employees in the workplace from things that could hinder safety and result in harm. In the United States, the regulatory duty of organizations to protect their workers from potential harm on the job is found in the **OSHA General Duty Clause**, which requires employers to manage the recognized and foreseeable hazards and dangers associated with workplace operations that could harm their employees. The General Duty Clause is as follows (adapted from osha.gov):

OSHA General Duty Clause

(a) Each employer —

(1) shall furnish to each of his employees employment and a place of employment which are free from recognized hazards that are causing or are likely to cause death or serious physical harm to his employees;

(2) shall comply with occupational safety and health standards promulgated under this Act [*Occupational Safety and Health Act, added by author for clarification*].

(b) Each employee shall comply with occupational safety and health standards and all rules, regulations, and orders issued pursuant to this Act which are applicable to his own actions and conduct.

Source: U.S. Department of Labor/OSHA

Thus, U.S. employers have the regulatory obligation or duty to provide as safe a workplace as possible for their employees. This is obviously a well-intentioned and even necessary regulatory obligation. However, as mentioned previously, a safe working environment should be not just a matter of compliance, but of intentional action on behalf of a company to protect its employees and assets.

The Ethics of Respect

Although **respect** is a common, everyday principle, many people do not always think about the importance of respect. Respect helps to establish minimal societal and individual standards governing human interaction, determining how we treat others and how we would like others to treat us in turn. Respect can be shown in many ways: respect for other people, respect for private property, respect for the rules and/or laws, respect for a company, and so on. What is considered respect varies from culture to culture. Interestingly, giving respect is not always a function of liking someone or agreeing with a principle. For example, a person may not like or agree with the president of their company, but they may still choose to show appropriate respect and deference to the person because of their position and status. Similarly, an employee may not agree with a given company policy, but they may choose to abide by the policy for reasons of respect for the person or company that created the policy. Of course, there are many reasons other than respect why a person may choose to follow the policy, but respect may be one of them.

In order to be effective, respect is ultimately the proverbial two-way street. In general, people tend to respect those who treat them with some level of respect. The same holds true for organizations and their employees: Organizations that treat their employees with respect may earn the respect and trust of the employee group. Conversely, employees who regard their leaders and coworkers with respect have a greater probability of earning the reciprocal respect of others. Employees may demonstrate respect for the organization and its leadership by following rules, policies, and procedures; treating others fairly; listening and learning; acknowledging the input of others; doing their jobs well on a consistent basis; being on time; taking their work seriously, and so on. Leaders in organizations may demonstrate respect for their employees by following the same rules and policies as the employees, not having double standards regarding treatment and discipline of employees, establishing reasonable and fair working practices, actively working to promote a safer workplace, encouraging employee involvement in decision making, listening to employees and actively soliciting their ideas and feedback, and in turn implementing appropriate employee ideas, paying reasonable wages commensurate with job expectations and worker experience, taking the effort to get to know their employees, and so on.

Arguably, perhaps, one of the best ways that a company can show it possesses and demonstrates respect for employees is through the establishment of a strong, consistent safety culture. Creating and promoting an active safety culture sends a clear message to the employee group that they and their well-being are regarded as being important to the company and that the company is working diligently to ensure their safety. Actions such as these help to foster a stronger organization ethical culture by demonstrating proactivity in safety, compassion, empathy, and genuine concern for the well-being of all workers and for the individual person by working to keep them safe as a function of respect.

The Ethics of Utilitarianism

It can be asserted accurately that a company working in the overall best interest of its employees is demonstrating the ethical principle of **utilitarianism**, which is the concept of doing what is perceived to be best for the greatest number of people. Utilitarian ethics states that a decision is ethical if it provides a greater amount of benefit than other possible alternative decisions for the greatest number of people. In some cases, a policy that is utilitarian in nature may be a very good idea and even necessary. For example, the OSHA General Duty Clause (discussed previously) could be said to be a policy of utilitarianism because it makes companies across the United States responsible for helping to protect their workers (a large number in total) from foreseeable hazards on the job. The clause provides a broad regulatory summary for workplace safety and health and provides accountability to companies and their leadership.

In other cases, a policy or decision that is designed to be utilitarian (for the greater good) may have a negative impact on a smaller group of people. Herein lies one of the difficulties of utilitarian ethics: Is it always the best decision to do what is good for the greatest number of people? What about people who may be harmed in some way by the decision or policy? Then things become more complex. For example, at the time of this writing a major U.S. airline has created a smaller subsidiary company (owned by the major airline) to perform ground handling services at some outstations that are served by some of its regional code-sharing airline partners. At many of these outstations, a prominent regional airline had operated with its own direct employees (station managers, supervisors, ramp agents, customer service agents, etc.) successfully for decades. However, the major airline has been incrementally replacing the regional airline staff at many of its outstations with staff of the ground handling subsidiary. The regional airline still operates flights in and out of these markets, but these are no longer handled by its direct employees. From a business standpoint, the decision makes good sense. It is less expensive to operate the stations with the new company, as employees of the ground handling company are paid significantly less than the regional airline employees were, making the decision cost-efficient. This decision was made in order to benefit the major airline (larger number of people) by saving the company money (net benefit). However, this decision has cost many employees (smaller number of people) of the regional airline their jobs. This issue created an understandable but nonetheless difficult situation with respect to the people who lost their jobs with the regional airline. This is not to imply that the major airline has done anything wrong or has acted with impropriety in making the decision to create and use the ground handling firm. Rather, the point is that doing what is best for the greatest number of people almost always has consequences for some on the other side of the decision and further complicates the concept of utilitarianism. It may be well intentioned and even noble to say that it is always the best decision to do the greatest good for the greatest number, but the reality is not that simple.

The Ethics of Rights

Yet another important ethical principle is found in the concept and reality of **rights**. The concept of rights and their applicability to people is perhaps one of the most powerful and influential forces that shape our lives and how we live them. Fundamentally, the establishment of rights seeks to create the minimal conditions governing decency in human life. Rights grant people certain basic privileges associated with being human and therefore possessing innate value that the rights seek to establish and protect. While rights have many different forms and functions, our concern here is with rights as an ethical and safety standard.

The roots of the theory and philosophy of ethical rights are attributed primarily to Immanuel Kant (1724–1804). Kant developed two basic philosophical formulations that in summary assert that all persons should be treated as free and equal to each other. Kant's two formulations are as follows (paraphrased):

1. A person should act only in a manner that they would like others to act in the same circumstances.
2. A person should always regard and treat other people with dignity and respect.

According to Kant, if all persons are equal (a point of tremendous debate and opinion in many cultures), then the rights of one person should not supersede or be placed above the rights of another. This view is rather idealistic, as equality is not universally present among people. In the context of safety, the rights and interests of an organization should include the rights and interests of its employees, including the right to a safe and healthy working environment, insofar as possible.

Under the **Occupational Safety and Health (OSH) Act of 1970**, which created OSHA, workers in the United States are afforded certain rights designed to keep them as safe as possible in the workplace. These same rights also seek to establish minimal workplace conditions intended to protect the health of workers as well. Because of this, there is a regulatory component to the ethics of safety rights. However (and back to our original point), safety must be motivated by more than just compliance—by a desire to keep people safe at work. The following section is a brief summary of some of the primary rights of employees and employer responsibilities for safety in the workplace as determined by OSHA. The information in the following sections is derived from various OSHA sources found online.

The Rights of Employees under OSHA and the OSH Act Under the auspices of OSHA and the OSH Act, employees are granted the right to a safe working environment provided by the employer, one that does not pose a risk of serious harm and is as free as possible from known hazards that could injure or kill. Employees have the right to receive specific training about job-related hazards, ways to prevent injury or illness, and the OSHA standards that are applicable to their workplace. Workers also have the right to review records of job-related illnesses and injuries and obtain a copy of their personal medical records from an employer. Employees can also report (anonymously, if they choose) to OSHA any known or potential safety problem in their workplace and request that OSHA conduct an inspection of their employer (which may or may not be granted, at the discretion of the governing OSHA office). Workers are also protected from retaliation and discrimination from their employer in the event that they do report a safety issue to OSHA. Employees have the right to learn the results of the inspection and to request a review if OSHA decides not to issue citations against the employer. Workers also have the right to request the employer to correct any hazards in the workplace, even if the hazards are not violations of OSHA standards.

Employer Responsibilities under OSHA and the OSH Act Under the auspices of OSHA and the OSH Act, most employers in the United States have specific responsibilities to ensure the safety and health of their employees. It is important to note that employer responsibilities under OSHA are in place to ensure that the rights of employees are honored. The following is a list of the primary responsibilities of employers to ensure workplace safety as required by OSHA.

- Provide a workplace free from serious recognized hazards and comply with standards, rules, and regulations issued under the OSH Act.
- Examine workplace conditions to ensure conformance to applicable OSHA standards.
- Ensure employees have and use safe tools and equipment and properly maintain this equipment.
- Use color codes, posters, labels, or signs to alert employees about potential hazards.
- Establish and update operating procedures and communicate to ensure employees follow safety and health requirements.
- Provide medical examinations and training when required by OSHA standards.

Source: U.S. Department of Labor/OSHA

- In a conspicuous location within the workplace, post the OSHA poster (or the state-plan equivalent) informing employees of their rights and responsibilities.
- Report any fatal accident or accident resulting in the hospitalization of three or more employees to the nearest OSHA office within eight hours of the time of the event.
- Keep records of work-related injuries and illnesses. Employers with 10 or fewer employees and employers in certain low-hazard industries are not required to do this.
- Provide employees, former employees, or their authorized representative(s) access to the Work-Related Injuries and Illnesses log (OSHA Form 300).
- Provide access to employee medical and exposure records to employees or their authorized representative(s).
- Provide the names of authorized employee representatives who may accompany an OSHA inspector during an inspection to the OSHA inspector.
- Employers may not discriminate against employees who exercise their rights under the OSH Act.
- Post any OSHA citations at or near the work area involved. Each citation must remain posted until the violation has been corrected, or for three working days, whichever is longer. Employers are required to post abatement verification documents or tags.
- Correct any OSHA standard violations by the deadline set in the OSHA citation and submit required abatement verification documentation.

Source: U.S. Department of Labor/OSHA

Employee Responsibilities under OSHA While much of the responsibility for safety in the workplace resides with employers, workers also have responsibilities under OSHA to help ensure safety on the job. Among the primary employee responsibilities for safety are:

- Follow all established procedures and policies of the organization.
- Keep up to date with health and safety regulations. This can be accomplished by requesting information from the employer or directly from OSHA, if desired. In actual practice, it is not always easy or practical for an employee to keep current with changing safety practices and regulations themselves, but getting information from the employer's safety department or personnel is often a viable means of obtaining information.
- Report unsafe conditions. Employees are expected to report potential and actual safety concerns to their supervisors as soon as possible after becoming aware of the issues. Employees can also report concerns directly to OSHA, should they choose to do so.

Source: U.S. Department of Labor/OSHA

Basic Economics and Safety

Beyond question, economics and safety have been and always will be inextricably linked; one cannot be separated from the other as each directly impacts the other continually. This factor ultimately affects decision making as it pertains to safety in virtually any company. Aviation accidents are incredibly expensive for reasons that extend well beyond the most significant impact, the loss of human life. Financial losses to an aviation company because of accidents (not just hull-loss aircraft accidents), incidents, employee injuries, aircraft ground damages, and other events are generally significant, each year cosing the industry (collectively) billions of dollars in losses.

A point that not many people are aware of or even stop to consider is this: The cost of preventing accidents is often less than the costs associated with accidents. However, the fact remains that convincing some aviation decision makers that safety is a good investment is not always an easy task. The argument for financing safety initiatives in an aviation company must be made in a way that is factual, compelling, persuasive, realistic. and economically sound. A safety argument that is strictly emotionally or ethically based is probably not going to get much in the way of serious consideration, much less appropriate funding by leaders who have a number of demands on their time and budgets. A balanced approach is needed to ensure success in aviation safety initiatives—one that considers the ethical, human-based issues and the often harsh realities around the availability and allocation of limited financial resources.

By way of definition, **economics** is the study of the allocation of **scarce resources** within a society or organization. "Scarce" refers to the fact that resources are finite and limited; there is only so much of any resource that is available. Economics is a powerful field of study that is highly influential and extremely useful. It is important to understand that virtually every significant decision made within an organization is related at least in part to economics. The decision for Delta Air Lines to acquire Northwest Airlines and become one giant air carrier was driven by economics. United Airlines and Continental Airlines have combined and are operating as another mega-carrier (United). When a corporation decides to acquire an aircraft to use as a time-saving business tool, economics is what ultimately drives the final decision. The corporate aircraft may provide a number of benefits to the company, not the least of which is significant time-saving capabilities versus traveling via the airlines. These are all examples of economics driving decisions.

Economics and Safety Decisions

As a statement of the reality of how the aviation industry tends to function, safety decisions are generally not made solely on the basis of ethical or moral dimensions, but rather on the basis of economic considerations. Often, a decision maker in aviation will ask questions such as: How much is a given safety enhancement, initiative, or program going to cost? What is the return on investment and the term for realization? What are the potential benefits? Are the potential benefits of the safety idea going to outweigh the costs? Is there a better use or more pressing need for the limited resources of the department/division/company rather than spending the money on safety? Where is the money going to come from to pay for the safety idea, especially if no additional funds are allocated in the annual budget to cover its costs? These are all examples of good, legitimate questions that could be asked by a decision maker, as availability of funding within any organization is finite. Therefore, economics does play a significant and necessary role in making decisions with respect to safety. Fiscal responsibility through budgetary compliance is very important to ensure the greatest opportunities for profitability and is thus a key responsibility of management in resource allocation and control. Any responsible manager should recognize the need to carefully weigh all possible courses of action before a decision is made, and decisions affecting safety are no different.

A useful tool used by many aviation organizations in economic decision making is **cost/ benefit analysis (CBA)**, sometimes known as benefit/cost analysis. As implied by the name, CBA is the practice of calculating and comparing the actual and projected costs of a decision against the actual and potential benefits of the decision. A valid CBA will look at both the short- and long-term costs and benefits to determine if the costs of a given decision will create benefits that may outweigh the costs. The FAA uses CBA regularly, as do a number of aviation companies.

Figure 1–5. Delta Air Lines acquired Northwest Airlines in 2008 to form a megacarrier.

According to the FAA, CBA is used to determine to determine whether or not a certain output (benefit) will be produced by a decision and, if so, how best to produce it. Because the objective of CBA is to compare benefits and costs, these must both be evaluated using the same unit of measure.

With regard to safety decisions, if the anticipated benefits of a decision are likely to outweigh the costs, it is likely the decision will be made to support the safety decision. In straightforward terms, the company may spend the money on the safety issue.

Balance between Safety and Economics

As stated previously, a balance between economic and ethical considerations is a key concept that must be present in any organization. It is a good and even noble thing for an organization to say that they are going to be serious about safety, but in reality, using money to "purchase" safety in not always a task easily completed. Stated another way, we know that we must have safety in aviation and that safety should be a high priority. However, safety expenditures are also governed by economics, and there is only so much money that can be given to fund a number of different activities within a company. In general, economics considers impact to the organizational bottom line, while ethics, when given its proper place in an organization, should consider the company, the employee group and the individual person. As discussed in Chapter 3, "Risk and Risk Management," consideration of risk is a factor impacting all aviation companies and plays a role in balancing safety and economics.

Basic Economic Concepts

Beyond argument, two of the biggest catalysts for change in the aviation industry are accidents and the economic impacts that accompany such events. Even a cursory consideration demonstrates that economics and accidents are inextricably linked, even inseparable. In short, accidents (flight- and ground operations–related) are extremely expensive in the aviation industry for a number of reasons. Loss of life, destruction of assets, rising insurance costs, the possibility of civil penalties through lawsuits, fines due to regulatory noncompliance, loss of reputation, and erosion of public trust are just a few of the many potential direct costs and indirect costs associated with accidents in aviation. **Direct costs** are those costs directly connected to the development and production of a specific product. In aviation, some examples of direct costs are aircraft acquisition, aircraft operating costs, insurance, fuel, maintenance, salaries of pilots and maintenance and ground handling personnel. In the case of an accident, direct costs would include loss of life and loss of assets (aircraft, etc.). **Indirect costs** are the costs that are present but not associated directly with the development and production of a specific product. Examples of indirect costs in aviation include technology, training, managing, and, in the case of an accident, litigation, lawsuit settlements, rising insurance costs, erosion of public confidence and perception, investigation, corrective actions, and fines.

Consider the following. As previously established, safety is directly related to and driven by economics, at least to a significant degree. Experience (our own and that of many others from the industry) and organizational reality clearly demonstrate that safety programs, initiatives, and related areas are, almost without exception, controlled or limited by available financial resources and the willingness of those who control the allocation of these resources to spend some of them to improve safety. After all, money spent on one thing is money that cannot be spent on something else. One of the difficulties encountered when trying to alter the safety culture of an organization is convincing leadership that safety is a worthwhile investment. In many companies, safety is treated as a non–revenue generating cost center. It costs money to fund safety programs, train and pay safety personnel, and so on. Due to its very nature, safety does not generate direct revenue for the company. Rather, the reality is that having a safety program or department causes the organization to spend money. Although preventing accidents may ultimately save an organization a significant amount of money, it is difficult to know exactly what was prevented, and it is therefore often hard to quantify the actual benefits of safety. This is an aspect of safety that is rather nebulous. Often an aviation decision maker may take the position that there needs to be a tangible reason to spend money on safety rather than on something else. Such a position involves taking consideration of opportunity cost, discussed in the following sections, along with one other applicable economic concept, diminishing returns.

Opportunity Cost

Opportunity cost is simply defined as follows: The real cost of something is what you give up in order obtain it. This concept includes not just how much money was spent to purchase something, but also the potential benefits that were sacrificed by not obtaining something else. As a simple example, a dollar spent on gasoline for a car is a dollar that cannot be spent on food. The spender received the benefit of the gasoline (transportation), but lost the potential benefit of the food that could have been purchased with that same dollar (sustenance). In the context of a company, money that is spent on safety is money that cannot be utilized elsewhere (or on safety, as the case may be). The opportunity of using the money (or other resource, such as time) on safety is lost if it was spent on something else, and vice versa. As mentioned previously, finances in any aviation organization are limited (especially during times of economic difficulty), and convincing a decision maker to spend money on safety initiatives that may or may not be in the budget is often a difficult task, as once the money is spent, it cannot be spent on something else. In this manner, opportunity cost may play a role in safety-related decisions. What it may really come down to is the priority, values, and goals of the decision maker. If the decision maker is motivated by improving safety, they may choose to allocate funding for that specific purpose. However, other internal and/or external forces may take precedence over those of the person trying to make the decision in influencing the final outcome.

Diminishing Returns

Another basic principle of economics that may be applied to safety decision making is the law of **diminishing returns**, also known as **diseconomy of scale**. The law of diminishing returns states that as equal quantities of a variable factor are increased and other factors remain constant, a point is reached beyond which the addition of one more unit of the variable factor will result in a diminishing rate of return. Stated more simply, the more you have of something, the smaller is the benefit of having even more; investing a quantity of money and/or other resources on something may not result in an additional benefit. In terms of safety, if spending additional resources to obtain a greater degree of safety (benefit) does not produce a positive potential or real benefit, then it may be difficult to convince a decision maker that more resources should be spent on safety if they believe that the resources may be better utilized elsewhere, especially in the absence of perceived tangible benefits. If the extra dollar spent on safety is not likely to produce at least a dollar's worth of benefit, then the leader may not spend the dollar on safety.

The Importance of Leadership in Establishing an Ethical Safety Culture

In this chapter, we have asserted the need for balance. We have briefly discussed the idea that safety should not be just a matter of regulatory compliance, but of taking genuine, active interest in protecting employees, assets, and the public from harm insofar as possible. We have also asserted that balance is needed between safety and economics in an organization. However, although we believe balance is essential, safety should not be given the proverbial backseat in an organization; it should be a part of the overall organizational culture. Making safety a part of the organizational culture may be accomplished through strong, balanced leadership. The following sections discuss a few basic elements that may be helpful for the leadership of an organization in supporting an ethical safety culture. Leadership is paramount in order for an ethical safety culture to develop and be maintained over time.

Leadership Support

In most organizations, it is the leadership that sets the tone governing the development of the internal culture. Ethics and safety, in order to be considered as serious and vital to the overall organization, must be embraced as important by the leadership. Thus, leadership is vital to both ethics and safety in a company. Communicating the importance of making ethics and safety priorities must take place through the vocal and financial support of viable programs designed to enhance the safety culture.

To go further, our contention is that the organizational culture, the ethical culture, and the safety culture should not be separate and distinct from each other. Rather, these three cultural elements should be blended into one that governs the internal nature and functioning of the organization; the organizational culture is the ethical culture is the safety culture. In other words, the organization should foster an internal culture embracing safety and ethics as vital, necessary parts of the overall organizational culture through active promotion and support by leadership. Ultimately, safety should be considered as an ethic by the organization.

As mentioned previously, making safety an important part of the organizational culture is in line with one of the key elements of a Safety Management System: Safety becomes intertwined as a vital and necessary part of the overall business plan of the organization; that is, safety becomes a key component of the organizational culture. In order for this to occur, the leadership must consistently support safety as a key organizational value. This support must not be just verbal, but financial. Employees need to see and hear their leaders consistently supporting safety through their actions, words, and decisions.

Put It in Writing!

Safety as a key organizational value and motivator needs to be in written form for all employees to see. In this context we are not so much referring to a written safety program (which, of course, is important and necessary); we are talking about an organization putting its support for safety as an internal ethical value in writing in the form of an organizational "commitment statement," so that it becomes an actual corporate value supported across the board by leadership. The commitment statement should always be signed by the CEO or the top executive of the organization. In doing so, the executive is demonstrating a commitment to safety and is also putting his reputation on the line by signing the written safety statement. Of course, an organizational value (even one in writing) is only as strong as the support for and demonstration of the value through the actions of leadership and employees. Nonetheless, putting safety in writing as an important organizational ethical value should be done in order to show commitment and gain trust.

Model Safety through Actions

Leaders need to back up their commitment to safety by living it out themselves through leading by their example. In other words, leaders should not say one thing and do another. Few things undermine the credibility of a leader faster than hypocrisy. When employees see and hear leaders violating safety rules, this harms the validity of the safety culture; the unspoken message given by leaders to employees is that it is okay for the rules to be broken. After all, the leader just did it! Leaders must consistently follow the rules, *especially* when they think no one is watching. Leaders should mentor employees in modeling positive safety behaviors.

Many aviation organizations, especially airlines, feature a "safety-first" mantra in their corporate values and mission statements. However, a safety-first mission is only validated through the consistent establishment of a strong safety culture in which the entire organization, from the top executives down to the frontline employees, is working to ensure that its actions really are aligned with putting safety first in everything. Employees will detect very quickly if the company says "safety first" but does not support this statement through action. This situation can actually be counterproductive to safety.

Accountability

Leadership should establish minimum expectations and standards for internal compliance with safety and ethical standards. Then, as in modeling any proper behavior, leaders need to hold themselves and the employee group accountable for maintaining the safety and ethical standards and bring appropriate corrective actions to bear as needed. Corrective actions are not automatically synonymous with disciplinary actions in this context, but may also include performing any action or changes needed to align personnel with established safety and ethical standards. In order to be fair and effective, application of accountability should apply not only to workers, but to the leadership as well.

Show Genuine Concern for Employees

Members of leadership and management in organizations are sometimes accused of being more concerned with the economic health of the organization than with the safety and health of employees (bottom line over front line). In some cases, an accusation of this nature is merited. It is imperative for the leadership of an organization to demonstrate consistent, genuine concern for the safety, health, and overall well-being of the employees in their charge. Actions of this nature help to establish trust among employees, which is ultimately beneficial to the organization. Satisfied employees have been shown to have higher morale and greater productivity on the job. Having an ethical organizational culture that values employees and their safety is very important.

Provide Training

In general, people cannot be expected to do something to standard if they do not know what the expectations are or if they have not received training. Leadership should establish appropriate training in safety and ethical standards for all employees in the organization, irrespective of position or job function. Training can be tailored to include the general organizational expectation and job-specific expectation, accounting for variances in department and positions. Employees should never be asked or required to do something for which they have not received appropriate training.

Chapter Summary

In this chapter, we have briefly discussed some of the concepts and ideas associated with the importance of ethics in safety. We have discussed the regulated nature of the aviation industry and have asserted the position that safety needs to be in place for ethical reasons, not just for regulatory compliance. We have discussed the importance of ethics in safety decisions, behaviors, and in an organization culture. We have also discussed the notion that a strong organizational culture should possess a solid ethical and safety culture as key components.

We have introduced the basic ethical concepts and theories of duty, egoism, respect, rights, and utilitarianism and covered some of the rights and responsibilities of both employees and employers under OSHA and the OSH Act of 1970. We also asserted the position that balance needs to be present between safety and economics in an organization. Some basic economic concepts and ideas associated with organizational ethics and safety were also covered. Finally, we discussed the importance of leadership in creating an ethical safety culture and touched on some elements that may assist leaders in supporting safety in their organizations.

In summary, ethics and safety can be viewed not just as good ideas, but as an essential way of life within an organization. It is our position that genuine safety in any organization goes beyond mere regulatory compliance: It is simply the economically and ethically responsible thing to do. Creating and managing a balance between fiscal responsibility and protection of people may prove to be beneficial to an organization in both the short and long term, and may ultimately have a positive impact on the economic health of an organization and the health and safety of its people.

Chapter Concept Questions

1. In your own words, define the following: ethics, business ethics, moral standards.

2. Define and describe the following in detail: organizational culture, ethical culture, safety culture.

3. What influence does the culture of an organization have on its ethical climate? On its safety culture? What general and/or specific ideas do you have that could improve the safety culture of many aviation companies?

4. In your view, what similarities exist between ethics and safety? How do ethics and safety fit together in an organization?

5. In your own words, summarize the OSHA General Duty Clause. List and discuss in detail at least three of the primary rights of employees under OSHA and the OSH Act of 1970.

6. Describe in detail some of the tensions that exist between safety and economics in virtually any aviation organization.

7. Define and describe each of the following: utilitarianism, egoism, duty, respect, rights.

8. What is opportunity cost? What is meant by diminishing returns? What does it mean to say that resources are scarce in an organization?

9. Name and describe at least three things that leaders can do to promote the development of an ethical safety culture within their organizational culture.

Chapter References

Banks, C. 2004. *Criminal Justice Ethics: Theory and Practice*. Thousand Oaks, CA: Sage Publications.

Censky, A. 2010. Tom Petters Gets 50 Years for Ponzi Scheme. *CNNMoney.com*. http://money.cnn.com/2010/04/08/news/economy/Tom_Petters/.

Economist.com. 2009. *Economics A–Z*. http://www.economist.com/research/economics/.

Economy Professor. 2009. Law of Diminishing Returns. http://www.economyprofessor.com/economictheories/law-of-diminishing-returns.php.

Federal Aviation Administration. 2006. Advisory Circular 120-59A. http://www.airweb.faa.gov/Regulatory_and_Guidance_Library/rgAdvisoryCircular.nsf/0/fd8e4c96f2eca30886257156006b3d07/$FILE/AC%20120-59a.pdf.

Federal Aviation Administration. 1998. Economic Analysis of Investment and Regulatory Decisions. http://www.faa.gov/regulations_policies/policy_guidance/benefit_cost/.

Fritzsche, D. 2005. *Business Ethics: A Global and Managerial Perspective* (2nd ed.). New York: McGraw-Hill.

Goetsch, D. L. 2008. *Occupational Safety and Health for Technologists, Engineers and Managers*. Upper Saddle River, NJ: Pearson Prentice Hall.

Healthline.com. 2009. Definition of Ethics. http://www.healthline.com/galecontent/ethics?utm_term=ethic&utm_medium=mw&utm_campaign=article.

Hosmer, L. T. 2006. *The Ethics of Management* (5th ed.). New York: McGraw-Hill Irwin.

Hughes, J. 2009. Southwest Air Agrees to $7.5 Million Fine, FAA Says. *Bloomberg.com*. http://www.bloomberg.com/apps/news?pid=20601103&sid=avFzkpTRRnHc&refer=us.

Krause, T. 2007. The Ethics of Safety. *EHS Today*. http://ehstoday.com/safety/best-practices/ehs_imp_67392/.

Merriam-Webster Online Dictionary. 2009. Ethics definition. http://www.merriam-webster.com/dictionary/ethics.

Northhouse, Peter. 2007. *Leadership Theory and Practice*. Thousand Oaks, CA: Sage Publications.

Occupational Safety and Health Administration. 2010. Employee Workplace Rights Document. http://osha.gov/Publications/osha3021.pdf.

Occupational Safety and Health Administration. 2010. General Duty Clause. http://osha.gov/pls/oshaweb/owadisp.show_document?p_id=3359&p_table=OSHACT.

Occupational Safety and Health Administration. 2010. Workers' Rights. http://osha.gov/as/opa/worker/rights.html.

Patankar, M. S., Brown, J. P., and Treadwell, M. D. 2005. *Safety Ethics: Cases from Aviation, Healthcare, and Occupational and Environmental Health*. Burlington, VT: Ashgate Publishing Company.

Roy, A. K., and Roy, L. C. 2004. The Importance of Teaching Ethics. *Business and Economic Review*, January, n.p.

Timm, D. 2000. Defining Ethics and Morality. *ASL Interpreting Resources*. http://asl_interpreting.tripod.com/ethics/dt1.htm.

Wood, Richard. 2003. *Aviation Safety Programs, A Management Handbook* (3rd ed.). Englewood, CO: Jeppesen-Sanderson, Inc.

2 Regulatory Oversight

Chapter Learning Objectives

After completing this chapter, the reader should be able to:

- Describe the primary functions of the Department of Transportation, Federal Aviation Administration, National Transportation Safety Board, Transportation Security Administration, Occupational Safety and Health Administration, and Environmental Protection Agency and the roles they play in providing regulation and oversight of aviation.
- Discuss the fundamentals of the Air Carrier Access Act and its applicability to the aviation industry.
- Understand some of the key programs run by the various regulatory agencies.

Key Concepts and Terms

Accumulation Start Date
Accumulation Time
Acutely Hazardous Waste
Administrative Actions
Administrator
Air Carrier Access Act (ACAA)
Aqueous Solution
Aviation and Transportation Security Act
Board of Inquiry
Certificate Actions
Characteristic of Corrosivity
Characteristic of Ignitability
Characteristic of Reactivity
Characteristic of Toxicity
Civil Penalties
Clean Air Act
Clean Water Act
Cockpit Voice Recorder (CVR)
Combustible Liquid
Company Materials
Complaint Resolution Officials (CRO)
Compressed Gas
Conditionally Exempt Small Quantity Generator (CESQG)
Corrosive
Crew Member Self-Defense Training (CMSDT)
Dangerous Goods
Department of Homeland Security (DHS)
Department of Transportation (DOT)
Department of Transportation Act of 1966
Direct Discharges
Empty Container
Enforcement
Federal Air Marshals
Federal Aviation Administration (FAA)
Final Report
Flammable Liquid
Flashpoint
Flight Data Recorder (FDR)
Full
Go-Team
Group Chairmen
Hazardous Waste
Hazardous Waste Manifest
Homeland Security Act of 2002
Ignitable
Impervious Surface
Large Quantity Generator (LQG)
National Explosives Detection Canine Team
National Transportation Safety Board (NTSB)
No Fly List
Notation Draft
Organic Matter
Oxidizer
Party System
Probable Cause

Public Hearing

Recordkeeping National Emphasis Program (NEP)

Resource Conservation and Recovery Act (RCRA)

Safety Recommendations

Satellite Accumulation Area

Security Inspectors

Site Specific Target (SST)

Small Quantity Generator (SQG)

Solid Waste

Storm Water Pollution Prevention Plans (SWPPPs)

Strikes

Toxic

Transportation Security Administration (TSA)

Transportation Security Officers

Treatment/Storage/Disposal (TSD)

Underwater Locator Beacon (ULB)

Watch Lists

Will Carry

Working Group

Introduction

As mentioned at the beginning of Chapter 1, aviation is among the most highly regulated industries, subject to a significant degree of legal and regulatory requirements, inspections, auditing practices, and many other types of oversight by a number of government agencies. Each regulatory body serves a very important function in ensuring that the aviation industry as a whole is safe and generally efficient. The Federal Aviation Administration (FAA), Transportation Security Administration (TSA), Environmental Protection Agency (EPA), Occupational Safety and Health Administration (OSHA), and state and local governing agencies provide a high degree of oversight. In addition, certain segments of the aviation industry, most notably air carriers, must provide proper accommodations for disabled passengers to ensure compliance with the Air Carrier Access Act (ACAA).

Regulatory requirements and compliance requirements abound in virtually every facet of an aviation organization. Knowing what each of these organizations does in the realm of aviation safety is important, as virtually all aviation industry personnel are impacted by regulations and compliance issues, and many of these individuals must interact directly with members of regulatory organizations at least occasionally and, in some cases, very frequently. The following sections provide a brief overview of some of the most prominent regulatory agencies that conduct oversight of the aviation industry and the basic functions that they serve. Safety is a key component in the actions and responsibilities of these agencies.

The U.S. Department of Transportation

The **Department of Transportation (DOT)** is an executive department of the U.S. government charged with providing oversight of a variety of major transportation functions that are of national importance. The DOT is the primary federal agency given the responsibility of developing and administering policies and programs designed to protect and enhance the safety, efficiency, and adequacy of the U.S. transportation system. The DOT was created on October 15, 1966, by the U.S. Congress through the **Department of Transportation Act of 1966**. The organization officially went into operation on April 1, 1967.

As noted on their website, the stated mission of the DOT is to "serve the United States by ensuring a fast, safe, efficient, accessible and convenient transportation system that meets our vital national interests and enhances the quality of life of the American people, today and into the future." In addition, the DOT's mission includes the development and coordination of policies designed to provide an efficient, cost-effective national transportation system, including consideration of the environment, national defense, and need.

The Department of Transportation consists of 11 individual operating agencies, each of which has a specific area of transportation responsibility:

- Federal Aviation Administration (FAA)
- Federal Highway Administration (FHA)
- Federal Motor Carrier Safety Administration (FMCSA)
- Federal Railroad Administration (FRA)
- Federal Transit Administration (FTA)
- Maritime Administration (MARAD)
- National Highway Traffic Safety Administration (NHTSA)
- Pipeline and Hazardous Materials Safety Administration (PHMSA)
- Research and Innovative Technologies Administration (RITA)
- Saint Lawrence Seaway Development Corporation (SLSDC)
- Surface Transportation Board (STB)

At present, the DOT employs approximately 60,000 people throughout the United States; it is responsible for keeping the traveling public safe and secure, increasing their mobility, and improving the transportation system to contribute to the nation's economic growth. The head of the DOT is the Secretary of Transportation, who is charged with the oversight of all operations of the DOT. The secretary serves as the principal advisor to the president on all matters relating to federal transportation programs and issues. The FAA oversees the regulation and safety of civil aviation, discussed in more detail in the next section.

In addition to FAA regulations, there are other DOT regulations that apply within the aviation industry. Among these are the Dangerous Goods and Hazardous Waste Transportation Regulations, found in 49 Code of Federal Regulations (CFR), Subchapter C. These regulations specify requirements for the safe packaging, labeling, and transport of dangerous goods and hazardous waste. There are two sides involved in each shipment of dangerous goods: the shipper and the carrier. Most airlines will likely be shippers of dangerous goods, particularly within their respective maintenance divisions. However, not all will be carriers. For example, while many commercial carriers (passenger and cargo) are "will carry" airlines, others are not or choose to carry only those hazardous materials related to the maintenance of their aircraft. A **will carry** airline is one that has chosen to carry select hazardous materials, otherwise known as **dangerous goods**, aboard their aircraft. Dangerous goods are defined as products meeting hazardous characteristics as defined by the Environmental Protection Agency (EPA) or described in the DOT or International Air Transport Association (IATA) Dangerous Goods Regulations. The section on the EPA later in this chapter provides more information on EPA's definitions of hazardous characteristics.

A will carry airline must develop a comprehensive program for the safe transportation of such goods and train all personnel who handle the goods in the details of the program, including receiving, handling, loading, and so on. For a carrier that chooses to carry hazardous materials only as **company materials**, or CoMat, the same requirements apply. Company materials are items that are owned or controlled by a company and are shipped within the organization in support of various parts of their overall operations. An example of CoMat would be shipping aircraft parts on a company aircraft from one location to another. The only difference is that the department receiving the goods does so from an internal customer, usually the maintenance or materiel department. While some may think not accepting such goods from the public would make it easier or demand less training and administration , there is no difference in practice for a carrier of dangerous goods: The carrier is considered a will carry, whether for CoMat or customer acceptance, or a "non-will carry" airline. The regulations are no different, and the same amount of training is required.

Although Title 49 CFR Parts 100–185 cover all modes of domestic shipment of dangerous goods, most carriers will follow the International Civil Aviation Organization (ICAO) Technical Instructions for dangerous goods transportation and the International Air Transport Association (IATA) Dangerous Goods Regulations. Products that cannot be classified under any of the listed dangerous goods of the IATA Dangerous Goods Table cannot be offered or accepted for shipment aboard an aircraft. The shipping of dangerous goods aboard an aircraft is an important consideration in maintaining safety, as there have been aircraft accidents as a result of the improper shipping and handling of such goods. One of the most recent examples is found in the May 1996 accident involving ValuJet 592, which crashed in the Florida Everglades after oxygen canisters in one of the cargo bins ignited shortly after take-off from Miami International Airport (MIA), causing a catastrophic fire. One hundred ten persons died in this accident; there were no survivors. This horrible accident was entirely preventable.

Federal Aviation Administration

The **Federal Aviation Administration (FAA)** is the agency of the DOT that serves as the regulatory body governing civil aviation interests and activities in the United States. The FAA is headquartered in Washington, D.C., and its primary authority possesses the title of **Administrator**. The administrator is appointed by the president, confirmed by Congress, and is charged with providing overall guidance, management, and leadership of the organization. The FAA has a number of regional and field offices in locations throughout the country. At the time of this writing, there were approximately 44,000 persons employed by the FAA.

The FAA has a significant number of responsibilities in providing oversight and regulation of the various facets of the aviation industry. In consequence, the FAA also possesses a rather high degree of authority in carrying out its duties. As noted on their website, the mission and vision of the FAA is to provide the safest, most efficient aerospace system in the world and to improve the safety and efficiency of flight. Safety is at the forefront of FAA activities; in fact, it is their primary responsibility. Virtually all the major functions of the FAA are rooted in safety, and this is the catalyst that drives most of its actions, including creation and revision of the Federal Aviation Regulations (FARs).

FAA Activities Related to Safety

The FAA has a number of primary functions and roles in support of promoting safety and efficiency in aviation. This section contains a brief overview of some of the primary functions of the FAA and their relation to safety in aviation.

Safety Regulation of Civil Aviation

The FAA is responsible for regulating all facets of civil aviation. It develops minimum standards and rules (the FARs) governing aircraft maintenance, operation, and manufacturing; air carrier certification and operation; airport operations, certification, and safety; and certification of airmen, dispatchers, maintenance technicians, air traffic controllers, and other specific positions requiring

FAA certification. A significant portion of the regulatory function of the FAA is **enforcement**, the practice of ensuring at least minimum compliance with FARs. As a function of enforcement, specific actions may be taken as appropriate against individual persons or organizations that have violated the FARs. Enforcement actions may take several different forms, depending on the type and degree of severity of the violations. These actions include **certificate actions, administrative actions**, and **civil penalties**. Each of these actions is discussed briefly in the next sections.

Certificate Actions Certificate actions are enforcement actions that may be taken against the certificate(s) of a person or organization. This includes type certificates, manufacturing production certificates, aircraft airworthiness certificates, airman certificates, air carrier operating certificates, air navigation facility certificates, and air agency certificates. FAR Sections 13.19(a) and 13.19(b) list the following regarding certificate actions:

> (a) Under section 609 of the Federal Aviation Act of 1958 (49 U.S.C. 1429), the Administrator may re-inspect any civil aircraft, aircraft engine, propeller, appliance, air navigation facility, or air agency, and may re-examine any civil airman. Under section 501(e) of the FA Act, any Certificate of Aircraft Registration may be suspended or revoked by the Administrator for any cause that renders the aircraft ineligible for registration.
>
> (b) If, as a result of such a re-inspection, re-examination, or other investigation made by the Administrator under Section 609 of the FA Act, the Administrator determines that the public interest and safety in air commerce requires it, the Administrator may issue an order amending, suspending, or revoking all or part of any type certificate, production certificate, airworthiness certificate, airman certificate, air carrier operating certificate, air navigation facility certificate, or air agency certificate.

Source: U.S. Department of Transportation/FAA

In general, FAA certificate actions are considered very serious, as they may inhibit the ability of a person or organization to carry out their aviation functions normally until the situation that caused the action to occur is resolved.

Administrative Actions According to FAA Advisory Circular 12-66B, administrative actions are a means of addressing violations, or alleged violations, that do not warrant the use of enforcement sanctions. More simply, administrative actions may be issued to a person or organization that either violated or allegedly violated FARs. The two basic types of FAA administrative actions are a *warning notice* and a *letter of correction.*

Civil Penalties As implied by the name, civil penalties are enforcement actions in the form of fines that may be issued by the FAA against a person or organization found to be in specific violation of FARs. Civil penalties vary in amount based on the severity of the violation. Some of these may be very large monetary fines assessed against an organization. Shortly before the time of this writing, Southwest Airlines agreed to pay a civil penalty of $7.5 million for operating some of their aircraft without required fuselage inspections. This represents the largest FAA-imposed fine paid by an airline in U.S. aviation history.

Airspace and Air Traffic Management

Part of the charge of the FAA is the continual safe and efficient management of navigable airspace in the United States. This includes the development and modification of air traffic regulations; airspace use; air traffic control system development and operation; provision and maintenance

In-Focus: Federal Aviation Administration

Dr. Don Arendt is the SMS program manager for the Federal Aviation Administration (FAA). His office is located in Washington, DC. He is considered to be a leading expert in the development of the framework for Safety Management Systems. In this brief excerpt from an interview conducted in late fall 2010, Dr. Arendt discusses his views on the role of the FAA as a regulatory agency in aviation.

It's easy to think of any regulatory agency from the basic standpoint of administrative law. We're all chartered by legislation, by the Congress, and the statutory system to provide regulations, basically, controls, controls on behavior, legal constraints, and provide implementation and enforcement strategies. But really the one overarching word I would use to describe the role of the FAA is leadership. I think that a regulatory agency shouldn't be just an organization that sits beside its industry and controls it from the side or keeps it from dancing over the precipice of disaster. I think it's really a leadership role; we should really be in front of the organization identifying what's important and imparting that to the industry. Really only then exercising control of the industry. The industry should really lead itself in terms of its business ventures; it should embrace the things that it would put in as controls in terms of part of their responsibility of safety. But I have to say the crux of the regulatory agency's real role is leadership, whether it's by force of law, audit, regulations, and legislation or if it's through collaboration through mutual effort in the safety area.

As far as the FAA's leadership role in Safety Management Systems, we need to supply goals, objectives, structure, mentorship, if you will. I think we also need to lead by example. If we stress that if senior and middle management of industry needs to be involved in safety efforts, then that means our own senior and middle managers and practitioners need to also be involved in safety efforts. After all, for the industry, it's certainly one of your major responsibilities. But frankly, the industry's major responsibility is providing a useful service to the public. It's a given that you need to be able to provide that safely. For a regulatory agency, that's our full time job. So there really isn't any reason that the management of regulators, the leaders in safety, shouldn't be fully embedded and immersed in safety. Again, there's that leadership role in structuring that mentorship in providing of goals and objectives for safety management. I see this as one of the main roles of the FAA as a regulator of aviation.

Source: U.S. Department of Transportation/FAA

of a variety of navigation facilities and aids for both civilian and military aircraft; and research, development, and management of the National Airspace System. The FAA operates an extensive network of air traffic control towers, air traffic control centers, and flight service stations (FSS). All of these functions are designed to make the airspace system in the United States as safe as possible, while also attempting to achieve efficiency.

Air Navigation Facilities

The FAA provides various aids for air navigation purposes across the country. These aids include of a number of electronic systems such as computer systems, non-precision and precision approach systems, data and voice communications equipment, radar, Global Positioning Systems, VORs, and other systems that provide navigational information to flight crews. The FAA operates and conducts maintenance on these systems and their support facilities to enhance aviation safety.

International Promotion of Civil Aviation

The FAA has long engaged in the practice of promoting civil aviation internationally, including promotion of aviation safety and encouragement of civil aviation. This is part of the reason the FAA has historically been viewed as serving in dual roles: aviation promotion as well as regulation.

Toward the goal of fostering international cooperation and standardization in aviation, the FAA negotiates and establishes bilateral aviation agreements with other nations; actively participates in international aviation conferences; exchanges all manner of aeronautical information with other countries; conducts certification of foreign maintenance organizations, maintenance technicians, and airmen; and a number of other activities in cooperation with the governments of participating nations. For international aviation to be as safe and efficient as possible, it is important to establish standards recognized around the world to ensure minimal differences among countries.

Research and Development

The FAA performs a great deal of research and development (R&D) across a variety of areas including air traffic control, aeromedical research, human factors, aircraft systems, air navigation, and other areas. Performing relevant research helps to ensure that new processes and appropriate technologies are in place to foster safety in aviation.

Inspections

As a regular feature of their work as aviation regulator, FAA personnel conduct a variety of inspections on a regular basis. The FAA performs inspections of air carriers, airports, maintenance repair organizations, fixed-base operators (FBOs), aviation manufacturers, commercial and general aviation aircraft, and a number of other types of aviation organizations and functions. One of the purposes of inspections is to ensure minimal compliance with the FARs. Inspections sometimes reveal hazardous situations or practices which may need to be corrected in order to prevent an undesired event such as an accident from occurring. Thus, inspection may play an important role in the safety mission of the FAA.

National Transportation Safety Board

The **National Transportation Safety Board (NTSB)** is an independent agency of the federal government responsible for investigating all civil aviation accidents in the United States, as well as significant railroad, highway, marine, and pipeline accidents. The NTSB is one of the world's leading accident investigation agencies. At the head of the NTSB is a five-member board. These individuals are nominated by the president for five-year terms and are confirmed by the U.S. Senate. Two of the members are chosen by the president to serve as the board chair and vice chair for two-year terms. The NTSB chair must also be confirmed by the Senate. The board presides over all the activities of the NTSB, providing oversight and leadership.

The NTSB manages an enormous database containing information and details about civil aviation accidents. It also conducts safety studies, evaluates other government agencies' transportation accident prevention programs, and reviews appeals of enforcement actions involving airman certificates as well as civil penalties imposed by the FAA. The work of the NTSB is focused on prevention of future accidents through determining the probable cause of accidents and issuing safety recommendations based on factual information gleaned from investigation of these events. The NTSB serves an extremely important function in working to make aviation safer.

Figure 2–1. The NTSB responds to the US Airways passenger jet crash into the Hudson River in 2009.

The NTSB is headquartered in Washington, D.C., and has regional offices in Chicago, IL; Dallas–Fort Worth, TX; Los Angeles, CA; Miami, FL; Seattle, WA; Parsippany, NJ; Anchorage, AK; Atlanta, GA; Denver, CO; and Ashburn, VA. In 2009, the NTSB had a total of 402 employees, 299 working out of NTSB headquarters in Washington, D.C., and 103 in their various regional offices. The NTSB features a number of transportation safety divisions, including aviation, maritime, highway, railroad, and pipeline/hazardous materials. Aviation is the largest division, conducting the majority of NTSB accident investigations and related activities.

Safety Recommendations

The most important aspect of NTSB activities is its **safety recommendations**. As implied by the name, safety recommendations are created and issued by the NTSB based on factual information gathered during the course of an investigation; they may address problems or areas of concern that do not pertain directly to what is determined to be the cause of an accident. Safety recommendations may be issued at any time during or after an investigation, and may even be issued very early in the process if something is discovered that may pose an immediate safety concern or hazard. Such safety concerns and areas of hazard must be addressed immediately after they are discovered, and therefore the NTSB often issues safety recommendations even before investigations are completed. The issuance of safety recommendations is a core principle of NTSB's mandate. These safety recommendations may have significant impacts on how aircraft are designed and manufactured and how air carriers operate their fleets, or they may address the role that FAA oversight (or lack thereof) and regulation may have played in an accident. Because U.S.-manufactured aircraft are popular and are operated around the world in a large number of countries, safety recommendations from the NTSB also carry international ramifications for safety in aviation. This factor is especially important given the increasing degree of globalization that is underway.

Safety recommendations, like virtually all NTSB activities, are designed to be preventive. Recommendations are issued to federal, state, and local government agencies, specific aviation industry companies (such as airlines and manufacturers), and other organizations having a vested interest or possible connection to the accident. These recommendations are the focal point of the NTSB's efforts to improve safety in the nation's transportation system. According to information on their website, the NTSB (at the time of this writing) has issued more than 12,000 safety recommendations in all the primary transportation modes to more than 2,200 organizations, and over 82 percent of these recommendations have been adopted by decision makers possessing the ability to bring about change. However, it is important to note that NTSB safety recommendations are not mandatory or binding on the organizations to which they are issued. The FAA and other organizations are required to acknowledge receipt of safety recommendations; however, the recommendations do not have to be adopted.

NTSB Aviation Investigations

Each year, the NTSB investigates a significant number of aviation accidents in the United States, including all air carrier, mid-air collisions, air taxi, and general aviation accidents resulting in fatalities. To date, the NTSB has investigated more than 124,000 individual aviation accidents.

The NTSB also participates in the investigation of accidents occurring in other nations that involve U.S.-based air carriers, a civil aircraft of U.S. registry, equipment manufactured or designed by U.S.-based companies (e.g., Boeing), or U.S. obligations under agreements with the International Civil Aviation Organization (ICAO). The NTSB also conducts safety studies on aviation issues from a systems perspective in order to go beyond individual accident investigations to determine problems and causal factors in accidents.

Because the exact time and location of an accident cannot be predicted, NTSB investigators are on call 24 hours a day, every day of the year. In this way, NTSB personnel are maintained in a continual state of readiness for expeditious dispatch to an accident site to begin the all-important work of investigation. As needed, these investigators travel throughout the country and even to different parts of the world in support of investigation activities.

The NTSB works toward the determination of the **probable cause** of accidents. As the term implies, probable cause considers all factors which are believed to have contributed to the occurrence of an accident under investigation. For aviation (as noted on their website), the NTSB determines probable cause for all U.S. civil aviation accidents and some public-use (government-operated) aircraft accidents, certain events involving the release of hazardous materials, and recurring accidents that may be systemic in nature or in potential impact.

Go-Team

An NTSB **go-team** consists of a group of individuals dispatched to an accident location to conduct the on-site investigation. An aviation go-team typically includes an investigator-in-charge (IIC) and a number of staff experts from up to 14 different specialties (discussed in the following section). The size of a go-team may vary from three or four to more than a dozen experts. Each staff expert leads a group of specialists from other government agencies and the aviation industry. NTSB staff experts are designated as **group chairmen** over their individual areas of responsibility in the entire investigation. These persons are experts in areas such as witness interviewing, aircraft systems and structures, aircraft maintenance, operations, meteorology, human factors, air traffic control, and other specific areas pertinent to the accident. The purpose of the go-team is to investigate an accident as expeditiously as possible at the scene, using a team of experienced individuals possessing specific expertise to determine the causal factors in the accident.

If available, and depending on the situation, go-teams will travel to an accident site on government aircraft, but sometimes travel via commercial airlines. Go-team members are on call every day of the year on a rotational basis. When on call, these individuals must be accessible 24 hours a day via telephone or pager. All go-team members carry flashlights, tape recorders, cameras, and extra tape and film used to record information at the accident site. They also use tools specific to their needs during the investigative process. A go-team is dispatched only to accidents that occur in the U.S., its territories, or international waters. In other places the investigator is the government in whose nation or territory the accident occurred. When a U.S. air carrier or aircraft of U.S. manufacture is involved, the investigating government will often be assisted by a specific representative of the NTSB, typically an investigator-in-charge. Although the time spent at a given site varies depending on the nature of the accident and other factors, a go-team will typically conclude its on-site work in around from 7 to 10 days, although the team will remain at a site as long as necessary to complete this phase of the investigation. Usually one of the five board members will accompany a go-team to an accident site to serve as a spokesperson and to assist in the investigation.

Investigator-in-Charge

The investigator-in-charge is an NTSB employee who is the primary authority at an accident site. The IIC is a senior air safety investigator possessing years of aviation industry and NTSB experience. This person is in charge of the accident site and manages all phases of major accident investigations. The responsibility of the IIC includes the supervision and control of the go-team and aviation industry representatives from the FAA, the airline(s) involved in the accident, airframe and engine manufacturer, pilot and flight attendant labor organizations, and others. The IIC also prepares accident reports and presents these at the public accident hearings. In addition, an IIC will develop safety recommendation proposals in conjunction with other team members, and manages NTSB staff and representatives of the investigating team in preparation for public hearings. The IIC also testifies before Congress on the findings and issues related to major aviation accidents.

Although the IIC is in charge at the accident site, if it been determined that an aviation accident is the result of a criminal or terrorist act, the Federal Bureau of Investigation (FBI) assumes authority over the investigation, and the NTSB provides support upon request.

NTSB Investigation Specialties

On a go-team, each staff expert is given responsibility over specific areas of the investigative process. Under direction of the IIC, each staff expert leads a **working group** in their respective area of expertise. As found on the NTSB website, the areas pertaining to aviation are classified as follows.

Human Factors/Performance: This involves the careful analysis of crew performance and all issues that were in place before the accident related to human factors such as fatigue, errors, use of medications, alcohol or drugs, crew members' medical history, crew training, crew workload, work environment, and design of equipment.

Operations: This area is concerned with the history of the flight involved in the accident and crew members' duties and related situations for as many days prior to the crash as is considered to be relevant.

Structures: This area includes the analysis and documentation of the wreckage of the aircraft and accident scene, and involves the calculation of impact angles to determine the aircraft's course and attitude before the accident.

Powerplants: Analysis of the aircraft engine(s), components, and propellers (as applicable).

Aircraft Systems: This involves analysis of the aircraft's instruments, flight controls, and hydraulic, electrical, pneumatic, and associated components and systems.

Air Traffic Control: This area includes the analysis and reconstruction of all air traffic services provided to the flight crew, including ATC radar data, instructions, altitude and heading assignments, and transcripts of radio transmissions between controllers and flight crew members.

Weather: This includes analysis of all applicable weather information and data from the National Weather Service and other pertinent sources for a broad geographic area surrounding the accident site.

Survival Factors: This area involves analysis and documentation of accident impact forces, injuries, evacuation, local emergency planning, and the efforts of crash–fire–rescue units.

Flight Data Recorder and Cockpit Voice Recorder

Commercial aircraft and some smaller private aircraft are required by the FAA to be equipped with two "black boxes" that record a great deal of information about a given flight. The term "black box" is a bit misleading, as these two devices are painted a high-visibility orange. These recorders are installed on aircraft to provide data on the events leading to an aircraft accident and to provide investigators with factual information about the aircraft's parameters and flight deck communications. These two devices are briefly discussed below.

The **cockpit voice recorder** (**CVR**) records flight crew conversation, radio transmissions, and other flight deck sounds. The **flight data recorder** (**FDR**) monitors aircraft parameters such as altitude, airspeed, flight control settings and movements, engine rpm, pitch, heading, and other measurements. The older units are analog and use quarter-inch magnetic tape for data recording and storage; newer units feature digital technology and memory chips. Both the CVR and FDR and are usually installed in the tail section of aircraft in order to be more likely to survive impact forces. The CVR and FDR are designed to withstand a significant amount of impact forces and still have usable data that can be retrieved for analysis. In case of an accident occurring over water, the CVR and FDR are equipped with an **underwater locator beacon** (**ULB**). The ULB is an internally installed radio beacon that automatically activates when the recorders are immersed, transmitting from depths up to 14,000 feet to make recovery from water more likely.

The CVR uses a cockpit-area microphone that is normally located on the overhead instrument panel between the captain and first officer positions. Sounds potentially important to the investigation such as conversation, engine noise, stall warning activation, landing gear extension and retraction, and other sounds are recorded so that significant parameters such as system failures, engine rpm, aircraft speed, communication issues, and other factors may be determined by investigators.

The FDR records a large number of functional parameters of the aircraft during the flight up to the time of the accident. Modern FDRs can record over 1,000 in-flight characteristics including throttle settings and changes, flight control settings and manipulation, airspeed, altitude, pitch, flap position, auto-pilot activation/deactivation, and a host of other parameters. The NTSB uses FDR data to generate a computer video re-creation of the accident flight. This is a powerful analytical tool, as it permits investigators to ascertain and observe the latter stages of the flight leading up to the accident.

Figure 2-2. An example of a Cockpit-Voice Recorder (CVR) and a Flight Data Recorder (FDR)

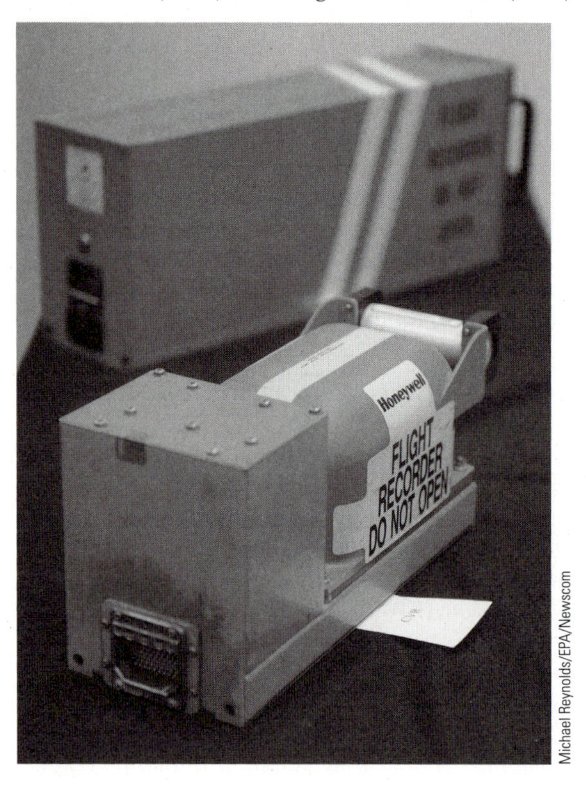

Michael Reynolds/EPA/Newscom

The CVR and FDR have proven to be extremely important tools in the accident investigation process, providing highly specific, factual information about the flight in question that may not have been obtainable otherwise. These devices are a part of the body of evidence used to help determine the probable cause of aviation accidents. As soon as possible after an aviation accident, both recorders are located, removed from the accident site, and transported to NTSB headquarters for analysis. Computers and audio equipment are used to recover the data stored on the recorders and translate it into a usable form.

As part of the analysis of accident data, a CVR committee is convened. Normally this committee will feature representatives from the NTSB, the aircraft operator (airline, etc.), the aircraft manufacturer, the FAA, and a member of the pilot union. The committee carefully reviews the CVR recording and develops a written transcript of audio retrieved from the recorder. In addition, air traffic control recordings are analyzed to determine the local standard time of one or more events during the accident timeline. These times are included in the CVR transcript, which lists the exact time of day (locally) for all events in the transcript. ATC recordings also provide investigators with information on radio transmissions that may have taken place between the crew of the aircraft involved and air traffic controllers. The CVR transcript may be released to the public at the time of an NTSB public hearing. If no public hearing is held, CVR transcripts may be made public when the majority of the factual reports become open to the public.

NTSB and Media Communications during an Investigation

When a major aircraft accident occurs, the NTSB works to engage in communicating some accident information to the media. What normally happens in the event of a major accident is a high degree of speculation by the media and the public as to what may have caused the accident to occur. This practice, although commonplace, is of questionable value because speculation is not always in line with the actual facts of a case. In order to try to minimize this type of response, NTSB policy is to provide factual information on the investigation by issuing press statements, holding press briefings and conferences, and having a board member meet with the press to brief them and answer questions in the days after an accident. To its credit, the NTSB does not engage in speculation about the possible causes of an accident; only confirmed, factual information of the investigation is discussed. The NTSB also has a public affairs officer who maintains contact with the media and communicates pertinent information.

The NTSB Party System

To assist in the investigation process, the NTSB employs the **party system.** The party system is the process by which the NTSB includes representatives from certain organizations that may have a vested interest or expertise in segments of an accident under investigation. At the discretion of the NTSB, these organizations are permitted to participate in the process and related activities as "parties" to the investigation. With the exception of the FAA (by law a party), the NTSB has absolute discretion over which organizations it designates and permits to serve as parties to the investigation. It is important to note that only organizations that will provide expertise to the investigation are designated are parties, and only individuals who provide demonstrated technical and/or specialized expertise are permitted to participate in an investigation as parties. Representatives from labor organizations, airlines, engine and aircraft manufacturers, regulatory agencies, and a number of other types of subject experts may be chosen as parties. Members of the media, attorneys, and members of insurance companies may not participate in the party process due to conflict of interest.

The party system is important to the NTSB and the investigation process as it provides experts in a variety of areas to assist in the investigation and the development of the factual reports of an accident. A significant amount of technical information is typically gathered through the party system, providing information pertinent to the investigation. The NTSB supervises and manages all investigation activities, and party members report directly to the appropriate NTSB members. Direct participation of aircraft operators, manufacturers, labor unions, and regulators as parties to NTSB investigations could actually be viewed as a form of crew resource management (CRM), as the party system could be viewed as an example of the NTSB using other available resources to supplement its own in carrying out its investigative work. This process also permits the NTSB to obtain information from the most reliable sources. An example: An airline is an operator of aircraft and understands the intricacies of aircraft operation. The airline, as a party, may provide valuable insights to the NTSB on aircraft operation or other areas where it may provide technical expertise.

As mentioned previously, information and data collected via the party system is used to generate factual reports. The individual investigative groups' chairs create a factual report in their respective areas of the investigation. Each of the parties working within the group is given the opportunity to review the report for clarity, accuracy, and completeness. The factual reports are eventually made available to the public, at the direction of the NTSB.

The Public Hearing

The NTSB may choose to conduct a **public hearing** on a specific major aviation accident. A public hearing is a scheduled public meeting in which the particulars and known factual information pertinent to an accident are discussed in detail. A public hearing is normally held during an investigation of a prominent accident where there is sustained public interest or significant safety issues involved. These meetings permit members of the public to learn about and observe the progress of the investigation to date. In general, public hearings usually take place within six months of an accident. However, in the case of more difficult or complex accidents and/or circumstances, a public hearing may be delayed. To avoid conflicts of interest, members of the media, family members of victims, attorneys, and insurance company representatives are not permitted to participate in public hearings or in any other part of the investigation.

During a public hearing, one of the five board members presides over the meeting. All of the investigators, party system members, and other experts are permitted to participate in the proceedings. A typical public hearing will also feature a number of subpoenaed witnesses and expert testimonies conducted under oath. This expert testimony provides information and viewpoints that may be helpful to the investigation and is used to ensure an accurate, complete, and well-documented factual record of information about an accident. Public hearings serve as a form of public accountability for the NTSB by demonstrating that the investigation is thorough, objective, and fair with respect to industry and all government agencies.

For all public hearings, a **Board of Inquiry** is created, consisting of senior NTSB staff members and chaired by a board member, who presides over the proceedings. This group is aided by a technical panel comprised of NTSB investigators and a number of specialists in areas such as aircraft performance, powerplants, CVR and FDR, aircraft systems and structures, operations, air traffic control, weather, and human factors.

After the conclusion of the public hearing, a written record of the accident investigation and hearing is placed into the NTSB public docket on the event. In addition, investigators work to obtain additional information and conduct further tests identified as important during the public hearing. Once the investigation is completed and all parties to the investigation have been provided the opportunity to review the factual record of the investigation, a technical review meeting of all involved parties is scheduled. The purpose of this meeting is to ensure accuracy and completeness in the investigation and that there is agreement among the parties that all investigative tasks have been adequately performed.

The Final Accident Report

After the public hearing, technical review meeting, and any additional testing and scrutiny of pertinent information are complete, NTSB personnel finalize their analysis of the facts of the investigation. Then the IIC submits an internal draft of a **final report** of the investigation to NTSB group chairs and supervisors for their review and comment. Then the report draft is given to the NTSB Office of Aviation Safety, Office of Research and Engineering, Office of Safety Recommendations, and the NTSB managing director for their input. Additional corrections or revisions to the report may be made at this stage. Then NTSB directors hold a closed meeting to discuss the specifics of the report in detail. The final draft of the report generated by the directors is called the **notation draft** and is presented to the five Safety Board members for their review, input, deliberation, and eventual adoption. This report contains detailed factual information and findings of the accident investigation, an analysis of the findings, a list of recommendations to prevent similar accidents in the future, and a statement of probable cause of the accident.

After the report is reviewed by the board, a public meeting is held in Washington, D.C., in which the notation draft is discussed and deliberated in detail. The IIC and the investigation group chairs present the final report draft before the board. Party investigation representatives are permitted to attend the meeting but may not comment or make presentations. At this meeting, board members may choose to do any of the following:

- Adopt the draft report, as is, as the final accident report
- Require additional investigation or revisions to the draft report
- Adopt the report with revisions discussed in the meeting

As soon as possible after the meeting, the NTSB Office of Public Affairs releases the Safety Board's conclusions, the statement of probable cause, and safety recommendations. The adopted final report on the accident is usually published approximately three weeks after the meeting.

This report is very extensive and may be several hundred pages in length. Copies of the final report are provided to families of accident victims, parties to the investigation, and the public.

NTSB Accident Investigation Conclusion

Here is a very important point to understand: Although the publication of the final report signifies the conclusion of an investigation, technically, NTSB investigations are never closed. Parties involved in an accident investigation can petition the NTSB to reconsider and possibly even revise the findings and probable cause of an accident if one of the following conditions is present: (1) New evidence pertaining to the accident is discovered which could alter the initial findings of the investigation, or (2) a party to the investigation believes that the NTSB made a mistake in the investigation or believes that the findings and conclusions of the investigation are erroneous. Petitions will only be accepted from parties that participated and made submissions during the investigation. Petitions to the NTSB to reconsider evidence may be made at any time after the conclusion of the meeting and the adoption of the final accident report.

Transportation Security Administration

The **Transportation Security Administration (TSA)** is the branch of the federal government charged with providing oversight and regulation of security in the various primary transportation modes. TSA conducts regulation and security activities in aviation, railway, maritime, pipeline, and other transportation modes.

Figure 2–3. TSA agents screen passengers at LAX.

Like most agencies of the federal government, the TSA is headquartered in Washington, D.C.; it employs over 50,000 people with a variety of job functions across the United States. The vast majority of these employees work under the title of **Transportation Security Officers**. These are the TSA agents who work as aviation passenger and baggage security screeners in commercial service airports. At the time of this writing, there were approximately 48,000 TSA security screeners in place at over 450 U.S. airports offering commercial air carrier service. On a typical day, over 2 million people are screened by TSA agents in support of aviation operations.

The Inception of the TSA

Although vulnerability in aviation security has been evident for a long time (based on past attacks against aviation), the creation of the TSA was precipitated by the tragic events of September 11, 2001 (9/11), when the United States was directly attacked by terrorists who were members of various Islamic extremist groups based in the Middle East and working at the behest of Osama bin Laden. With respect to the aviation industry and to the overall psychological impact of 9/11, arguably one of the most shocking aspects of the attacks was that large transport-category aircraft were rather easily hijacked and subsequently used as missiles (in effect) in targeting the World Trade Center (WTC) in New York and the Pentagon near Washington, D.C. The basic sequence of the attacks was as follows: First, at 0846 local time, American Airlines Flight 11 (a Boeing 767) was deliberately flown into the North Tower of the WTC. Next, at 0903, United Airlines Flight 175 (a Boeing 767) was crashed into the South Tower. Then at 0937, American Airlines Flight 77 (a Boeing 757) was flown into the west side of the Pentagon. Finally, United Airlines Flight 93 (a Boeing 757) crashed in a field in Pennsylvania. United 93 is believed to have been on its way to attack either the White House or the U.S. Capitol when some of the passengers fought back against the terrorists in an effort to regain control of the aircraft. All onboard were killed in the crash.

According to information in the 9/11 Commission Report, despite the fact that the federal government, the media, and the public had received warnings that Islamic extremists intended to indiscriminately kill large numbers of Americans, the United States was largely unprepared for such an attack. The events of 9/11 revealed in dramatic fashion some of the weaknesses present in the American aviation security system. The 9/11 hijackers were organized in groups of four or five on each aircraft. They were armed with simple weapons such as box cutters, small knives, and cans of pepper spray or mace. Each terrorist (and their weapons) had passed successfully through the security checkpoints at the airports where the flights had originated that morning. In total, nearly 3,000 persons lost their lives in the events of 9/11; most of these were persons who perished inside the WTC when they were unable to escape from the burning structures, which eventually collapsed. Many first responders such as firefighters and police officers also died while attempting to rescue people trapped inside the WTC. In the end, 9/11 brought irrevocable change to the aviation industry in a number of ways: operationally, economically, and in the area of security.

As a result of 9/11 and its aftermath, the TSA was created on November 19, 2001, through an act of the U.S. Congress. The legislation was named the **Aviation and Transportation Security Act** and was signed into law by President George W. Bush. This act created the TSA, and placed the primary responsibility for transportation security with this new agency. Before the advent of the TSA, responsibility for security in civil aviation resided with the FAA. Thus, the inception of the TSA resulted in a shift in transportation security responsibility in the federal government. Initially, the TSA was placed within the DOT, but was shifted into the **Department of Homeland Security (DHS)** in March 2003. The DHS was created by the passage of the **Homeland Security Act of 2002** by the Congress to coordinate responses and identify terrorism threats against the United States, that is, the "homeland." The mission of the DHS is to prevent terrorist attacks within the United States, reduce the vulnerability of the country to terrorism, minimize the damage of attacks, and assist in the recovery from terrorist attacks. The TSA functions as one of the DHS entities working to prevent attacks against transportation.

TSA Initiatives and Programs in Aviation

As mentioned previously, the TSA was created as a result of 9/11 through the passage of the Aviation and Transportation Security Act to strengthen the security of transportation systems in the United States. Under the provisions of the act, the TSA was given three primary mandates:

1. Responsibility for security in all primary modes of transportation.
2. Recruit, assess, hire, train, and deploy security screeners for 450 commercial service airports in 12 months.
3. Conduct 100 percent screening of all checked baggage for explosives by December 31, 2002.

The stated mission and vision of the TSA is as follows:

> **Mission:** The Transportation Security Administration protects the nation's transportation systems to ensure freedom of movement for people and commerce.
>
> **Vision:** The Transportation Security Administration will continuously set the standard for excellence in transportation security through its people, processes, and technology.

Source: U.S. Department of Homeland Security/Transportation Safety Administration

In working toward the support of its mission and goals in aviation, the TSA has a number of programs and initiatives intended to promote security and make aviation safer. The following sections briefly discuss some of the TSA's principal security activities.

Federal Air Marshals

Federal Air Marshals are the primary law enforcement group within the TSA. They deploy on flights around the world and in the United States. These individuals are highly trained law enforcement professionals who fly aboard certain commercial flights. While on duty, they are armed and are trained to blend in with other passengers so as not to bring any attention to themselves that could permit others to identify them as law enforcement officials. The point is that they should not be identified as Air Marshals unless it becomes necessary in the execution of their duties.

Air Marshals must be able to function independently without the benefit of backup. They tend to be highly skilled in the use of handguns and must maintain firearms proficiency. They are trained in investigative techniques, criminal/terrorist behavior recognition, aircraft-specific tactics, and close-proximity self-defense techniques. Collectively, Air Marshals fly millions of miles each year on a large number of flights. After 9/11, there was resurgence in the Air Marshal program as the federal government actively recruited more qualified individuals into the program.

TSA Security Inspectors

The TSA employs a number of **security inspectors.** In support of aviation security, these individuals frequent airports, passenger terminals, ramp areas, and so on, providing enforcement of regulations and guidelines. At the time of this writing, the TSA had over 100 inspectors working out of 18 field offices throughout the country. They frequently work with personnel from airlines, airports, and other groups on security-related issues. They also conduct frequent inspections of airports and terminals, looking for security threats, gaps, and violations.

Federal Flight Deck Officer Program

The Federal Flight Deck Officer (FFDO) program authorizes select and qualified aircraft flight crew members to use firearms and other self-defense measures to defend their aircraft, fellow crew members, and passengers against acts of criminal violence and against attempts to gain control of an aircraft by use of terror or force. These crew members may include the captain, first officer, flight engineer, navigator, or other authorized flight crew member aboard the aircraft. Pilots of cargo aircraft are also eligible to participate in the program. The program is quite selective, and not all persons who apply are accepted into the FFDO program. It is important to note that the main purpose of the FFDO program is to protect the aircraft through the protection of the flight deck. The weapons carried by FFDO qualified flight crew members are not permitted to be removed from the flight deck unless they are secured by a qualified crew member.

Federal Flight Deck Officers (FFDOs) are trained by the Federal Air Marshal Service in firearms, use of force, legal issues, defensive tactics, survival psychology, and program standard operating procedures (SOPs). The FFDO program is strictly voluntary, and flight crew members in the program are not eligible for compensation by the government for FFDO duties.

Crew Member Self-Defense

The **Crew Member Self-Defense Training** (CMSDT) program provides basic self-defense training to airline crew members who volunteer to participate. A provision of the FAA Reauthorization Act of 2003 is that scheduled air carriers providing passenger service are required to conduct basic security training for flight and cabin crew members (pilots and flight attendants) to prepare them for possible threats that may occur on an aircraft. This act also requires the TSA to create an advanced self-defense course to train flight and cabin crew members to defend against potential attackers. Like the FFDO program, the CMSDT program is conducted by the Federal Air Marshal Service and is made available at no cost to participating crew members.

Canine Explosives Detection

The TSA also uses dogs in its transportation security efforts. The **National Explosives Detection Canine Team** utilizes highly trained law enforcement personnel and dogs to locate explosives and other dangerous items that may pose a threat to aviation interests. In addition, these units of dogs and handlers are used to rule out the presence of hazards in airport buildings, baggage, cargo, packages, and vehicles. The dogs may also be used to detect or rule out dangerous items on aircraft. At present, hundreds or these dog/handler teams are present throughout the nation and are called upon for duty as necessary.

Participating law enforcement officers from various parts of the country are paired up with a dog at the Canine Team training course, conducted at Lackland Air Force Base in San Antonio, Texas. The dogs in the program are specifically bred for detection work by TSA's "puppy program," also located at Lackland. The teams complete an intensive ten-week training course to learn to locate and identify a significant variety of dangerous materials. The training includes search techniques for aircraft, baggage, cargo, vehicles, and buildings. The teams are also trained in procedures for the dog to "alert" the handler when dangerous materials are detected.

Air Cargo Security

The TSA is also responsible for the security of cargo carried aboard aircraft, focusing especially on cargo carried aboard passenger aircraft. A key congressional mandate of the 9/11 Act (signed into law in August 2003) was that air carriers screen at least 50 percent of cargo flown on passenger aircraft by February 3, 2009. The TSA issued security directives to all air carriers informing them of the need to comply with this mandate. These security directives went into effect on February 1, 2009. Each air carrier is required to submit to the TSA data each month on how much cargo it has screened. According to information on its website, the TSA has met the cargo screening mandate of the 9/11 Act, and at present 50 percent of cargo flown on passenger flights is screened.

Figure 2–4. TSA canine checks baggage for explosives.

Although these cargo screening initiatives are good, the numbers show that half of air cargo is not subject to screening at present. However, the TSA states that 100 percent of air cargo present on 96 percent of the flights originating in the United States is screened and that 85 percent of passengers flying daily from U.S. airports are on aircraft on which all air cargo has been screened. In addition, the TSA requires all airlines operating narrow-body (single aisle) passenger aircraft from U.S. airports to screen 100 percent of cargo transported on them. This has been in effect since October 2008. The TSA also conducts **strikes**, which are surprise security inspections of air cargo at selected air carrier facilities and airports. Also in 2008, the TSA eliminated all exemptions to screening of air cargo for the first time and also increased the amount of air cargo subject to mandatory screening.

Secure Flight

Secure Flight is a behind-the-scenes TSA program intended to enhance the security of domestic and international commercial air travel through the use of **watch lists**, also known as the **No Fly List**. A watch list is a large database containing the names of known terrorists, persons suspected of terrorist activity, and persons having or suspected of having connections to terrorist groups. Watch lists are highly controversial due to the fact that not everyone appearing on a watch list actually has ties to terrorism; some critics view the use of watch lists as a violation of civil liberties.

Under the Secure Flight program, passengers will be required to provide the following Secure Flight Passenger Data (SFPD) to each airline when making travel reservations:

- The passenger's name as it appears on a government-issued I.D. (driver's license, etc.)
- Date of birth

- Gender
- Redress Number (if available). This is basically a case number assigned to a person seeking resolution of problem experienced during the screening at an airport or when crossing a U.S. border. Problems might involve delayed or denied airline boarding, denied or delayed entry into or from the United States at a port of entry or border, or being regularly subjected to secondary security screening.

Next, the airline will transmit the passenger information to Secure Flight, where it will be compared to the watch list. This is to prevent individuals on the list from boarding an aircraft and also to identify individuals on the Selectee List to receive additional security screening. After matching passenger information against watch lists, Secure Flight then transmits the results back to the airlines. Dating back to August 15, 2009, for domestic flights and October 31, 2009, for international flights, airline employees must request and obtain passengers' full name, date of birth, gender, and Redress Number (if available). By collecting additional data on passengers, the TSA hopes to improve travel for all airline passengers, including any person who may have been have been misidentified previously.

The Occupational Safety and Health Administration (OSHA)

The Occupational Safety and Health Act was signed into law by President Richard M. Nixon on December 29, 1970. Under this act, the Occupational Safety and Health Administration (OSHA) was created as an agency of the U.S. Department of Labor, with oversight given to the Deputy Assistant Secretary of Labor. OSHA regulations cover most private-sector workplaces across the country, but states are also allowed to develop their own plans provided that they are at least as stringent as the federal regulations and cover the public sector as well. Twenty-two states and U.S. territories have developed and operate plans covering both the public and private sectors, and four operate public-only plans (federal OSHA still governs the private sector in these states).

With each change of administration, there is also some level of change within OSHA. In recent years, OSHA claims to have changed its fundamental operating paradigm from one of command and control to one which provides employers a real choice between a partnership and the traditional enforcement relationship. According to the OSHA website describing the "New OSHA," there are six principles for protecting America's workforce:

- OSHA's purpose is to save lives, prevent workplace injuries and illnesses, and protect the health of all of America's workers. This includes efforts to protect small groups of workers who are unorganized but who are particularly vulnerable or who face special hazards.
- Whenever possible, OSHA will seek and expect implementation of hazard control strategies based on primary prevention, that is, strategies focused on fixing the underlying causes of problems or reducing hazardous exposures at their source.
- OSHA will initiate public–private partnerships to identify and encourage the spread of industry best practices to solve national problems.
- Employer commitment and meaningful employee participation and involvement in safety and health are key ingredients in effective programs.
- All safety and health services, resources, rules, and information must be readily accessible and understandable to employees, employers, and OSHA's staff.
- OSHA intends to be a performance-oriented, data-driven organization that places the highest premium on real results rather than activities and processes. OSHA's programs must be judged according to their success at eliminating hazards and reducing injuries and illness.

Source: U.S. Department of Labor/OSHA

Title 29 CFR Part 1910 contains the majority of the regulations and requirements that involve OSHA within the aviation industry. These are the standards for general industry. Some of the most frequently cited standards, such as Hazard Communication, Powered Industrial Trucks, Electrical Wiring, Machine Guarding, and Fall Protection, are found in most aviation operations. In fact, these are consistently the most frequently cited problems within the aviation industry (North American Industrial Classification System [NAICS] code 481).

An OSHA inspection of a facility is either programmed or unprogrammed, or a follow-up. An unprogrammed inspection can result from imminent danger, fatality/catastrophe, complaints, or referrals. A programmed facility inspection will come from a facility qualifying for the OSHA **Site Specific Target (SST)** list or one of the National, Regional, or Local Emphasis Programs (NEP/REP/LEP). These programs target high-risk hazards and industries. OSHA currently has seven NEPs, focusing on amputations, lead, crystalline silica, shipbreaking, trenching/excavation, petroleum refinery process safety management (PSM), and combustible dust. There are approximately 140 REPs and LEPs.

The SST list (main programmed inspection plan) consists of facilities across the Unites States and its territories identified in the prior year's Bureau of Labor Statistics (BLS) survey as having very high incident rates. Specifically, for 2009, the SST Primary Inspection List includes facilities (non-manufacturing) from the 2008 BLS survey with a days away, restriction, or transfer (DART) rate at or above 15.0, or a days away from work injury or illness (DAFWII) rate at or above 8.0. About 500 sites are involved.

In addition, establishments that received notice but did not participate in the prior year's BLS survey will be included on the Primary Inspection List. Prior to 2009, OSHA also included establishments in high-rate industries (including air transportation) with lower individual corporate incident rates. Now, OSHA will implement a **Recordkeeping National Emphasis Program (NEP)** in working toward this objective. The new program will be used to verify the accuracy of the OSHA Form 300 (also known as the First Report of Injury/Illness) for selected low-rate establishments in high-rate industries.

The SST Secondary List may be used by an OSHA area office once the Primary List has been completed. For non-manufacturing, this list consists of those establishments with a DART rate at or above 6.0, but less than 15.0, or a DAFWII of 4.0 or greater, but less than 13.0. Injury and illness rates are calculated as the number of cases (DART, DAFWII, etc.) divided by the number of hours worked by all employees of the operation or facility, multiplied by 200,000 (equivalent of 100 full-time employees working a 2,000-hour year). More information about the SST program can be found in the current OSHA directive on the OSHA website, www.osha.gov.

OSHA and BLS define an incident rate as the number of injuries and/or illnesses per 100 full-time workers. An incidence rate for injuries can be calculated from the following formula:

OSHA Injury Equation

$$\frac{\textit{No. injuries and illnesses} \times 200{,}000}{\textit{Hours worked}} = \textit{Incident rate}$$

For example, ABC Company has experienced 23 DART (days away, restricted duty, or transfer) injuries over a given period (works for any given period). The company's actual hours worked during the period was 851,875. Their DART rate would be calculated as follows:

$$\frac{23\ \textit{injuries/illnesses} \times 200{,}000}{851{,}875\ \textit{hours}} = 5.4$$

The incident case (IC) can be the overall incidents, OSHA recordable, DART (days away, restriction, transfer), or LWD (lost workday) cases. The 200,000 hours in the formula represent the equivalent of 100 full-time employees and provide the standard base for incident rates.

OSHA typically conducts about 38,000 inspections per year, with an additional 58,000 conducted by state-run agencies in states with such programs. Penalties range from $0 to $70,000 per violation. Repeat and willful violations typically carry the highest monetary penalties.

Although some would argue otherwise, OSHA's intent is not to penalize employers, but rather to ensure the safety and well-being of employees. The administration realizes that many organizations have limited or no resources to implement or even understand how to achieve compliance. For this reason, OSHA provides compliance assistance in variety of ways. General information is provided on the OSHA website at www.osha.gov. Included on the website are the regulations and, for many regulations, interpretation letters. There are also web-based training tools called "eTools," posters, guides, fact sheets, and other publications such as OSHA's e-newsletter, *Quick Takes*. There are many print publications that can be ordered online or by phone and a 24-hour assistance number that can be called by employers or employees for workplace safety and health information. A Spanish version of OSHA's website is also available at www.osha.gov.

Assistance can also be found through one or more of OSHA's cooperative programs, by which the employer or labor group can work cooperatively with the agency in a prevention capacity. One of these is the Alliance Program, consisting of a group or groups committed to safety and health. These could include businesses, trade or professional organizations such as the National Safety Council (NSC) or American Society of Safety Engineers (ASSE), unions, and educational institutions working with OSHA. OSHA and the group(s) pool their resources and share information with employers, signing formal agreements with goals to address safety challenges. One of the most prevalent of these has been the partnership between 13 airlines and OSHA in an Ergonomics Alliance signed by each partner in 2002 to address baggage handling. Similar to the Alliance Program, the OSHA Strategic Partnership Program (OSPP) works with the same types of groups, but originates at a specific worksite.

OSHA's On-Site Consultation Program provides free and confidential assistance to small and medium-sized organizations in all states, with priority given to high-hazard worksites. These services are provided completely separate from enforcement and will not result in penalties or citations. This program helps an organization to identify and correct hazards at the worksite. It also recognizes small employers that operate exemplary safety and health management systems through the Safety and Health Achievement Recognition Program (SHARP). Active SHARP sites are exempt from OSHA's programmed inspections.

The OSHA Voluntary Protection Program (VPP) establishes a cooperative relationship between OSHA, management, and labor at worksites that have implemented comprehensive health and safety management systems exceeding minimum federal and state standards. The worksite is recognized at different levels of achievement upon successful application and an onsite evaluation by a team of OSHA safety and health experts. The different levels or programs in which the site may qualify are Star, Merit, and Star Demonstration. A Star site has comprehensive, successful safety and health management systems and has achieved injury and illness rates at or below the national average for its industry. These sites demonstrate their ability to control workplace hazards and are reevaluated every three to five years, with incident rates reviewed annually. A Merit site is a step down from Star. It is a site that has good safety and health management systems, but needs some improvement to become Star quality. These sites show potential and commitment to achieve Star and must do so within three years unless a second term is approved by the Assistant Secretary of

Labor. Onsite evaluations occur every 18–24 months. A Star Demonstration site is a worksite that is promising, but still needs considerable work and demonstrated success to reach Merit or Star.

Details of many of the OSHA programs and regulations that apply to the aviation industry are covered in Chapter 7, "Ground Safety," which will also demonstrate how many of these programs and regulations apply within different types of aviation organizations as well as within the different departments of larger organizations (major, regional, and cargo airlines).

The Environmental Protection Agency (EPA)

The Environmental Protection Agency (EPA) officially opened on December 2, 1970, in Washington, D.C., signed into legislation by Richard Nixon. Earlier that year, on New Year's Day, Nixon had signed a bill known as the National Environmental Policy Act (NEPA). NEPA was later described by Wisconsin Senator Gaylord Nelson as "the most important piece of legislation in our history." The result of the generation of NEPA and the EPA was an explosion of environmental activity in the press, by activists, and involving many affected cities and industries. What was extremely popular for some was abhorred by others. Also during this year, the first Earth Day was celebrated, with more than 20 million Americans coming out for peaceful demonstrations in support of environmental reform.

Today's EPA has about 17,000 employees, about half of whom are engineers. Their mission is to protect human health and the environment. Their duty is to conduct research, provide education to the public, and enforce environmental regulation. There are three types of enforcement actions the EPA may use: civil administrative actions, civil judicial actions, and criminal actions. Civil administration actions do not involve the court system, but may be informal, conducted via phone call or letter for minor violations (typically recordkeeping in nature), or formal, which could include immediate orders of compliance that could carry a penalty of up to $27,500 per day for noncompliance or suspension of the facility's permit. Civil judicial actions involve the courts and include penalties equal to that of administrative actions. A criminal action may be issued based on the severity of the violation. Seven criminal acts have been identified and are subject to criminal action and carry penalties. Six of the seven criminal acts impose a penalty of up to $50,000 per day and up to five years in jail. They are:

- Transporting waste to a non-permitted facility.
- Treating, storing, or disposing of waste without a permit or in violation of a permit condition or interim status standard.
- Omitting significant information from or making false statement in a label, manifest, report, permit, or interim status standard.
- Generating, storing, treating, or disposing of hazardous waste without complying with the Resource Conservation and Recovery Act's (RCRA's) recordkeeping and reporting requirements.
- Transporting waste without a manifest.
- Exporting a waste without the consent of the receiving country.
- Knowing that the transportation, treatment, storage, disposal, or export of any hazardous waste can place a person in imminent danger of death or serious bodily injury.

This last criminal act can result in a penalty of up to $250,000 or 15 years in prison for an individual, or a $1 million fine for a corporation.

As with many government agencies, however, the EPA offers an abundance of information and provides resources for small businesses and grants for environmental education, emissions reduction projects, and other "green" programs. It also provides for many partnership programs. One of the most well-known of these is perhaps the Energy Star program, which partners the EPA with thousands of organizations across the country to increase sales of energy-efficient products and provides tax credits or rebates for many businesses and homeowners.

The following sections provide an overview for some of the regulatory acts and standards that are applicable to many aviation organizations.

The Clean Air Act (CAA)

The **Clean Air Act** defines the EPA's responsibilities for protecting and improving the nation's air quality. Major changes to the law were included as the Clean Air Act Amendments of 1990 enacted by Congress. The Clean Air Act was incorporated into the U.S. Code as Title 42 (Public Health), Chapter 85. Under the Clean Air Act, the EPA sets limits on certain air pollutants, including limits on how much can be in the air anywhere in the United States. The CAA gives the EPA authority to limit emissions of air pollutants coming from sources such as chemical plants, utilities, and steel mills. Individual states must provide air pollution laws at least as stringent as those set by the EPA. For larger organizations, maintenance requirements and limitations may sometimes make it difficult to stay within the limits imposed by the CAA. For those entities, it is possible to purchase credits from other organizations that have them available.

Clean Water Act

The purpose of the Federal Water Pollution Control Act, otherwise known as the **Clean Water Act**, is to restore and maintain the chemical, physical, and biological properties of the nation's waters. The National Pollutant Discharge Elimination System (NPDES) controls the direct discharge into navigable waters. **Direct discharges**, or "point source" discharges, are those from sources such as pipes and sewers. This is most applicable in aviation within the aircraft maintenance function. NPDES permits issued by a state environmental agency or the EPA contain industry-specific limits and establish pollutant monitoring and reporting requirements. (The EPA has authorized 40 states to administer the NPDES program.) A facility that intends to discharge into the nation's waters must obtain a permit before initiating the discharge. The permit applicant must provide quantitative analytical data identifying the types of pollutants present in the facility's effluent (liquid waste stream). The permit will then provide conditions and effluent limitations under which a facility may make a discharge.

Storm water discharges or discharges from building or surface runoff are also regulated by the act. As part of storm water permitting, facilities are often required to implement pollution prevention plans. **Storm water pollution prevention plans (SWPPPs)** should identify potential sources of pollution that may reasonably be expected to affect the quality of storm water discharges associated with industrial activity at the facility. The plan should also describe and ensure the practices and procedures implemented to reduce the pollutants in storm water discharges. A term often used as a rule of thumb for the facility is "only rain down the drain."

Waste Generation

The **Resource Conservation and Recovery Act (RCRA)** of 1976 governs the regulation of solid and hazardous waste. Under the act, all organizations and facilities that generate, transport, treat, store, or dispose of hazardous waste are required to provide information about their activities to state environmental agencies, which is then transmitted to EPA. Generators of hazardous wastes are categorized into three classes: Conditionally Exempt Small Quantity Generators, Small Quantity Generators, or Large Quantity Generators.

Conditionally Exempt Small Quantity Generators (CESQG) include facilities that produce less than 100 kg of hazardous waste, or less than 1 kg of acutely hazardous waste, per calendar month. A CESQG may only accumulate less than 1,000 kg of hazardous waste, 1 kg of acutely hazardous waste, or 100 kg of spill residue from acutely hazardous waste at any one time. **Small Quantity Generators** (SQG) include facilities that generate between 100 kg and 1,000 kg of hazardous waste per calendar month. **Large Quantity Generators** (LQG) include facilities that generate more than 1,000 kg of hazardous waste per calendar month, or more than 1 kg of acutely hazardous waste per calendar month. The state environmental agency imposes a fee for each generator based on its classification and amount of waste produced.

Waste Characterization

Most aviation organizations will produce wastes as by-products of operations, usually as part of a maintenance function. Because of this, there are a number of reasons why it is important for the organization to determine the characteristics of each waste. First, although many materials are mandated by the aircraft manufacturer for maintenance purposes, knowing the characteristics of the products can help to minimize hazardous waste streams through careful product selection. Second, some wastes with similar characteristics may recyclable, thus reducing the waste stream charged to the organization. Another important reason is to understand the proper methods for storage and disposal of the waste to maximize protection of employees, assets, and the environment.

A material may be categorized as a hazardous waste either by meeting the criteria for a listed waste or by exhibiting one or more hazardous waste characteristics. A hazardous waste determination may be made using the knowledge of the process and using available documentation (such as material safety data sheets) or through sampling and laboratory analysis. A waste will be considered hazardous if it meets any one or more of the following criteria:

Characteristic of ignitability. A solid waste exhibits the **characteristic of ignitability** and is assigned the EPA Hazardous Waste Number D001 if a representative sample of the waste has any of the following properties:

- It is a liquid, other than an **aqueous solution** (a solution in which water is the solvent), containing less than 24 percent alcohol by volume and has a flash point less than 60°C (140°F), as determined by a Pensky-Martens Closed Cup Tester, using the test method specified in ASTM Standard D93-79 or D93-80 (incorporated by reference, see §260.11), or a Setaflash Closed Cup Tester, using the test method specified in ASTM Standard D3278-78 (incorporated by reference, see §260.11).
- It is not a liquid and is capable, under standard temperature and pressure, of causing fire through friction, absorption of moisture, or spontaneous chemical changes and, when ignited, burns so vigorously and persistently that it creates a hazard.
- It is an ignitable compressed gas. The term **compressed gas** shall designate any material or mixture having in the container an absolute pressure exceeding 40 pounds per square inch (p.s.i.) at 70°F or, regardless of the pressure at 70°F, having an absolute pressure exceeding 104 p.s.i. at 130°F; or any liquid flammable material having a vapor pressure exceeding 40 p.s.i. absolute at 100°F as determined by ASTM Test D323. Refer to §261.21 for information describing ignitability of a compressed gas.

- It is an oxidizer. An **oxidizer** for the purpose of this section is a substance such as a chlorate, permanganate, inorganic peroxide, or a nitrate that yields oxygen readily to stimulate the combustion of **organic matter** (of, relating to, or derived from living organisms such as wood or manure).
- An organic compound containing the bivalent -O-O- structure (consisting of two oxygen molecules bound together) and which may be considered a derivative of hydrogen peroxide where one or more of the hydrogen atoms have been replaced by organic radicals must be classed as an organic peroxide (exceptions may apply, as specified in §261.21).

Characteristic of Corrosivity. A solid waste exhibits the **characteristic of corrosivity** and is assigned the EPA Hazardous Waste Number D002 if a representative sample of the waste has either of the following properties:

- It is aqueous and has a pH less than or equal to 2 or greater than or equal to 12.5, as determined by a pH meter using Method 9040C in "Test Methods for Evaluating Solid Waste, Physical/ Chemical Methods," EPA Publication SW-846.
- It is a liquid and corrodes steel at a rate greater than 6.35 mm (0.250 inch) per year at a test temperature of 55°C (130°F) as determined by Method 1110A in "Test Methods for Evaluating Solid Waste, Physical/Chemical Methods," EPA Publication SW-846.

Characteristic of Reactivity. A solid waste exhibits the **characteristic of reactivity** and is assigned the EPA Hazardous Waste Number D003 if a representative sample of the waste has any of the following properties:

- It is normally unstable and readily undergoes violent change without detonating.
- It reacts violently with water.
- It forms potentially explosive mixtures with water.
- When mixed with water, it generates toxic gases, vapors, or fumes in a quantity sufficient to present a danger to human health or the environment.
- It is a cyanide- or sulfide-bearing waste which, when exposed to pH conditions between 2 and 12.5, can generate toxic gases, vapors, or fumes in a quantity sufficient to present a danger to human health or the environment.
- It is capable of detonation or explosive reaction if it is subjected to a strong initiating source or if heated under confinement.
- It is readily capable of detonation or explosive decomposition or reaction at standard temperature and pressure.
- It is a forbidden explosive as defined in 49 CFR 173.54, or is a Division 1.1, 1.2, or 1.3 explosive as defined in Title 49 CFR (Transportation) 173.50 and 173.53.

Characteristic of Toxicity. A solid waste (except manufactured gas plant waste) exhibits the **characteristic of toxicity** and is assigned an EPA Hazardous Waste Number as specified in §261.24 Table 1 if, using the Toxicity Characteristic Leaching Procedure, test Method 1311 in "Test Methods for Evaluating Solid Waste, Physical/Chemical Methods," EPA Publication SW-846, the extract from a representative sample of the waste contains any of the contaminants listed in §261.24 Table 1 at the concentration equal to or greater than the respective value given in that table. Where the waste contains less than 0.5 percent filterable solids, the waste itself, after filtering using the methodology outlined in Method 1311, is considered to be the extract for the purpose of this section.

The following are typical hazardous waste streams generated from aviation operations:

- *Waste Paint–Related Material:* generated as a result of residual paints and thinners from painting aircraft parts or GSE. This waste would typically have an ignitability characteristic.
- *Waste Aerosol Cans:* generated as a result of puncturing aerosol cans or disposing of cans with product in them.
- *Waste Jet Fuel:* usually generated as a result of contaminated fuel.
- *Expired Shelf Life Material:* generated as a result of time-controlled chemicals expiring. Paint, resins, epoxies, and other such chemical products that are applied to the aircraft are typically time-controlled products that become waste upon expiration.

Definitions and Terms

Accumulation Time: The amount of time starting when a container (up to 55 gallons for most hazardous wastes or 1 quart for acute hazardous or P-listed wastes) is full and ending when the container is removed from the site.

Accumulation Start Date: The day that a container is full (up to 55 gallons for most hazardous wastes or up to 1 quart for acute hazardous or P-listed wastes). The accumulation start date is required to be clearly marked on all hazardous waste containers. (*Note:* Some states have slightly different interpretations of accumulation start dates.)

Combustible Liquid (OSHA Definition): A liquid having a flash point at or above 37.8°C (100°F) and below 93.3°C (200°F). Combustible liquids are divided into two classes. A Class II combustible liquid includes those liquids with flash points at or above 37.8°C (100°F) and below 60°C (140°F). A Class III combustible liquid includes those with flashpoints at or above 60°C (140°F).

Conditionally Exempt Small Quantity Generator (CESQG): a facility that generates no more than 220 pounds of hazardous waste in any one calendar month and stores not more than 2,200 pounds on-site at any time.

Corrosive: A characteristic of hazardous waste defined as a waste with a pH less than 2.0 or greater than 12.5.

Empty Container: Drums that last contained a hazardous waste (*except acutely hazardous or P-listed wastes*) are considered empty only if:

- all material has been removed as much as possible by common removal practices (pouring, pumping, aspirating, or squeezing); and
- no more than one inch of residue remains on the bottom of the drum or other container; or
- no more than 3 percent by weight of the container's total capacity remains if the container is less than or equal to 110 gallons; or
- no more than 0.3 percent by weight of the container's total capacity remains if the container is greater than 110 gallons by size.

Drums, containers, or inner liners that last contained an **acutely hazardous waste** (or a chemical product which becomes an acute waste on disposal) are considered empty only if the drum container or inner liner has been triple rinsed using a solvent capable of removing the product or waste material; or, in the case of the container, the inner liner which prevented contact between the product and the container has been removed and disposed of as a hazardous waste (where the liner itself is not triple rinsed). Refer to the Chemical List section of the manual to identify acute hazardous wastes under RCRA.

Flammable Liquid: Flammable liquid is defined slightly differently by the DOT, EPA, and OSHA. (These are the basic forms of these definitions where some of the parameters and exclusions may have been removed due to limited relevancy in the aviation industry).

> *DOT Definition:* A liquid having a flash point of not more than 60°C (140°F).
>
> *EPA Definition:* The EPA uses the term "ignitable characteristics" for waste regulations rather than "flammable liquid." Materials with ignitable characteristics are used for those with flash points lower than 60°C (140°F).
>
> *OSHA Definition:* A liquid having a flash point below 37.8°C (100°F). This is the same definition used by the National Fire Protection Agency (NFPA) in the *NFPA 30: Standard for Flammable and Combustible Liquids.*

Flash point: The minimum temperature at which a liquid gives off vapor in a sufficient concentration to form an ignitable mixture in air near the surface of the liquid.

Full: As it relates to a hazardous waste satellite accumulation drum, the accumulation drum's contents have reached the most extruded portion of the drum's upper ring (approximately four inches from the top of the drum).

Hazardous Waste: A solid waste, which exhibits the characteristics of:

- Ignitability (EPA Hazardous Waste Number D001),
- Corrosivity (EPA Hazardous Waste Number D002),
- Reactivity (EPA Hazardous Waste Number D003), and
- Toxicity (EPA Hazardous Waste Numbers D004 through D0043).

As defined in 40 CFR 261 Subpart C, or is a listed waste found in 40 CFR 261(D):

- *F list:* List of wastes from nonspecific sources (EPA Waste Numbers F001 through F039)
- *K list:* List of wastes from specific sources (EPA Waste Numbers K001 through K172)
- *P list:* List of acute hazardous discarded commercial chemicals (EPA Waste Numbers P001 through P205)
- *U List:* List of discarded commercial chemical products (EPA Waste Numbers U001 through U411)

Hazardous Waste Manifest: The shipping document EPA form 8700-22 and, if necessary, EPA form 8700-22A, originated and signed by the generator in accordance with the instruction included in the appendix to 40 CFR Part 262.

Ignitable: A characteristic of hazardous waste defined as a waste with a flash point less than 140°F.

Impervious Surface: A surface that will not absorb any moisture, such as a non-porous texture that is impenetrable.

Large Quantity Generator (LQG): A generator which generates more than 1,000 kg (2,200 pounds) of hazardous waste in a calendar month or accumulates more than 6,000 kg (13,200 pounds) of hazardous waste on site at any time.

Satellite Accumulation Area: A designated area at the point where hazardous waste is generated, which is under the control of an operator. Up to 55 gallons of hazardous waste (or 1 quart of acute hazardous waste) may be accumulated in a satellite accumulation area without starting the accumulation time clock. The accumulation time starts when the container is full. The hazardous waste container must be moved to the accumulation area within 3 days.

Small Quantity Generator (SQG): A generator which generates less than 1,000 kg of hazardous waste in a calendar month, but more than 100 kg (220 pounds) and the total quantity of hazardous waste accumulated on site never exceeds 6,000 kg (13,200 pounds).

Solid Waste: Any material that has fulfilled its intended purpose and is being discarded, has been abandoned, or is inherently waste-like. Solid waste includes both hazardous and non-hazardous waste and may take the form of a solid, semi-solid, liquid, or gas.

Toxic: Having the quality of being hazardous to human health. For waste purposes, a waste that will leach any of the chemicals at a concentration greater than the Regulatory Level provided in the Maximum Concentration of Contaminants for the Toxicity Characteristics Table found in 40 CFR 261 and is considered a toxic characteristic hazardous waste. This may be determined using a Toxicity Characteristic Leaching Procedure (TCLP) test.

Treatment/Storage/Disposal (TSD): The processes involved with converting a hazardous substance to a less dangerous one that is then later moved to a designated EPA site for hazardous waste.

Air Carrier Access Act (ACAA)

Another area of oversight and compliance impacting aviation is the need to provide accessibility to aircraft and facilities for passengers with disabilities. Each day, a large number of persons with disabilities travel via the airline industry and it is very important that these passengers be appropriately accommodated to ensure their safety, security, and ease of accessibility to transportation by air. The ACAA is applicable only to air carriers that provide regularly scheduled service revenue flights to the public.

The requirement for air carriers to provide reasonable access to air transportation for passengers with disabilities is regulated under the **Air Carrier Access Act (ACAA)**. The ACAA prohibits air carriers from discriminating against passengers on the basis of disability in air transportation and requires carriers to accommodate the needs of passengers with disabilities. This is an important piece of legislation in ensuring reasonable access to airline travel for persons with disabilities.

The ACAA was enacted by the U.S. Congress and signed into law by President Ronald Reagan in 1986. This act was originally coded as Section 404(c) of the Federal Aviation Act (49 USC 1374(c)). Since the inception of the act, the DOT has changed the coding of the ACAA to Title 14 CFR Part 382. The ACAA was originally applied to U.S. domestic airlines. However, in 2000, the ACAA was amended by Congress to require that foreign air carriers comply with U.S. standards for disabled passenger accessibility. Thus, foreign air carriers operating flights to and from destinations in the United States are required to abide by the ACAA in providing access and accommodation to passengers with disabilities. As an example, an international air carrier operating flights between London Heathrow (LHR) and John F. Kennedy International Airport (JFK) in New York is required to follow the ACAA for disabled passengers on each flight segment between these city pairs (LHR-JFK-LHR and vice versa). Over the years, the ACAA has been amended ten times, including areas of concern such as the use of passenger lifting devices for enplaning/deplaning passengers with disabilities on and off aircraft where level-entry boarding is not available; providing seating accommodations for passengers with disabilities; reimbursement of passengers for loss or damage to wheelchairs; modifications to policies and practices to ensure nondiscrimination; implementation of airline terminal accessibility standards; and technical changes to terminology and dates of compliance. The most recent amendment of the ACAA took place in May 2008, with implementation occurring on May 13, 2009. This amendment includes issues such as revised provisions concerning passengers who use medical oxygen and deaf, hard-of-hearing, or deaf–blind passengers.

Major Provisions of the Air Carrier Access Act

Because air carriers must comply with the ACAA, it is important that present and future employees of the airline industry possess at least a fundamental working knowledge of the requirements of the ACAA. This is especially true for those working in customer service positions, who may have direct interaction with passengers with disabilities. The following paragraphs briefly discuss some of main requirements and principles of the ACAA.

The ACAA prohibits discrimination in air transportation by domestic and foreign air carriers against passengers with physical or mental disabilities. In other words, an airline may not refuse transportation to a person on the basis of their disability. However, airlines reserve the right to exclude anyone who would jeopardize the safety of a flight. If an airline excludes a disabled person from flying on the basis of safety, the air carrier must provide the person a written explanation of the reason for the exclusion. In addition, airlines are not permitted to limit the number of disabled passengers on a given flight.

Also, air carriers may not require that advance notice be given that a passenger with a disability will be traveling. In certain instances, air carriers may require up to 48 hours' advance notification for specific passenger accommodations that require a degree of preparation time, such as connection of a respirator, passengers traveling with an emotional support or psychiatric service animal, and other needs. In addition, airlines are not permitted to require passengers with disabilities to travel with a personal care attendant. Rather, a person can be specified as being a safety attendant for a passenger. A safety assistant would help the passenger in exiting the aircraft in an emergency or would assist with establishing communication with airline personnel for required safety briefings. A safety attendant could be a volunteer passenger, a person chosen by the passenger, or off-duty airline employees who voluntarily elect to serve in this capacity.

Basic Aircraft and Facility Requirements In order to ensure reasonable accommodation of passengers with disabilities, there are certain requirements for aircraft and terminal facilities that must be followed for airlines and airports to be ACAA compliant. Under the ACAA, any "new" aircraft (defined as those ordered after April 5, 1990, or delivered after April 5, 1992) having 30 or more seats are required to have movable aisle armrests on half the aisle seats. This constitutes the vast majority of the U.S. commercial airline fleet. Also, newer wide-body aircraft (two passenger aisles, such as 767, 777, A330, etc.) are required to have lavatories accessible to passengers with disabilities. In addition, passengers with disabilities may be accompanied onboard aircraft by trained service animals, such as dogs. The basic distinction is that the animal must be specifically trained and must be shown to be an actual service animal providing assistance to the passenger. Reasonable accommodation of the service animal must be made by the airline onboard the aircraft. Also, any airport facilities owned and/or operated by airlines are required to satisfy the same accessibility standards that apply to federally funded airports.

With respect to wheelchairs, new aircraft with 100 or more seats must have priority storage space for a passenger's folding wheelchair in the cabin of the aircraft. Wheelchairs and other assistive devices must be given priority over other items for storage in the aircraft baggage compartment. Airlines must accept battery-powered wheelchairs, and contain the batteries in hazardous materials packages provided by the air carrier. Airlines may not charge for packaging for batteries. Also, any aircraft containing more than 60 seats and an accessible lavatory must have an onboard wheelchair, regardless of when the aircraft was ordered and/or delivered to the carrier. For any aircraft with more than 60 seats that does not have an accessible lavatory, the air carrier must place an onboard wheelchair on the aircraft if a disabled passenger provides the airline 48 hours' notice that they are able to use a non-accessible lavatory, but require a wheelchair to reach the lavatory.

Enplaning and Deplaning of Passengers with Disabilities Under the ACAA, airlines are required to provide assistance for enplaning and deplaning passengers with disabilities and to assist them with making connecting flights. In locations where there is no passenger loading bridge available, airlines often use some form of mechanized passenger lift device. There are a variety of different types of these devices available for purchase or lease. At U.S. airports having 10,000 or more annual passenger enplanements and where level-entry boarding is not available, boarding assistance for disabled passengers must be provided by the airline through the use of a lift or ramp. This requirement applies to aircraft with 19 or more seats, with a few exceptions. In addition, all air carriers serving U.S. airports are required to have agreements in place with the airport to provide, operate, and maintain passenger lifts and ramps used to meet the boarding requirements of the ACAA. Under the new revisions effective in May 2009, this requirement also applies to foreign airlines operating to and from the United States. Airlines are permitted to require passengers needing lift assistance to check in one hour before the anticipated check-in time for the flight. Finally, even when level-entry boarding is not required, airlines must provide whatever is needed to assist passengers with disabilities in enplaning and deplaning aircraft. As stated in the 2009 ACAA amendments:

> *. . . boarding and deplaning assistance using lifts is not required at smaller U.S. airports and foreign airports, or when severe weather or unexpected mechanical breakdowns prevent the use of a lift. In those circumstances, airlines must still provide enplaning and deplaning assistance by other available means, such as by placing the passenger in a boarding chair and carrying him or her up the boarding stairs unless the design of the aircraft (e.g., the Fairchild Metro, the Jetstream 31 and 32, the Beech 1900 (C and D models) and the Embraer EMB-120) makes this impossible. The only limitation on the means of providing this assistance is that hand-carrying by carrier personnel as defined in that section is prohibited, except in situations of an emergency evacuation where no other timely means of assistance is available.*

Source: U.S. Department of Transportation/FAA

Other Airline Requirements Under the ACAA, airlines must provide specific training for their employees and contractors who are in positions involving direct contact with members of the traveling public. Some of this training specifically focuses on the handling of passengers with disabilities to ensure their safety, security, proper accommodation, and accessibility to air transportation. Some of the areas covered in this training must include the use of equipment by the carrier for enplaning and deplaning that protect the safety and dignity of passengers; awareness and appropriate responses to passengers with disabilities, including persons with physical, sensory, mental, and emotional disabilities; recognition of requests for communication accommodation from individuals who are hearing and/or visually impaired, including the use of writing notes or taking care to enunciate clearly, use of Braille cards (if available), or reading information to a passenger upon request, or communicating through an interpreter. Employees in passenger contact positions must complete refresher training at least once every three years.

Airlines are also required to designate and train certain employees to be **Complaint Resolution Officials (CRO)**. A CRO assists passengers with disabilities in a variety of ways, including addressing any concerns or complaints that such a passenger might have. A CRO must be on duty in every location wherever and whenever an air carrier is conducting operations. A CRO is considered to be the airline's on-site expert in assisting disabled passengers, and as such these individuals must be familiar with the ACAA and their own company policies. A CRO must have the authority to address, resolve, and decide on behalf of the carrier with respect to issues involving the handling of disabled passengers. However, this authority does not supersede the pilot-in-command authority of an aircraft captain with respect to making a decision about the safety of a flight. If a complaint or problem is not resolved through a CRO to their satisfaction, passengers with disabilities have the right under the ACAA to file a complaint against an airline with the DOT, or in some cases to file a lawsuit in federal court.

All CROs complete initial training to designate them as a CRO. Each CRO is then required to complete annual recurrent training concerning their duties and any regulatory changes in the accommodation of disabled passengers.

Chapter Summary

In this chapter, the primary functions of the Department of Transportation (DOT), Federal Aviation Administration (FAA), National Transportation Safety Board (NTSB), Transportation Security Administration (TSA), Occupational Safety and Health Administration (OSHA), and Environmental Protection Agency (EPA) were described in general terms, including a brief summary of each agency's history and key responsibilities. Although each of these impacts the aviation industry at varying levels, the FAA (an agency within the DOT) is still the most prominently visible agency.

The role each agency plays in regulation and oversight of aviation was described in general terms. The roles of many individual positions within some of these agencies were also described to provide a more focused understanding of some of the key responsibilities within the applicable agency. Where high-level programs were relevant, such as OSHA's Recordkeeping National Emphasis Program or the EPA's Clean Air Act, the program was noted within the text and briefly described, along with a reference to the specific regulatory agency and standard.

Finally, this chapter discussed the fundamentals of the Air Carrier Access Act and its applicability to the aviation industry. It described the need to provide accessibility to aircraft and facilities for passengers with disabilities and the many amendments made to the act over the years to ensure that discrimination does not exist with this sector of the population.

Chapter Concept Questions

1. Describe, in your own words, the function of enforcement as it relates to the FAA.
____
____
____

2. Which government agency regulates dangerous goods that are to be transported via aircraft?

    a. OSHA

    b. FAA

    c. DOT

    d. EPA

3. CRO stands for:

    a. Credit Reporting Official

    b. Compliance Resource Office

    c. Complaint Resolution Officials

    d. Claim Response Office

4. The law that requires airlines to provide "reasonable accommodations" is called the ________________ ________________.

5. The aircraft maintenance department for ABC Airlines experienced 30 injuries last year. The operation worked 475,000 hours. What is the overall incident rate for the operation? ________________

6. The FDR records flight crew conversation, radio transmissions, and other flight deck sounds. True or False

7. The mission of the ________________ ________________ ________________ is to protect the nation's transportation systems to ensure freedom of movement for people and commerce.

8. Briefly describe OSHA's Voluntary Protection Program and how it may (or may not) benefit an organization.
____
____
____

9. By environmental definition, jet fuel removed from an aircraft that cannot be reused is considered a:

    a. Solid waste

    b. Liquid waste

    c. Storm water waste

    d. Non-regulated waste

Chapter References

Accessible Journeys. 2009. Air Carrier Access Act. http://www.disabilitytravel.com/airlines/air_carrier_act.htm#summary.

ADA Home Page. 2010. http://www.ada.gov/.

Aero Society Channel. 2010. Black Boxes to Start Tweeting? http://www.aerosocietychannel.com/2010/09/black-boxes-to-start-tweeting/.

Aviation Consumer Protection and Enforcement. 2010. Part 382: Passengers with Disabilities. Rule in effect through May 12, 2009. http://airconsumer.dot.gov/rules/rules.htm.

Aviation Consumer Protection and Enforcement. 2010. Part 382: Passengers with Disabilities. Rule in effect beginning May 13, 2009. http://airconsumer.dot.gov/rules/rules.htm.

AvStop.com. 2010. Cockpit Voice Recorder Legislation Opposed by CAPA. http://avstop.com/news_march_2010/cockpit_voice_recorder_legislation_opposed_by_capa.htm.

Bloomberg.com. 2009. Southwest Air Agrees to $7.5 Million Fine, FAA says. http://www.bloomberg.com/apps/news?pid=20601103&sid=avFzkpTRRnHc&refer=us.

Department of Homeland Security. 2008. Homeland Security Act of 2002. http://www.dhs.gov/xabout/laws/law_regulation_rule_0011.shtm.

Department of Justice. 2005. A Guide to Disability Rights Laws. http://www.ada.gov/cguide.htm#anchor63814

Department of Labor. 1996. *The New OSHA: Reinventing Safety and Health*. OSHA Office of Training and Education.

Department of Transportation. 2009. Aviation Consumer Protection and Enforcement, Part 382: Passengers with Disabilities. http://airconsumer.dot.gov/rules/rules.htm.

Economist.com. 2009. Economics A–Z. http://www.economist.com/research/economics/.

Economy Professor. 2009. Law of Diminishing Returns. http://www.economyprofessor.com/economictheories/law-of-diminishing-returns.php.

Environmental Protection Agency (EPA). 2010. About EPA. http://www.epa.gov/aboutepa/index.html.

Environmental Protection Agency (EPA). 2010. Partnership Programs: List of Programs. http://www.epa.gov/partners/programs/index.htm#air.

Federal Aviation Administration. 2006. Advisory Circular 120-59A. http://www.airweb.faa.gov/Regulatory_and_Guidance_Library/rgAdvisoryCircular.nsf/0/fd8e4c96f2eca30886257156006b3d07/$FILE/AC%20120-59a.pdf.

Fritzsche, D. 2005. *Business Ethics: A Global and Managerial Perspective* (2nd ed.). New York: McGraw-Hill.

Lawrence, H. W. 2004. *Aviation and the Role of Government*. Dubuque, IA: Kendall Hunt Publishing.

Lebow, C. C., Sarsfield, L. P., Stanley, W., Ettedqui, E., and Henning, G. 2000. *Safety in the Skies: Personnel and Parties in NTSB Aviation Accident Investigations*. RAND Corporation.

National Council on Disability. 2004. Position Paper on Amending the Air Carrier Access Act to Allow for Private Right of Action. http://www.ncd.gov/newsroom/publications/2004/aircarrier.htm.

National Transportation Safety Board. 2009. Cockpit Voice Recorders (CVR) and Flight Data Recorders (FDR). http://www.ntsb.gov/aviation/CVR_FDR.htm.

National Transportation Safety Board. 2009. The Importance of an Independent Safety Board. http://www.ntsb.gov/speeches/sumwalt/UAAprese.pdf.

National Transportation Safety Board. 2009. What Is the National Transportation Safety Board? http://www.ntsb.gov/Abt_NTSB/What-Is-The-NTSB.pdf.

9/11 Commission Report. http://www.9-11commission.gov/report/911Report.pdf.

Transportation Security Administration. 2010. Our Programs, Law Enforcement. http://www.tsa.gov/lawenforcement/programs/indexshtm.

Transportation Security Administration. 2010. What We Do. http://www.tsa.gov/what_we_do/index.shtm.

Transportation Security Administration. 2010. Who We Are. http://www.tsa.gov/who_we_are/index.shtm.

Wensveen, J. G. 2007. *Air Transportation: A Management Perspective* (6th ed.). Burlington, VT: Ashgate Publishing Company.

Wood, R. 2003. *Aviation Safety Programs: A Management Handbook* (3rd ed.). Englewood, CO: Jeppesen-Sanderson, Inc.

Risk and Risk Management

Chapter Learning Objectives

After completing this chapter, the reader should be able to:

- Understand the term "risk" and the difference between risks and hazards.
- Explain the risk assessment process by addressing probability (frequency and likelihood) and severity.
- Classify and prioritize risks.
- Use a risk matrix to score risk and demonstrate risk reduction techniques.
- Understand the difference between a typical risk management system and the FAA's System Safety Risk Management approach.
- Understand why risk management is important to an organization, basic risk management principles, and the risk management process itself.

Key Concepts and Terms

Administrative Controls
Control
Elimination
Engineering Controls
Frequency
Hazard
Hazard Analysis
Job Hazard Analysis
Likelihood
Personal Protective Equipment (PPE)
Probability
Qualitative Assessment
Residual Risk
Risk
Risk Analysis
Risk Assessment
Risk Reduction
Risk Statement
Safe
Scope of Exposure
Severity
System Safety Process
TEAM
Unidentified Risk

Introduction

Risk can be simply defined as the probability and severity of harm. Merriam-Webster provides an alternative definition: "someone or something that creates or suggests a hazard." A combination of these two definitions perhaps better describes its meaning. Risk can be exemplified as the chance of winning in a hand of cards, the chance an insurance company takes in covering an individual, the chance an investor takes in the stock market, or the chance a person takes with the decision to shortcut a safety provision. For each, the *likelihood* of the event occurring in combination of the *frequency* of occurrence and the *severity* or potential outcome of the event is an important consideration.

The fact is that everyone takes risks on a daily basis. Certain risks are present in everyday life as we pull out of our driveways, walk down the street, or play sports. Risks are also present in the workplace; many cannot be avoided, but perhaps can be minimized. The word "**safe**" can be defined as being free from harm or risk. But how can a task or operation be considered safe if there is still risk involved? Where the risk cannot be completely eliminated, the answer is often a matter of (sometimes subjective) opinion. In reality, a task or operation can only be considered safe if the risks involved are considered acceptable. For instance, an aviation organization may have a policy that an employee must seek assistance or use available equipment for lifting baggage or freight in excess of 50 pounds. The organization has accepted that the risk of lifting under 50 pounds by an individual is acceptable, while the risk above that limit is not. Certainly an injury could occur while lifting less than 50 pounds, but the risks are less (and therefore acceptable), and there may be procedures or methods that have been implemented and trained for to minimize those risks as well. **Figure 3–1** demonstrates the types of risks that every organization faces.

Many assessed risks may be considered "unacceptable" by an organization. Of these, some can be eliminated or controlled (mitigated). Both of these terms will be explained in detail later in this chapter. The risk remaining after risk reduction or system safety efforts have been employed is the **residual risk**. Residual risk includes those risks that are considered "acceptable" as well as those that have yet to be identified. In other words, after protective measures have been taken, the residual risk is that risk which remains. An example of an acceptable risk might be the noise

Figure 3–1. Types of Risk.

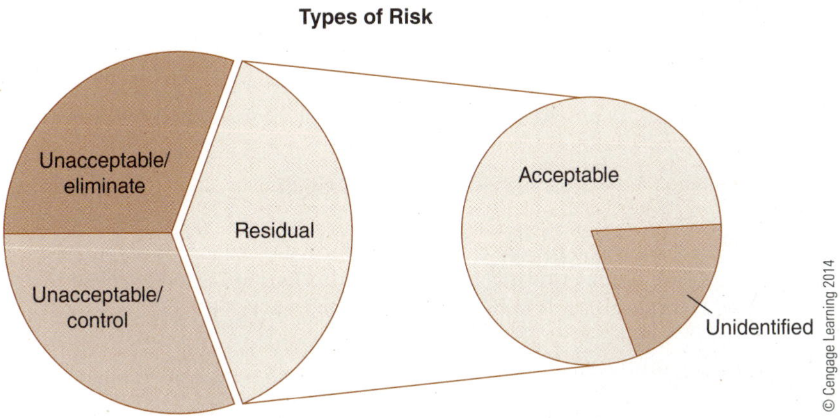

exposure of ramp personnel. In this case, it is not practical "engineer out" the exposure to noise. The risk is considered "acceptable" only after minimizing the risk through proper training, the use of personal protective equipment (PPE), and perhaps minimizing the length of exposure. An **unidentified risk** is one that has not been determined, perhaps overlooked or unrealized. Some unidentified risk is realized only after an unwanted or unplanned event has occurred, making it more evident.

A **hazard** is the source of a problem or danger that poses a *risk* of injury or damage to property or the environment. Hazards can be categorized into basic types, including physical, environmental, chemical, and psycho-social, and can have a range of threat level, from negligible to immediately dangerous to life and health (IDLH). Each of these terms will be used and explained extensively throughout this chapter as they relate to risk management in the aviation industry. At the conclusion of this chapter, the reader should understand why risk management is important to the organization, the basic risk management principles, and the risk management process itself.

Risk Management: An Overview

Risks associated with an organization certainly have an effect on the "bottom line," either positive or negative. That is why it's important to manage those risks. If only for the financial and judicial survival of the organization, its leadership has a moral and ethical obligation to manage risk and ensure no unnecessary (or unacceptable) risks are taken. Risk decisions should be made at the appropriate level within the organization to establish clear accountability. Thus, those held accountable for the success or failure of any given process should be included in the risk decision process. Leadership at a level that makes financial decisions impacting risk reduction should be part of the process as well. This could be as simple as providing lifting equipment to reduce the stress of manual lifting for employees, or it could mean providing a professional safety staff for advice on the risk management process. The best and most cost-effective time to approach the risk management process is during the planning phase of the organization, of a new process, or of a change to a standard operating procedure.

Of course, there is a balance that must be realized. The organization is in business to make a profit. While there can be little or no compromise on some risk decisions, others will require justification in the form of return on investment. This is not to say that a judgment should be made to disregard safety to ensure a profit, but that credibility and justification are earned by the safety professional or other individuals by demonstrating that the potential (and probability) for loss exceeds the cost of the control. A **control** is a form of mitigating a risk. There are four types of controls:

- **Elimination** (including substitution): removing the process, equipment, or associated product that presents the risk, or providing an alternative.
- **Engineering controls:** designing or modifying a process, a piece of equipment, and so on to reduce the source of risk exposure.
- **Administrative controls:** altering the way a process is completed through a change in standard operating procedures for completion, time controls for work (work rotation), or other rules, including training. Time and rotation controls reduce the amount of time an employee is exposed to a risk.
- **Personal protective equipment (PPE):** equipment worn by an employee to reduce or eliminate certain exposures such as chemicals or noise.

The order of these is the hierarchy in which they should be chosen as well. In other words, PPE should be the last resort. Obviously, not all hazards can be eliminated or engineered out, but a concerted effort should be made to reduce the risks at higher levels before resorting to the use of PPE.

The Risk Management Process

As mentioned in the previous section, making risk management decisions during the inception of the organization or process, or when there is a change to a process or piece of equipment, is very important. However, risk management is a *process* for a reason. As opposed to a program that has a beginning and an end, a process is continuous. As such, the risk management process should be designed into the organization to detect, assess, and control risks on a continuous basis while enhancing operational performance and effectiveness.

There are a few stages of an organization's life where risks can be reviewed and managed. They include inception, expansion, operational change, risk observation, and risk history. As an organization is developed or expands its operations, it is critical to include key decision makers and knowledge experts in the planning phase. These decision makers may include executive staff, financial leaders, safety and/or industrial engineers, and outside industry experts in an effort to determine what equipment will be necessary to conduct business safely, while weighing any budgetary concerns. The planning process will often include invitations to several vendors to display the efficiencies and safety features of their products. Financial leaders will want to understand the risk factors associated with using or not using, as may be the case, certain equipment that determines overall cost factors (for example, engineered features that would eliminate or reduce manual handling—where risk of injury resides) that will affect purchasing decisions. Some cost elements in this stage of decision making, such as training and administrative factors, including procedure development and oversight (including labor and behavioral factors), are often neglected.

An operational change in the organization is similar to the inception or expansion stage in that there is an opportunity for planning that includes the same decision makers. Usually, an operational change results from a problem that has occurred in the current system, where a change has been ordered or technology or other developments have provided for more efficient production. One advantage of an operational change over a change in the inception stage is that the organization has a history of experience from which to draw (including production, injury, and damage).

Risk observation and risk history are related by the fact that both provide evidence of which areas within the organization contribute to losses (injury and damage). As part of the risk management process, it is critical to review these factors to determine whether:

- a change in the operation is necessary
- safety controls work as designed or if they are bypassed
- there is a lack of accountability
- insufficient procedures exists

Decision making for these stages usually occurs at the operational management level and may include the safety manager/coordinator with input from the safety committee.

For each of these stages, identifying hazards is the first step of the process. Where operational tasks are concerned, an operational and **hazard analysis** (also referred to as a **job hazard analysis** or JHA) is an important part of the process. It includes a detailed breakdown of each of the jobs or tasks expected within each operation to determine where hazards exist, using historical data, experts, regulations, and logic to determine and prioritize risk. More information on the JHA process is detailed in the Accident Prevention Strategies section of Chapter 5.

The FAA's System Safety Process

As with the traditional risk management process, the FAA's **System Safety Process** is a proactive approach of identifying, assessing, and eliminating or controlling safety-related hazards to an acceptable level to prevent injuries and accidents. System Safety is a formal and flexible process that includes eight steps designed to proactively search for opportunities to improve the process at each step. The eight steps are identified and explained below:

1. *Define Objectives:* Define the objectives of the system or process, typically documented in the operating specifications and the business plan for the organization.
2. *System Description:* A description of the system being reviewed (procedures, tools, materials, equipment, facilities, etc.). This also includes descriptions of data/records available.
3. *Hazard Identification: Identify Hazards and Consequences*: Potential hazards may be identified from a number of sources and listed and grouped by functional equivalence for analysis. Inclusion of the consequence (undesired event) resulting from the hazard scenarios is required prior to the risk analysis. Hazard scenarios may address who, what, where, when, why, and how.
4. *Risk Analysis: Analyze Hazards and Identify Risks:* Risk analysis is the process whereby hazards are characterized according to their likelihood and severity. (*Note:* The Risk Analysis and Assessment section of this chapter will differ from the System Safety risk analysis process by including the term "frequency" as a measure of how often the exposure exists, along with likelihood.) Risk analysis looks at hazards to determine what can happen when. Either a qualitative or quantitative analysis can be conducted. Some type of a risk assessment matrix is normally used to determine the level of risk. This will be explained in more detail in the Risk Analysis and Assessment section.
5. *Risk Assessment: Consolidate and Prioritize Risks:* For System Safety, risk assessment is defined as the process of combining the impacts of risk elements discovered in risk analysis and comparing them against some acceptability criteria. Risk assessment can include the consolidation of risks into risk sets that can be jointly mitigated, combined, and then used in decision making.
6. *Decision Making: Develop Action Plans*: A prioritized list of risks is reviewed to determine how to address each risk, beginning with the one of highest priority. The four options that may be chosen for a risk are transfer, eliminate, accept, or mitigate (TEAM).
7. *Validations and Control: Evaluate Results of Action Plan for Further Action:* Validation of the effectiveness of actions taken and the current status of each prioritized risk. If the residual risk is acceptable, the modification is documented along with the rationale for accepting the residual risk. If it is unacceptable, an alternate action plan may be needed, or an additional modification may be necessary.
8. *Modify System/Process (if needed):* If the change or mitigating action does not produce the intended effect, a determination must be made as to why. The System Safety Process will need to be reexamined at the hazard identification step.

Source: U.S. Department of Transportation/FAA

The FAA has System Safety processes for many areas, including procedures, tools, materials, equipment, and facilities, included in the System Safety Handbook (search at www.faa.gov). The FAA also offers System Safety classes open to aviation professionals for a nominal fee.

Risk Assessment and Analysis

A **risk analysis** is a proactive approach to identifying the most probable threats of loss to an organization and analyzing the most likely effects of loss to the organization (system or procedure) from these threats.

Risk assessment is the process of evaluating existing controls and assessing their adequacy to the potential threats of loss to the organization. In other words, it involves making a determination as to how safe a job task or situation is by determining the probability (how often the unwanted event could occur) and severity (the potential consequences or seriousness of an unwanted event) of loss from exposure to the hazards involved. While there is a level of subjectivity in the determination of each, the amount of quality historical data will increase the validity of the assessment. Although no risk assessment is 100 percent accurate, this process is adequate and necessary for the decision makers with regard to managing risks or, more appropriately stated, determining acceptable risks for the organization.

The process by which a risk assessment is conducted based on recognized hazards (by employees, management, safety committees, etc.) is considered a **qualitative assessment**. The process by which internal and/or external data is used to analyze risk is considered a quantitative assessment. A combination of the two can be used as well. An element of quantitative risk analysis is often favored by management, since the risk assessment can be expressed and emphasized using metrics or charts.

Publications vary in risk assessment terminology (some use "consequence" or "degree of consequence" as opposed to "severity"), but our intent is to keep the terminology simple and understandable. Some risk assessment models can use complicated formulas and assumptions. While these may appear professional and impressive (and may sometimes be more appropriate in high-level staff settings), a simplified approach (see **Figure 3–2** for example) will allow for more individuals to be involved in the process. Having front-line employees, safety committee members, supervisors, and others involved in the risk assessment process provides more "buy-in" for the process and is another measure of employee involvement, which speaks well for the organization and gains favor with OSHA. In addition, no one will understand the risk better than those who interact with it daily. The process should be led by person(s) who are knowledgeable and skilled in the development of risk assessments. Employees participating in risk assessment exercises should be trained to understand the process and terminology.

As mentioned earlier, there are various models that can be used for a risk assessment. The model used in this text will include the terms frequency (F), likelihood (L), and severity (S), used in a scale to rate each on a risk assessment matrix. The model used is simply an example and can be modified to meet the needs of different types of organizations and operations. For instance, where likelihood on a ramp operation may include how often an employee conducts a task, the likelihood scale for chemical or noise exposures might be modified to include different levels or percentages of exposure. For each (F, L, S), the scale will be ranked from 1 through 5, with 5 being the highest element of risk for the term (see Tables 3–1, 3–2, and 3–3 for examples).

Frequency (or exposure opportunity) is how often the work activity that produces the hazard occurs. Frequency is influenced by the scope of exposure and how often the exposure exists. **Scope of exposure** is a variable that may be customized to meet the needs of the assessment. For example, for general workplace safety, the number of employees at risk may define the scope; when determining damage risk, the number of aircraft and/or tugs on the ramp may be applicable. From these, clear parameters can be established for frequency.

Likelihood (or the chance of occurrence) is the most subjective element of the process in that it is often misunderstood. Persons involved in the assessment process should consider likelihood in terms of the life of the system or their time working within the system rather than the likelihood of an incident occurring every time a hazard exists. Employees may be exposed to hazards several

Figure 3–2. Qualitative Risk Assessment Matrix.

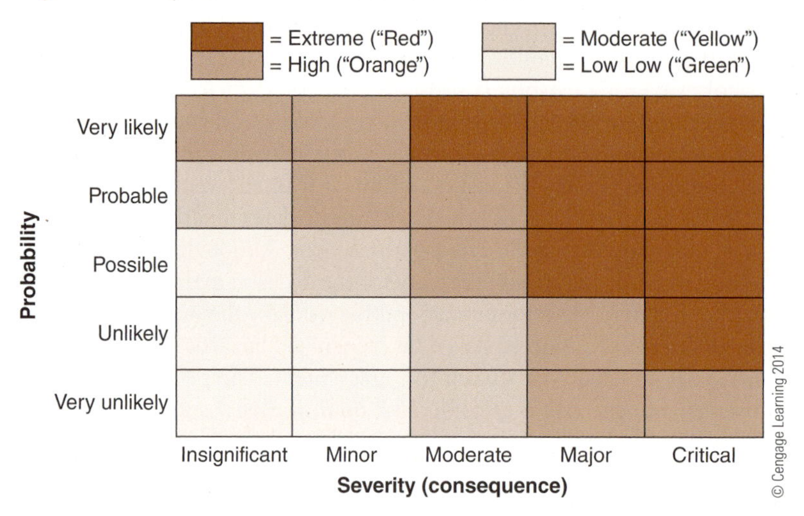

times without incident. One method to reduce the subjectivity of this element is by reviewing the system's history. If an incident has occurred previously, the likelihood of reoccurrence is higher.

The combination of frequency and likelihood is recognized as the **probability** or opportunity for occurrence of an unwanted event. **Severity** is the degree of harm for which the unwanted event has potential. For this element, the worst-case scenario (within reason) is considered. The product of all three elements (F, L, and S) is used to quantify the risk: $F \times L \times S = \text{Risk}$. **Figure 3–3** is an illustration of the product of frequency and likelihood equalling risk.

Before measurement of risk can occur, the risk analysis should identify and prioritize the risks to be assessed. A clear and concise description of the risk, or **risk statement**, must be developed for each process or procedure being assessed (the concern). The risk statement is a single phrase or sentence that expresses the condition (hazard) and consequences and is used to evaluate the likelihood and

Figure 3–3. Example Risk Criteria.

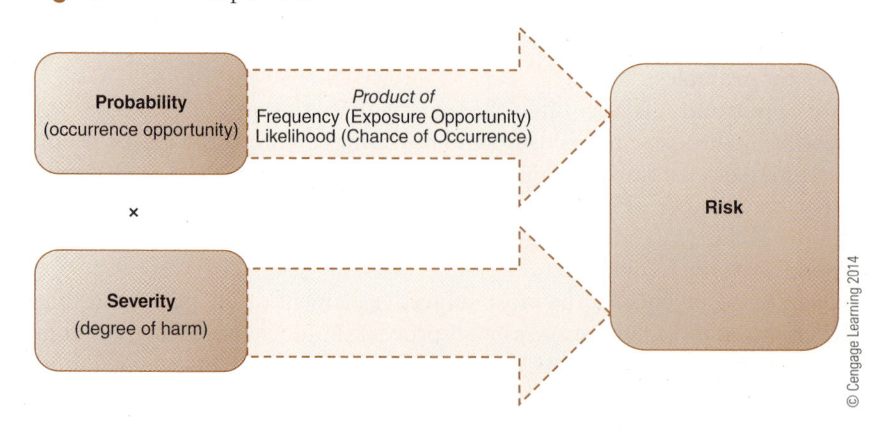

severity of the risk (example risk statement: "Allowing the placement of large bags onto the second/top shelf of a bag cart exposes baggage handlers to back and shoulder injury while lifting/lowering").

From the example risk statement, one can assume there is the likelihood that a baggage handler could incur an injury to the back or shoulder as a result of lifting and/or lowering. To continue the example, it is now necessary to determine how often (the frequency) a baggage handler employee is exposed to the hazard, how likely an injury is to occur (when following the established procedures), and how severe the potential consequences could be if an injury occurred. To quantify the frequency, likelihood, and severity associated with the risk statement, a risk matrix can be used. The following text is an example of a simple risk matrix that can be used at any level within the organization (management, safety committee, etc.):

Table 3-1. Example Frequency Scale (How often the activity is performed during the employee's shift)

1	The activity is not normally performed during the shift.
2	The activity is performed equal to or more than once per shift.
3	The activity is performed equal to or more than once per hour.
4	The activity is performed equal to or more than once per 5 minutes.
5	The activity is performed equal to or more than once every minute.

Table 3-2. Example Likelihood Scale (The chance that the concern could occur)

1	Highly unlikely
2	Not likely
3	Possible
4	Probable
5	Likely

Table 3-3. Example Severity Scale (The potential degree of harm, within reason, should the concern event occur)

1	First aid or minor property damage.
2	Medical treatment, but no lost time from work, light to moderate damage to equipment.
3	Moderate medical treatment and/or lost time from work, moderate to heavy damage to equipment.
4	Heavy medical treatment, hospitalization, lost time from work, some permanent impairment, heavy to severe damage to equipment.
5	Catastrophic injury (including major permanent impairment) or damage to equipment.

To develop a risk index or number to quantify the risk for the risk statement, the individual job tasks must be broken down and analyzed. An example, including the original risk statement, is provided below. There are a few assumptions given, including that this is for a small regional airline and the number of employees exposed is an average of 22 per shift. The task, when averaged to include downtime, is about once every 3–4 minutes, and the severity is moderate medical treatment that may result in lost time from work.

Example Risk Assessment

Risk Statement: Allowing the placement of large bags onto the second/top shelf of a bag cart exposes baggage handlers to back and shoulder injury while lifting/lowering.

The concerns are listed (with the concern split for lifting and lowering, which will be explained later in the Risk Reduction section of this chapter) along with the existing controls. The number of employees (see **Figure 3–4**) does not factor into the index. The frequency, likelihood, and severity are determined using the scales and the given information. For this model, a "1" indicates "green" or good or slight risk for likelihood and severity (frequency is not rated as good or bad). A "2" or "3" indicates "yellow" or caution (moderate risk), and a "4" or "5" indicates "red" or a severe or catastrophic risk. The product of these is taken and added together to determine the risk index. For example, in the first row of the example risk assessment (see Figure 3–3), you multiply the ratings for frequency, likelihood, and severity (5 × 4 × 3) to get a product of 60. Do this again with the second row (60) and add the two products to determine the risk index

Figure 3–4. Example Risk Assessment before new controls are implemented.

= Extreme ("Red")
= High ("Orange")
= Moderate ("Yellow")
= Low Low ("Green")

Concerns	Existing Controls	Number of Employees Exposed (during shift)	Frequency (rate frequency of activity)	Likelihood (rate likelihood of concern)	Severity (rate severity of concern)	Risk
Strain injury to shoulder or back by lifting large and heavy bags to second/top shelf of bag cart	Current procedure does not specify a load or size limit for lifting bags to the second/top shelf of the bag cart. There is no procedure in place for lifting methods.	22	5	4	3	60
Strain injury to shoulder or back by lowering large and heavy bags to second/top shelf of bag cart	Current procedure does not specify a load or size limit for lowering bags to the second/top shelf of the bag cart. There is no procedure in place for lifting methods.	22	5	4	3	60
					Risk index	120

(in this case, 120). While the individual elements may be high (in the red), the product of the concern is the more telling factor. For each concern, the product may also help to determine if the risk is acceptable. If the product of the concern is between 1 and 24, there is a negligible or slight risk (indicated as green) associated with the activity. If the product is between 25 and 63, there is a moderate risk (indicated as yellow) associated with the activity, but it may be a generally acceptable risk, (This is not to suggest that the risk should not be evaluated for reduction.) If the risk is 64 or higher, there is a severe or catastrophic risk (indicated as red), which should not be considered acceptable. The risk index number is the sum of the risk from the concerns assessed. For this example, the risk index is 120. This number means little unless and until a risk reduction exercise is performed. At that point, the risk index created from the process changes is compared to the baseline as a percentage of risk reduction (which will be provided in the Risk Reduction section of this chapter).

The next part of the assessment is the analysis of existing controls. The same concerns and existing controls are used. This part of the analysis looks at the environment that can lead to the injury, a specific description of the exposure to each employee, and the consequence of the current environment. Photos are useful tools for helping those involved in the process to better understand the environment. This part of the example is shown in **Figure 3–5.**

It was mentioned earlier in this chapter that the risk assessment process is a good way to promote employee involvement. A risk assessment exercise can be used by a safety committee or safety focus group to demonstrate the organization's commitment to safety and its awareness of the role of frontline employees in the safety process. By using a simple worksheet such as those in the example above, the team can provide input to a facilitator discussing the level of risk for each element and using the consensus to determine the risk index. If the risk is deemed acceptable by management, there may be no need to continue with the concern, or the concern may be moved down on the priority list. If the risk is unacceptable, the team can continue with a risk reduction exercise.

Figure 3–5. Example System Analysis.

Concern	Existing Controls	Analysis of Existing Controls		
		Environment	Capability	Motivation
Strain injury to shoulder or back by lifting large and heavy bags to second/ top shelf of bag cart	Current procedure does not specify a load or size limit for lifting bags to the second/top shelf of the bag cart. There is no procedure in place for lifting methods.	Requires lifting heavy and large bags at or above shoulder level, extending the arms and shoulders to an end-range motion.	Exposure is approximately once every 3–4 minutes when averaged with downtime.	Current environment could cause acute or cumulative injury.
Strain injury to shoulder or back by lowering large and heavy bags to second/top shelf of bag cart	Current procedure does not specify a load or size limit for lowering bags to the second/top shelf of the bag cart. There is no procedure in place for lifting methods.	Requires reaching at or above shoulder level to pull bags off of shelf which also creates force on arms, shoulders, and back when heavy bags drop.	Exposure is approximately once every 3–4 minutes when averaged with downtime.	Current environment could cause acute or cumulative injury.

Risk Reduction

Risk management is the responsibility of the organization's management team. When a risk is identified and considered unacceptable, the risk management function should determine which measures to take to mitigate those risks to an acceptable level. First, the organization should take the stance that it will not accept any unnecessary risk. It should make such decisions at an appropriate level within the organization and accept risks only when the benefits outweigh the costs. The FAA System Safety section of this chapter discusses the TEAM approach to reducing risk. **TEAM** stands for transfer, eliminate, accept, and mitigate. The transferring of risk is a method by which the organization does not accept the risk for its employees, but decides to contract out the work to another organization that may be better suited or equipped to complete the task. Elimination is simply eliminating the process or the circumstance that requires the process to exist. The organization should only accept a risk when its frequency, likelihood, and severity are all low. Mitigation means taking action to reduce the frequency, likelihood, or severity of risk.

If an unacceptable risk cannot be transferred, eliminated, or accepted (as is), the organization must mitigate the risk. **Risk reduction** is the process by which an organization mitigates risk through a change in the process, procedures, equipment, behavior, and so on. The goal is to analyze each of these to determine if there is a corresponding element that exists that may reduce the frequency, likelihood, and/or severity. Continuing with the baggage handler example, the risk assessment completed previously would serve as a baseline assessment. An example of a subsequent system analysis and risk assessment is shown in **Figures 3–6** and 3–7; it includes changes made to the organization's policy and procedures.

For this example, the new controls established were changing company policy by not allowing the heavier bags on the second shelf and providing safe work methods for handling the bags that will be placed on the second shelf. Training will be necessary to teach baggage handlers about the new policy and procedures along with a demonstration of how to properly conduct the task.

Figure 3–6. Example Subsequent System Analysis.

Concern	New Controls	Analysis of New Controls		
		Engineering	Training and Education	Behavioral
Strain injury to shoulder or back by lifting large and heavy bags to second/top shelf of bag cart	Policy has been established to not allow a bag of more than approximately 40 lbs to be placed on the second/top shelf. All bags placed on the second/top shelf must be lifted with two hands, bag close to the body. The shelf should be used as leverage to place and push the bag onto the shelf. Observations to be conducted to ensure compliance.		Training for each baggage handler on the new policy and procedures established. Demonstration of the proper methods provided.	Weekly observations conducted with appropriate feedback given to the employee to encourage safe behavior and reduce at-risk behavior. Progressive discipline provided for those who do not comply.
Strain injury to shoulder or back by lowering large and heavy bags to second/top shelf of bag cart	Policy has been established to not allow a bag of more than approximately 40 lbs to be placed on the second/top shelf. All bags placed on the second/top shelf must be removed by leveraging the bag on the edge of the shelf, which will help to determine weight. The bag must then be grasped firmly by both hands and pulled close to the body to lower with a steady motion. Observations to be conducted to ensure compliance		Training for each baggage handler on the new policy and procedures established. Demonstration of the proper methods provided.	Weekly observations conducted with appropriate feedback given to the employee to encourage safe behavior and reduce at-risk behavior. Progressive discipline provided for those who do not comply.

Figure 3–7. Example Subsequent Risk Assessment.

= Extreme ("Red")
= High ("Orange")
= Moderate ("Yellow")
= Low Low ("Green")

Concerns	New Controls	Number of Employees Exposed	Frequency (rate frequency of activity)	Likelihood (rate likelihood of concern)	Severity (rate severity of concern)	Risk
Strain injury to shoulder or back by lifting large and heavy bags to second/top shelf of bag cart	Policy has been established to not allow a bag of more than approximately 40 lbs to be placed on the second/top shelf. All bags placed on the second/top shelf must be lifted with two hands, bag close to the body. The shelf should be used as leverage to place and push the bag onto the shelf. Observations to be conducted to ensure compliance.	22	5	2	3	30
Strain injury to shoulder or back by lowering large and heavy bags to second/top shelf of bag cart	Policy has been established to not allow a bag of more than approximately 40 lbs to be placed on the second/top shelf. All bags placed on the second/top shelf must be removed by leveraging the bag on the edge of the shelf, which will help to determine weight. The bag must then be grasped firmly by both hands and pulled close to the body to lower with a steady motion. Observations to be conducted to ensure compliance.	22	5	2	3	30
					Risk index	60

Risk reduction = 50%

Observations are included to reinforce safe behaviors while reducing at-risk behaviors through feedback and coaching.

In the example subsequent risk assessment, the number of employees exposed has not changed; neither has the frequency of exposure to the heavy bags or the potential for placing them on the second shelf. However, the likelihood of injury occurring has been reduced due to the change in policy, procedures, and subsequent observations. The severity will remain the same, since an injury occurrence would likely be just as severe as before. The product for each concern is 30, which provides a sum of 60 for the risk index. This demonstrates a 50 percent risk reduction from the baseline assessment.

It will often take some time to complete the tasks involved in developing new controls. Establishing new procedures and providing training, especially for larger organizations, can be quite a task. However, completion of a risk reduction exercise that demonstrates a significant reduction of risk is a big motivator to management to ensure the process is taken seriously and given due attention and priority. Making sure the status of each exercise is regularly communicated to the assessment team will promote increased morale and buy-in by those involved.

Case Study: Airline Near-Miss Ground Event

Event Overview

In spring 2011, a near-miss event occurred involving an aircraft of a major airline and a company hired by the carrier to provide aircraft ground handling services on the ramp. An aircraft pushback unit inadvertently rolled into a tow bar attached to the nose gear of the aircraft prior to the pushback of the aircraft from the gate for departure. The pushback unit was being operated by a Lead Ramp Agent of the ground handling service to the airline. At the ground handling company in question, Lead Agents are responsible for all ramp operations in the area where they are working.

The Lead Agent approached the aircraft while driving the pushback unit into position to hook up the tow bar to the unit in preparation to push the aircraft for departure. The Lead stopped the unit when he noticed that one of his ramp crew members was having difficulty attaching the tow bar to the aircraft. The Lead placed the pushback unit gear lever into neutral, left the engine running, set the parking brake, and got off the pushback to assist the ramp agent with the connection of the tow bar. After connecting the tow bar to the aircraft, the Lead turned around to see the pushback unit rolling toward the aircraft on its own. He then climbed on the unit to apply the brakes, but it was too late, as the pushback unit made contact with the tow bar before it could be stopped. The flight crew onboard the aircraft reported feeling the aircraft shudder as a result of the unit rolling into the attached tow bar.

The aircraft and tow bar were inspected by local qualified maintenance technicians, who determined that no damage had been done to the aircraft or the tow bar. It was also determined that the pushback unit had not been damaged either. The event was determined to be a near-miss, as no damage or injuries occurred. However, there was a significant risk of damage to the aircraft and injury to the two agents. If the tow bar had not been attached to the nose gear, the unit would have likely rolled into the aircraft and inflicted damage. Moreover, the two employees could easily have been injured if they had been struck by the pushback unit.

According to the airline's station manager, the Lead Agent was going to proceed with the pushback of the aircraft using the same pushback unit that was involved in the event. The pushback unit had not been inspected apart from the front external area that impacted the tow bar. At the insistence of the station manager, the pushback unit was taken out of service for inspection immediately following the event.

Investigation and Findings

- The Lead Agent and the Ramp Agent were both interviewed by local leadership of the ground handling company immediately after the event. Written statements from both agents were also obtained.
- During the initial interview, the Lead Agent reported feeling rushed to get the pushback of the aircraft completed due to approaching thunderstorms moving into the area.
- The ground handling company conducted a formal internal investigation meeting the day after the event to discuss the incident in detail.
- A board of inquiry between leadership of the airline, the ground handling company, and the employees involved in the event was conducted two days after the event occurred.
- The pushback unit was taken out of service to be inspected by ground equipment maintenance technicians employed by the ground handling company. The airline's investigation revealed that the pushback unit's parking brake was broken and inoperative at the time of the event. This is

significant because the unit was in use with an inoperative parking brake and because, though the agent did set the brake before exiting the unit to assist the agent with the tow bar, it was not functional and, as a result, permitted the unit to roll forward on its own. A pre-operational inspection of the pushback unit might have discovered this problem, but this did not occur.

- Preventive maintenance records from the pushback unit were provided to the airline by the ground handling company. The most recent preventive maintenance inspection on the unit had been performed four weeks prior to the event in question.
- The investigation also revealed that the Lead Agent violated the policies of both the airline and his own company, as he left the engine of the pushback unit running when he exited the vehicle. He also did not install wheel chocks after exiting the unit to help his fellow agent.
- The ground handling company created and distributed "Read and Sign" memos informing their employees of the need to turn engines off before exiting vehicles and that drivers are not to exit pushback units until after a unit is connected to the tow bar.
- A new parking brake and handle were installed on the pushback. The unit was kept out of service until after the repair to the parking brake was tested and found to be functioning normally.
- Of particular concern was that after the pushback rolled and hit the tow bar, the Lead Agent intended to use the unit for the pushback of aircraft. This demonstrates a willingness to accept an *unacceptable* level of risk on the part of the Lead Agent.

Corrective Actions and Recommendations

- The ground handling company created a number of Read and Sign memos reminding employees of the following:
  - Pushback units are not to be positioned until the tow bar is connected to the aircraft and a guide person signals for the pushback unit operator to approach the aircraft to connect the tow bar.
  - When the driver leaves the cab of any ground support equipment, the ignition is to be turned off, without exception.
  - Whenever possible, pushback units shall be hooked up upon arrival rather than just prior to departure.
  - Complete proper pre-use vehicle inspections are to be performed, without exception.
  - Lock out/tag out retraining for local employees was conducted three weeks after the event, as the pushback unit should have been removed from service until the parking brake was repaired.
- The airline also recommended that all Read and Signs be distributed to all of their stations serving the airline to ensure consistency and standardization. This was accomplished.
- *Policy/Practice.* The investigation also found that while the ground handling company requires their stations to perform daily inspections of all ground support equipment, they do not require written documentation of these inspections. Although written documentation does not create a flawless system of ensuring inspections actually are being performed, a written record does nonetheless help to provide a means of tracking and establishing responsibility and accountability. Therefore, the airline recommended that the ground handling company create a standardized system for logging daily ground support equipment inspections and that the results should be maintained for a period of at least 12 months. The records would also be made available to the airline upon request. At the time of this writing, this process is being established by the ground handler.

- The airline also recommended that the ground handling company leadership ensure that a consistent lock-out/tag-out process is in place and is followed consistently at all of their stations. At the time of this writing, this process is being established by the ground handler.
- The airline requested that the local ground handling leadership discuss this event at the next monthly joint safety meeting between them and the airline and to communicate with all their stations about this event to raise awareness.

Case Conclusion

The Lead Agent failed to follow established procedures when he exited the pushback unit and left the engine running and did not chock the wheels. More significantly, he did not ensure that a full operational pre-use inspection was conducted of the pushback unit. As a result, the faulty parking brake on the unit was not discovered, and the pushback unit rolled into the tow bar and would have impacted the aircraft had it not contacted the tow bar. This was a near-miss event that could have easily produced injuries to the persons involved and damage to

Figure 3–8. Event Risk Assessment.

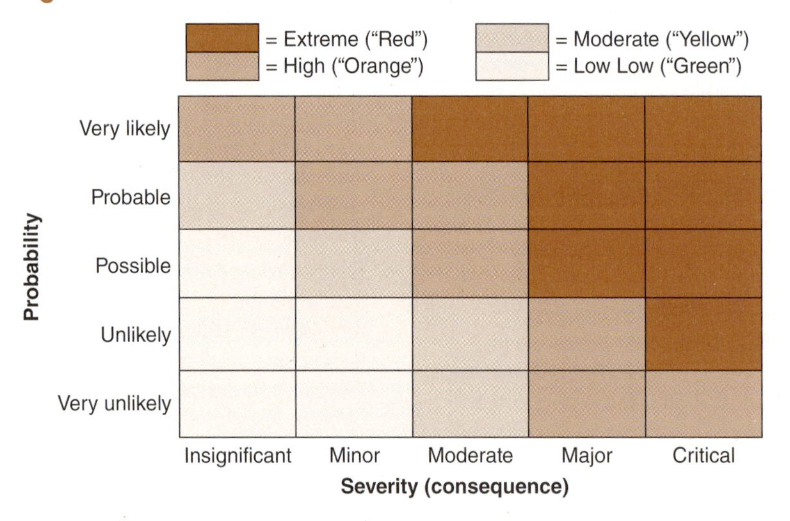

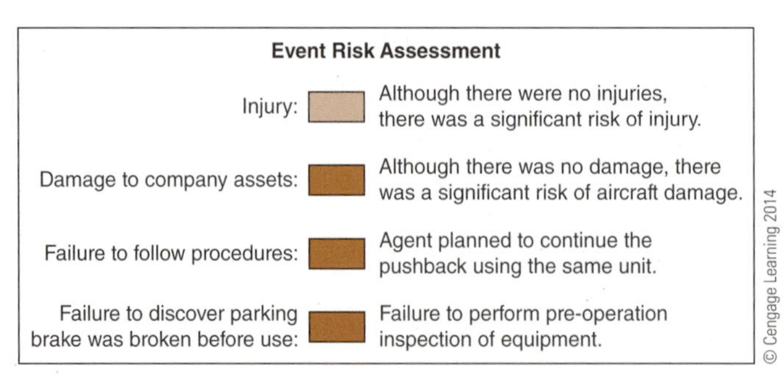

the aircraft. The Lead Agent also demonstrated a willingness to accept an unacceptable level of risk, as he was going to use the same unit to push the aircraft for departure.

As is the case in many safety events, there were several root and contributory causes in this event. Airlines and other aviation companies should be concerned with proactively identifying and mitigating the hazards involved in events of this nature so that they can work to prevent similar occurrences, and avoid undesired events involving aircraft and injury to teammates, passengers, and business partner employees.

Events of this nature necessitate looking beneath the surface evidence to discover the latent hazards and risks before they are given the chance to develop into an issue that contributes to an occurrence. The key is to detect and mitigate potential and actual problems before they cause an event to occur. Regarding the event discussed in this report, following basic procedures would likely have prevented this near-miss event from occurring.

The Lead Agent in the event was suspended pending the results of a mandatory drug and alcohol test administered soon after the event. The test results were negative. The Lead Agent was permitted to return to duty after receiving additional counseling and completing retraining on pushback procedures.

Figure 3–8 is a copy of the risk assessment that was performed by the airline's safety department during the investigation using a sample risk assessment matrix.

Chapter Summary

In this chapter, the terms "risk" and "hazard" were defined and their key differences described. The very notion that risks are a part of everyday life, whether at work or at play, is often something taken for granted or overlooked as a result of complacency. The key consideration for being considered "safe" is the determination of which risks are considered acceptable, whether as a company or as an individual.

The basic concepts of risk analysis and risk assessment were demonstrated, and risk reduction techniques were described. A risk analysis, as defined in this chapter, is the process whereby hazards are characterized according to their likelihood and severity. A risk assessment is a tool used to combine the impacts of the risk elements discovered during the risk analysis and comparing them against some acceptability criteria (a range or numbering system). This will provide a measurement criterion for determining if an alternate solution (engineering, administrative, etc.) provides a lesser risk or if the risk is too great to consider at all. The FAA's TEAM approach was described as a risk reduction approach as well. These are proactive measures for reducing risk within an organization/operation, with the expectation that injuries, damage, and so on will also be reduced.

Including various personnel, such as safety committee members, supervisors, and others, encourages cross-functional participation and is a good tool for employee empowerment. As mentioned in other chapters of this text, employee empowerment breeds good morale and attitude among employees and provides them with a sense of ownership of the process. Thus, the risk management process can certainly prove to be a positive for both the employees and the organization.

Chapter Concept Questions

1. Describe how a risk can be considered "acceptable."
2. Probability can be observed as the ______________ and ______________, the product of which is the occurrence opportunity.
3. Explain who should be involved in the risk assessment process and why.
4. ______________ is the process by which an organization can mitigate risk through a change in the process, procedures, equipment, behavior, and so on.
5. Describe the FAA System Safety TEAM approach.

Chapter References

Eaton, G. H., and Little, D. E. 2011. Risk: Assessing and Mitigating to Deliver Sustainable Safety Performance. *Professional Safety Magazine*, July.

Federal Aviation Administration. 2000. *FAA System Safety Handbook*, Chapter 3: Principles of System Safety.

Loghry, J. D., and Veach, C. B. 2009. Enterprise Risk Assessments. *Professional Safety Magazine*, February.

National Safety Council. 2007. *Aviation Ground Operation Safety Handbook* (6th ed.). Itasca, IL: National Safety Council.

Peterson, D. 1989. *Techniques of Safety Management: A Systems Approach*. Goshen, NY: Aloray, Inc.

U.S. Air Force System Safety Handbook, Chapter 3: Risk Management. 2000. Kirtland AFB, NM: Air Force Safety Agency. http://www.system-safety.org/Documents/AF_System-Safety-HNDBK.pdf.

4 Introduction to Safety Management Systems

Chapter Learning Objectives

After completing this chapter, the reader should be able to:

- Explain the basics of Safety Management Systems (SMS) and their function.
- Discuss some key components, features, and the framework of ICAO and FAA SMS requirements.
- Describe the basics of how SMS can transform the safety culture of an organization.
- Explain some of the regulatory and standardization issues driving the development and implementation of SMS globally.
- Discuss the importance of including SMS in the overall business plan of an organization.
- Explain the basics of the eight components of safety management.
- Describe the four "pillars" of SMS.
- Discuss the basics of the levels of SMS (Level 0–4).
- Describe how SMS can move an organization from being reactive to proactive to predictive.
- Explain the basics of SMS applicability in business aviation, including the IS-BAO code.
- Discuss the importance of including aviation stakeholders in the SMS process.

Key Concepts and Terms

Acceptable Level of Safety (ALoS)
Detailed Gap Analysis
Gap Analysis
ICAO SMS Features:
 Systematic
 Proactive
 Explicit
Implementation Plan
International Business Aviation Council (IBAC)
International Organization for Standardization (ISO)
International Standard for Business Aircraft Operation (IS-BAO)
National Business Aviation Association (NBAA)
Organization Safety Responses and Transition:
 Reactive
 Proactive
 Predictive
Safety Management System (SMS)
Stakeholders
State Safety Program (SSP)
Toolbox

Introduction

With respect to the modern safety movement in aviation and some other industries, an important trend has gradually emerged: The status quo is becoming increasingly unacceptable. Doing things the way they have been done for years or even decades in an organization while giving only token attention to safety for reasons of compliance is simply not an advisable practice. Increasing issues of regulatory compliance, employee awareness, constant economic pressures, the increasing availability of information and related technology, and the influence of international aviation organizations have caused many aviation companies to focus on new ways of doing business in order to control costs, remain competitive, ensure compliance, and improve safety. With these changes, the relatively new position that safety is a good business investment has also emerged. Accidents, incidents, and employee injuries are extremely expensive, annually costing the aviation industry enormous sums of money, time, and other finite resources. The old way of doing things is simply not going to work if aviation companies are going to remain viable and competitive. There is too much at stake to give only minor amounts of attention and resources to safety initiatives. As discussed in Chapter 3, "Risk and Risk Management," the risks of ignoring or giving only cursory attention to safety are too great for a company to be complacent. An active approach to integrating safety into the overall business plan and culture of an organization is essential.

In the strongest sense of the word, aviation is a truly *global* industry, connecting the world by providing for the rapid transport of people, goods, and services, and is one of the most important features of the militaries of many nations. Aviation has long been recognized as vital to business and commerce, and as a necessary element in growing and protecting local, state, national, and international economies. Aviation provides convenience and expediency that simply cannot be found in other modes of transportation. Further, as is now generally acknowledged, a significant feature of aviation is that it is characteristically safe. The accident rate in the U.S. and global air transport sector has declined appreciably since the 1950s, though it has leveled off somewhat in the last couple of decades. Statistically speaking, aviation is a remarkably safe mode of transportation.

However, as inherently safe as aviation is, the areas of risk in our industry remain significant. Despite the positive overall safety record, aviation, like any industry, remains imperfect and contains some areas of significant risk. An example: As a categorized industry, air transportation has one of the highest injury/incidence rates with respect to OSHA-recordable injuries. Most of these injuries are associated with various positions such as ramp workers, maintenance technicians, and flight attendants. Another example: It is estimated that each year tens of thousands of transport category aircraft are dispatched into revenue service in the U.S. with at least one maintenance error. Fortunately, most of these errors are not of a catastrophic nature. However, a harsh reality in aviation is that many of the problems present in the system are the seemingly "small" issues that are "beneath the surface" and thus unknown to much of the leadership of an organization (see **Figure 4–1**). Because certain issues are perceived by some to be small and inconsequential, many are left unreported and uninvestigated. Over time, the continued occurrence of such small issues can accumulate and contribute to much larger and more serious incidents and even accidents.

Errors in aviation maintenance are just one example of occurrences that would fit into the iceberg concept shown in Figure 4–1. The majority of small events remain below the "waterline" unseen, yet potentially problematic in terms of safety. Thus, even though aviation is inherently safe by and large, there are still a number of hazards and areas of risk that need to be addressed. Hazards within any aviation organization that pose an unacceptable level of risk to safety need to be identified and handled through an active process of hazard identification and risk mitigation.

Accepting as status quo the fact that aviation is "safe" can in and of itself be risky, as this perception can lead to complacency, a proven enemy of safety. If an aviation organization believes that all is well in the area of safety, this belief could cause its members to be less than proactive in looking for ways to further improve safety. In order to be truly effective in both the short and long term, safety must be made an organizational priority within aviation companies. This is a point that has been and will continue be made throughout this book.

Figure 4–1. Simple example of an "iceberg" of safety issues in organizations.

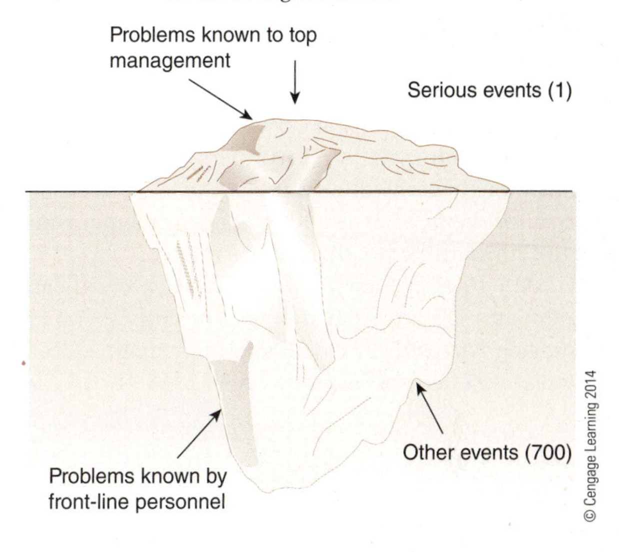

In Chapter 1, we discussed the importance of an ethical organizational culture that includes a strong safety focus as a primary feature. Changing an organization's culture to focus on safety as a key element often takes deliberate and consistent action on the part of the decision makers and leadership of the organization. The focal point of this chapter is on an important and timely change taking place in a number of aviation organizations around the world. This change represents a move toward altering the existing culture at many aviation companies in such a way that safety is integrated into the overall business plan of the organization. This is being accomplished through the adoption and implementation of Safety Management Systems (SMS). This chapter will introduce the reader to some of the basic concepts and features of SMS as identified by the International Civil Aviation Organization (ICAO) and the FAA.

Safety Management Systems

A **Safety Management System** (**SMS**) is a system composed of policies, processes, procedures, and practices adopted and implemented to enhance the overall safety of an organization. Fundamentally, SMS is the application of general business management practices to the active management of organizational safety. It includes specific practices such as gap analysis; hazard identification and analysis; risk mitigation; incident and accident investigation; active reporting of potential and real safety problems and actual events; data collection, analysis, and dissemination; corrective action; and continual monitoring and improvement of the safety system to determine areas of success and those needing modification. SMS is therefore an active process that represents a significant move from the traditional reactive safety culture present in many organizations to a deliberate, proactive safety process.

At present, SMS is becoming a safety standard in many sectors of the global aviation industry. SMS has been officially recognized by the FAA, the Joint Planning and Development Office (JPDO), ICAO, and the civil aviation authorities of many nations, as well as a number of aviation companies as the next progression in aviation safety. In fact, ICAO's Standards and Recommended Practices (SARPS) specify that an SMS be incorporated into the governing safety regulations for operators of non-commercial aircraft having over 12,500 pounds maximum takeoff weight (MTOW) or turbojet aircraft. At a minimum, the SMS for commercial and non-commercial aircraft operators must actively identify safety hazards and areas of risk, continuously monitor and assess the SMS to determine effectiveness, work toward the continuous improvement of the SMS, and implement any corrective actions necessary to obtain and maintain minimum safety performance.

SMS is not currently required for U.S. certificate holders; however, under Public Law 111-216, signed into law on August 1, 2010, the FAA was required to produce a rule by August 1, 2012, requiring Part 121 air carriers to implement SMS in their organizations. The following section of the law was derived from the FAA's SMS "Read Me First" document.

House Resolution 5900 was signed into Public Law 111-216, on August 1, 2010.

Section 215 requires that:

A. *Federal Aviation Administration shall conduct a rulemaking proceeding to require all part 121 air carriers to implement a safety management system.*

B. *In conducting the rulemaking, the Administrator shall consider, at a minimum, including each of the following as a part of the safety management system:*

   (1) An aviation safety action program.
   (2) A flight operational quality assurance program.
   (3) A line operations safety audit.
   (4) An advanced qualification program.

C. *The Administrator shall issue—*

   (1) Notice: A Notice of Proposed Rulemaking (NPRM), not later than 90 days after the date of enactment of the Act, and
   (2) Regulation: Not later than 24 months after the date of enactment of the Act, a final rule.

Source: U.S. Department of Transportation/FAA

SMS principles are also a standard for safety management in fields other than aviation, such as occupational safety and health, medicine, security, and environmental protection. Safety Management Systems integrate contemporary risk management and assurance concepts into a system that is replicable, proactive, and preventive by actively seeking areas of hazard that may cause safety-related problems for an organization, mitigating the associated risks, and continuously monitoring the overall system to ensure safety improvement and efficiency. On its website, the FAA states the following:

By recognizing the organization's role in accident prevention, SMSs provide to both certificate holders and FAA:

- *A structured means of safety risk management decision making*
- *A means of demonstrating safety management capability before system failures occur*
- *Increased confidence in risk controls through structured safety assurance processes*
- *An effective interface for knowledge sharing between regulator and certificate holder*
- *A safety promotion framework to support a sound safety culture*

Source: U.S. Department of Transportation/FAA

SMS and the Company Business Plan

By design, an SMS should be integrated within the organization's overall business plan and operations. By adopting an SMS as a primary element of its business plan, an organization can identify the actual and potential areas of risk specific to their operations and implement responses appropriate to each area of concern throughout the company. It is important to understand that safety management is not restricted to one or more specific functions of an aviation organization. Rather, an SMS should be integrated into each operational activity (flight, ground and maintenance operations, cargo, dispatch, flight attendants, etc.) that contains hazards that could directly impact the organization. An SMS may also be integrated into non-operational departments of an organization such as human resources, logistics and procurement, and finance. Thus, SMS is ultimately a top-to-bottom business approach toward organizational safety enhancement and protection.

In order to ensure effectiveness, an SMS needs to be specific to the type of organization in which it will be applied and its various modes of operation. For example, a corporate flight department has a different focus, operations, and goals than an airline, and the SMS of each organization must be reflective of and specific to how each operates. In other words, an SMS is scalable to fit the size, needs, and scope of an organization. Ideally, an SMS should enhance safety while not compromising operational efficiency; these two elements can work together for the overall benefit of the company. An SMS should also provide the organization and its personnel with the appropriate tools for managing their areas of risk and efficiently allocating resources.

SMS as a Safety Toolbox

The International Civil Aviation Organization (ICAO) asserts that an SMS can be viewed as an organizational **toolbox** containing the items an aviation company needs to identify, manage, and control the risks to safety posed by the hazards present in their various operations. The SMS itself is not the tool; rather, the SMS contains the tools needed to perform hazard identification and risk management. An SMS must be matched in an appropriate manner to each organization.

By functioning as an organizational safety toolbox, an SMS should be designed to ensure that the proper safety tools commensurate with the specifics of the organization are available, that tools and tasks are matched, and that all tools are readily available and accessible to the employees.

SMS and Safety Culture

As discussed in Chapter 1 and in the Introduction to this chapter, adopting a Safety Management System may represent a means of changing the culture of an organization to one that has a strong safety focus. However, before we go any further on this point, consider the following.

Whether intentional or unintentional, any aviation organization (regardless of size) has a prevailing safety culture that develops over time. Even in the absence of a deliberate safety program or focus, a safety culture develops naturally and will come to characterize the organization's working conditions, policies, expectations, and concern for employee safety. This type of unintentional culture is generally unfocused, and safety may be given only cursory attention at best. Further, even within an aviation company that has a specific safety program, the safety culture is not automatically strong, consistent, or deliberate. It is often difficult to convince people that they need to do things differently than they have been doing for years, especially if change means that more time, effort, and money need to be spent on safety. This is one of the hard facts about organizational safety cultures. Establishing a strong safety culture happens through consistent, deliberate effort, not by leaving things to develop on their own.

Therefore, an SMS can enhance the safety culture of an aviation company significantly. Just as a weak safety culture will make it difficult for an SMS to achieve the desired results, an SMS may enable a positive safety culture to develop and flourish in an aviation company. In fact, it can be asserted that the adoption of an SMS will foster the development of an intentional proactive and preventive safety culture by promoting safety; identifying and addressing hazards and risks; and showing real concern for the company, its employees, and its assets. This may help create an environment that is conducive to continual assessment and education, and to openly sharing information with employees, stakeholders, and even other aviation organizations with similar operations—that is, airlines sharing safety information to the mutual benefit of all.

The creation of a positive safety culture has been directly identified as a necessary primary element for developing a successful SMS. In their various SMS publications, ICAO regularly addresses the need for senior leadership to support safety management. The FAA has also directly expressed the need for the active promotion of a positive safety culture in their SMS document entitled *Aviation Safety (AVS) Safety Management Requirements*, reproduced here as follows (FAA Order VS 8000.367, p. B-8):

7. Safety Promotion.
   a. Safety Culture. Top management must promote the growth of a positive safety culture demonstrated by, but not limited to:
      (1) publication to all employees of senior management's stated commitment to safety;
      (2) communication of safety responsibilities with the organization's personnel to make each employee part of the safety process;
      (3) clear and regular communications of safety policy, goals, objectives, standards and performance to all employees of the organization;
      (4) an effective employee reporting system that provides confidentiality and de-identification as appropriate (as described in Chapter 6, Section c);
      (5) use of a safety information system that provides an accessible, efficient means to retrieve information; and
      (6) allocation of resources to implement and maintain the SMS.

Source: U.S. Department of Transportation/FAA

How SMS Addresses the Organization's Role in Safety

With respect to aviation safety, organizational responses to significant events (accidents, incidents, etc.) can be grouped into three basic types: reactive, proactive, and predictive (see **Figure 4–2**). The traditional approach of most aviation organizations (including the FAA) to accidents and incidents has been largely **reactive**: Changes are made only when and after something bad has occurred, such as an aircraft accident. Although it is necessary and important to react to accidents and incidents in terms of rescue, investigation, and fact-finding, this approach does little to enhance safety until after the fact because it is post-event and therefore not forward looking.

The reactive phase can nonetheless yield important data to aid in determining what caused an event and what can be done to try and prevent the next such event. The reactive mode is considered to be an integral part of the overall safety management process, helping to determine what, why, and how an event happened.

Figure 4-2. The Role of SMS in Organizational Safety Transition.

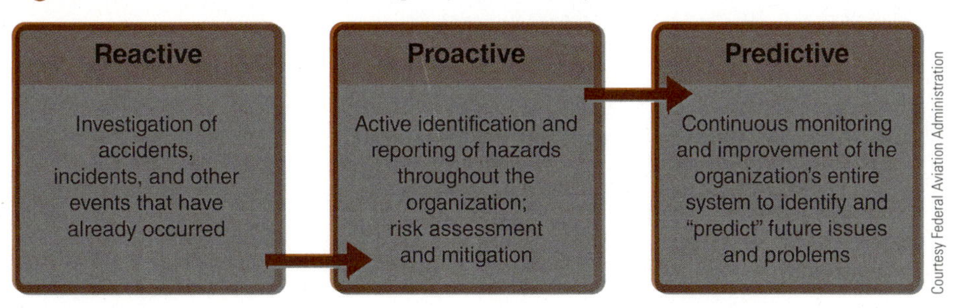

A **proactive** approach, as the name implies, works to identify issues that are or may be of concern before they become problems. The proactive approach is based on the belief that the probability of system failures in an organization can be minimized or reduced by identifying safety hazards and risks and taking the necessary actions to mitigate risk to an acceptable level. Examples of actions within a proactive system include mandatory and voluntary safety reporting systems, conducting safety audits, trend analysis of data, hazard identification, and risk mitigation. An organization's activities and operations are observed and analyzed constantly to identify potential problems ahead of time.

The **predictive** approach is forward thinking in that it is used to identify and anticipate problems and issues well before the fact, based on how the organization functions. Operational data are continually captured, as events occur. Prediction is based on the concept that safety management can be best accomplished by trying to find potential and actual problems before they can become real-time incidents rather than reacting when they occur. Therefore, the predictive approach to safety aggressively seeks safety information that may be indicative of active, emerging, or latent safety risks from a variety of sources within the organization. The predictive method captures the status of system performance as it happens in real-time daily operations to identify potential future problems and address them appropriately. Being predictive allows an organization to better anticipate where and when problems and safety issues are likely to occur, based on historical performance.

By design, an SMS requires that an aviation organization conduct a thorough overview of its entire operation to assess its current safety status. Properly implemented and managed, an SMS fosters the continuous improvement of safety through active identification and prediction of hazards and areas of risk. Information and data pertinent to hazards and risk will typically be obtained through reports of personnel in the organization. This information should then be used to further assess and manage areas of risk affecting the company, which also involves the continual monitoring of the SMS itself to ensure effectiveness. Modifications to the SMS should be made as need becomes apparent through continuous monitoring. One of the intentions of SMS is to assist in encouraging the safety culture of aviation organizations to undergo a "metamorphosis" of sorts, changing from a state of reactive response to proactive response, and ultimately to predictive response in order to anticipate safety issues and concerns before they become problems. Continuous improvement of the overall safety of the organization is essentially the goal of SMS.

In-Focus: Safety Management Systems

Bunty Ramakrishna is the Safety Management System (SMS) Program Manager for a large airline in the southern U.S. She is considered to be an expert in airline safety management system development and has guided her company into a mature and advanced phase of SMS development. In this brief excerpt from an interview conducted in late fall 2010, Bunty discusses the impact of SMS on her company.

Traditionally, air transportation has safety organizations managing and maintaining their safety programs, interacting with the operating divisions as needed. The development of a comprehensive SMS business model at our company has enabled the operation to own and control the safety implication specific to their division, enabling the operating division to interact with the safety organization as needed. This helps the safety organization have a higher degree of oversight, improving accountability and responsibility of the operating division. In our company, SMS allows the divisional leadership to have a better insight of their safety performance, empowering them to influence the outcome and contribute to a sustainable safety in the operation in the long term.

SMS bridges the gap between all operating divisions in our company. Top management address how each dimension of the organization is doing and manages the operation in a safe manner. So how do they do this? All the divisions come with their addressed safety issues, use the processes they have built to identify the hazards, assess the risk using the SMS risk management technique to the table. They then collaborate and unify to address the issues of high risk. Of course this has to be documented, tracked; risks rated, assessed, mitigated, and finally the data must be collected for use in aggregate analysis. To summarize, all they are doing is identify their hazards, channel this information through the process, and put it through the machine of SMS. This brings visibility and connects top down and bottom up. Now all departments are working toward the same goal collectively in unified manner, which helps to eliminate the traditional silo approach. It's a unified strategy to address safety. That's what an SMS does, at least in our company.

SMS enables a company to progress from being reactive to proactive to predictive, meaning you are able to anticipate problems and hazards based on what you have learned previously. You have accumulated all this data and have determined what the structure needs to look like to bridge the gap; now all you need to figure out is how to analyze the data proactively and predict before an incident or an accident can occur. Collecting and looking at safety data, proactively, and managing data are critical to a good SMS. It's a long-term process and a company cannot achieve this overnight. SMS slowly shrinks the reactive tendencies of a company while expanding the proactive. It impacts the operation and company from a cultural point of view.

One of the positive things we have had at our company is we believe SMS is the future of global aviation and the leadership believe it is the right thing to do. It is commitment of the management at the highest level that brings it all together. We were able to bring all the senior leadership to the table, commit to SMS and now we have achieved greater interdepartmental cooperation. This has resulted in quicker decisions being made on cross-divisional issues, addressing issues that pose a greater and higher risk, and allocating resources where needed. Risk management tells you where you can best allocate your business resources and SMS has enabled us to do this.

SMS Standardization and Regulatory Requirements

As mentioned previously in this chapter, ICAO mandated that a basic SMS be put into practice by both commercial and non-commercial operators of aircraft over 12,500 pounds maximum MTOW and/or turbojet aircraft. Furthermore, ICAO requires SMS for safety risk management in other aviation areas such as aircraft maintenance, air operations, air traffic control, airports, flight training, and aircraft design and manufacturing. According to information found on their website, the **National Business Aviation Association (NBAA)** pointed out that at the time of this writing, the 189 member nations of ICAO are in the process of creating regulations and guidelines to meet the ICAO requirements for SMS. The FAA has developed Advisory Circular (AC) 192-20A (2010), which provides specific guidelines for aviation organizations seeking to implement an SMS. AC 192-20A is considered the governing document for SMS development in the U.S. As of the time of this writing, the FAA is in the process of creating specific SMS regulations, and, as described above, in accordance with P.L. 111-216, it is anticipated that the SMS requirements for Part 121 certficated airlines will be put into effect by August 1, 2012.

ICAO has created and published SMS requirements for their member nations around the world, mandating that ICAO member states establish what is known as a **State Safety Program (SSP)**. Notification to member nations of the SSP guidelines and training was made on November 18, 2008, in the form of a letter distributed by ICAO. The following passage is taken from the SSP document on the ICAO website:

> *Under an SSP, safety rulemaking is based on comprehensive analyses of the State's aviation system; safety policies are developed based on hazard identification and safety risk management; and safety oversight is focused towards the areas of significant safety concerns or higher safety risks. An SSP thus provides the means to combine prescriptive and performance-based approaches to safety rulemaking, policy development and oversight by States.*

Source: ICAO

One purpose of an SSP is to enable member states to achieve an **Acceptable Level of Safety (ALoS)** within the civil aviation system of their respective nations. ALoS helps to ensure that minimum acceptable levels of aviation safety performance are achieved by participating ICAO member nations.

Because international promotion, cooperation, and standardization of aviation is a primary function of ICAO, the organization is interested in the various civil aviation authorities of their member nations working together to standardize, collaborate, and implement SMS and SSP requirements. International cooperation and standardization of SMS efforts should prove beneficial to the continued globalization of the aviation industry by ensuring that the aviation regulatory agencies of the world are developing and implementing SMS protocols in a similar fashion, promoting a standardized format from one nation to the next. This also includes sharing of SMS best practices, lessons learned, and program results among nations and among the various aviation regulatory bodies and operators. ICAO points out that individual aviation organizations may also benefit from SMS standardization, since many companies operate internationally and carry multiple operating certificates in a number of different countries.

In AC 192-20A, the FAA describes the importance of SMS standardization and harmonization, pointing out that many of their SMS precepts are similar to internationally recognized standards for quality control management, environmental protection programs, and occupational safety and health management. For example, the standards for SMS are written very much in line with the **International Organization for Standardization (ISO)** standards for quality management (ISO 9000) and management of environmental protection (ISO 14000). By design, the SMS standards were developed to parallel ISO 14001. The FAA SMS standard was produced after an exhaustive review of SMS information and systems in other nations. Information and management standards from the British Standards Institute (BSI), the American National Standards Institute (ANSI), the International Air Transport Association (IATA), the International Business Aviation Council, and other organizations were reviewed by the FAA in the development of their SMS standard.

Safety Management Components and Responsibilities

The process of safety management involves a number of different features and components, each of which is considered both fundamental and essential to the safety management process. As applied toward the development of SMS, ICAO has identified eight essential safety management elements. These are discussed briefly in the following section.

Eight Components of Safety Management

1. *Commitment of Senior Leadership to Safety Management* An organization's leadership must fully and consistently support safety management in order for it to be successful. Part of the support needs to be financial, as safety initiatives require funding. However, just as important as providing financial support, senior leadership must also support the SMS consistently through their words and actions. This element was discussed at length in Chapter 3, "Risk and Risk Management."
2. *Effective and Consistent Reporting of Safety Issues* Quite simply, that which remains unknown cannot be changed or prevented from occurring. In order to be effective, the management of safety requires information and data in the form of reports completed and submitted by employees of an organization. Most of these reports should be of a voluntary nature, except those involving a definitive and/or immediate safety threat, which requires mandatory reporting. The data gathered from the submitted reports can then be used to conduct trend analysis in order to indentify the most common types of issues and to rank-order events and hazards according to the risk(s) they pose.
3. *Continual Monitoring* Proper safety management necessitates monitoring of the entire process within the organization to determine effectiveness, areas of concern, successes, failures, and so on. Monitoring should lead to revisions and modifications of the safety processes as needed. In short, an organization should never get to the point that members are comfortable and satisfied with respect to safety; effort must be made to ensure quality safety management.
4. *Investigation of Accidents and Other Events* As soon as possible after a safety-related event has occurred, the events should be thoroughly investigated to determine all causal factors and to develop effective, specific correction actions. The focal point of the investigation should not be to assign blame but to find out exactly what happened to cause the event so that future events have a greater probability of being prevented. All accidents, incidents, near-misses (as much as possible) should be investigated using a standardized investigation process throughout the organization. Again, prevention of future events is the goal of investigation.
5. *Sharing Safety Lessons Learned and Best Practices* Safety is one area of the aviation industry that should not be competitive. Rather, sharing safety-related information both within the organization and with other entities is potentially beneficial to the industry as a whole. For example, an airline sharing safety information with other airlines and organizations that may benefit is a best practice that may produce benefit to all.

6. *Provision of Safety Training for Operational Personnel* People cannot be expected to know and follow that which they do not understand. Providing specific, comprehensive safety training will help to ensure that employees know the expectations and policies related to safety that they are expected to follow.
7. *Effective Implementation of Standard Operating Procedures (SOPs)* Everything related to safety management and programs needs to be put in writing in the form of organizational SOPs. This information needs to be written clearly, revised as necessary, and maintained in such a manner that it is continuously accessible by all employees. When revisions are made to safety procedures, manuals and other SOPs need to be updated as soon as possible.
8. *Continuous Improvement of the Overall Level of Safety* In order to ensure the greatest potential for effectiveness and long-term success, the management of safety must be a continual, ongoing activity. Modifications and changes within the organization must be made as problems are identified.

SMS Components and Elements

With respect to the framework of an SMS, both the FAA and ICAO have identified four primary components that must be present. Each of these contains a number of elements specific to each component. These components and their accompanying elements, derived from the FAA and ICAO, are discussed in the following section.

The Four Pillars of SMS

Component: *Safety Policy*

The SMS safety policy contains all of the particulars, methods, processes, guidelines, expectations, and features of the SMS. The safety policy also demonstrates the commitment of the senior leadership to continually improve safety in the organization.

Elements of the Safety Policy:

- Establish management commitment to SMS for safety improvement
- Appoint specific safety personnel
- Establishment of specific safety objectives and management of the objectives
- Outlining and documentation of specific SMS methods, processes, practices, and the structure needed to achieve safety performance goals
- Establish and foster transparency through safety management
    - Voluntary hazard and event reporting system
    - Establish accountability of management and all employees under SMS
- Establishment and coordination of emergency response planning

Component: *Safety Risk Management*

This component determines the need for risk identification, based on the assessment of acceptable and unacceptable risk in an organization.

Elements of Safety Risk Management:

- Identification of hazards
- Risk assessment and mitigation, including:
    - Identification and assessment of risk(s)
    - Analysis of risk(s)
    - Control of the identified risk(s)

Component: *Safety Assurance*
This component provides for assessing the effectiveness of risk management strategies and a continual improvement of the SMS in the organization.

Elements of Safety Assurance:

- Monitoring and measurement of safety performance through:
  - Internal and external audits
  - Employee safety reporting
  - Data and trend analysis
- Assessment of the overall system
- Continuous improvement of the SMS
- Compliance with all SMS requirements and ICAO/FAA standards, policies, and directives
- Management of change through analysis regarding processes and opportunities to improve safety and minimize risk

Component: *Safety Promotion*
This component provides for training, communication, promotion, and so on in order to foster the development of an excellent safety culture within the organization.

Elements of Safety Promotion:

- Provide effective SMS training and education for employees
- Continual creation and promotion of a positive, strong safety culture
- Promotion of safety communication and awareness throughout the organization
- Share information on safety lessons learned, successes, failures, and so on, internally and externally

SMS Levels and Phased Implementation

Under both ICAO and FAA guidelines, SMS follows a phased implementation strategy within an organization. The FAA SMS process features five levels (0–4), as illustrated in **Figure 4–3.** These levels are briefly described below.

Level 0: This is where an organization commits to developing and implementing an SMS. The top management expresses its support (in writing) for SMS, and the organization begins to develop the initial basic strategy for creating their SMS and their goals and objectives.

Level 1: The gap analysis phase. The organization's operating departments (flight operations, maintenance, etc.) begin the process of conducting a **detailed gap analysis** and the development of their **implementation plan.** The detailed gap analysis and the implementation plan are described in more detail below.

Level 2: The reactive phase. The organization develops and implements a basic safety risk management strategy, including active hazard identification, reporting, risk mitigation, investigation of safety events and issues, and gathering and processing of safety data. Known problems are addressed and preventive measures are put in place. Undesired safety issues and events are responded to as they occur; hence the name "reactive."

Figure 4–3. SMS Levels.

Level 3: The proactive phase. Safety risk management is applied to the initial design of systems, processes, organizations, products, policies, development of operational procedures, and planned changes to the operation. This process includes careful analysis of systems and tasks involved, identification of potential hazards in these functions, and development of risk controls. Risk management is actively used to analyze, document, and track safety functions. Because the organization is now using the processes to look ahead, the term "proactive" is used. However, although proactive processes have been implemented, their performance has not yet been proven.

Level 4: The continuous improvement phase. Safety processes have been in place and their effectiveness has been verified. The safety assurance process, including continuous monitoring and safety risk management, are functioning. The main objective of a successful SMS is to achieve and maintain the continuous improvement status throughout the life of the organization. At this point, the SMS is implemented fully into the business plan of the organization.

ICAO SMS Features

According to ICAO, an SMS is characterized by three distinct features: It is systematic, proactive, and explicit. As applicable to an SMS, **systematic** means that safety management programs and initiatives are applied throughout the aviation organization as part of their overall operational plan. The systematic plan ensures consistency of the SMS across the organization. It is important to note here that the goal of the SMS is not immediate, radical change, but rather consistent improvement of safety performance over time.

An SMS is **proactive** because the primary focus of an SMS is the identification and effective management of hazards and areas of risk before problems can occur. An SMS is a deliberate process of improving the safety of an organization through effective risk management. The proactive nature of SMS is also seen in the fact that an SMS is a continuous process of monitoring and seeking ways to further improve safety performance.

A key component of the proactive nature of the SMS is found in the process of gap analysis, which is the examination of the system of the organization and its existing safety processes as compared to what is needed for an SMS to be fully established and operational. Gap analysis is important for identifying the status of the organizational structure needed to operate an SMS and ensuring that the existing structures of the organization are utilized to develop the SMS.

There are two main objectives of gap analysis: (1) identification and response to any mismatches or "gaps" in the interactions between the various components of the organization's system (areas of risk); and (2) identification of any additional resources needed to address the areas of risk (close the gaps) and help the persons responsible accomplish their responsibilities. The completion of a detailed gap analysis is a primary component of the initial SMS implementation process in an organization (SMS Level 1). A detailed **gap analysis** involves the completion of a comprehensive checklist for each operating department or division within the organization. The FAA has produced a detailed gap analysis checklist that is in used by a number of U.S. airlines. The completion of the detailed gap analysis should be accompanied by the completion of an implementation plan, which is a written framework that addresses how the organization plans to close each gap identified. Both the detailed gap analysis and implementation plan are completed as required components of SMS Level 1.

Finally, an SMS is **explicit**, meaning that all safety management processes are thorough, comprehensive, consistent, documented, readily available to employees, and transparent throughout the organization. The SMS is a written document and process that is clear, carefully outlined, and revised as often as necessary. The explicit nature of SMS also necessitates that all employees receive SMS training, as mentioned previously.

SMS in Business Aviation

The global movement toward the adoption of SMS is not restricted to the world's air carriers; it is also present in business and corporate aviation. Through SMS, organizations operating business aircraft can receive all the same SMS benefits as the air carriers. The NBAA supports and advocates that all business aircraft flight departments create and adopt their own SMS tailored to the size and specifics of their operation.

To assist operators of business aircraft in the development of an SMS, the **International Business Aviation Council (IBAC)**, the NBAA, and other IBAC member associations developed the **International Standard for Business Aircraft Operation (IS-BAO)**. IS-BAO is a code of best practices created to assist corporate flight departments around the world achieve and maintain high levels of professionalism and safety. The primary feature of IS-BAO is an SMS tool for all business aircraft operators to use in developing their own SMS. This SMS tool is scalable in order to accommodate the entire range of corporate flight departments, from very small to very large. The IS-BAO tool satisfies ICAO SARPS for SMS.

The Stakeholders of SMS

The aviation industry contains many **stakeholders**. For the purposes of a simple definition, an aviation stakeholder is any individual or group that may have a vested interest in, or may be directly or indirectly impacted by, the decisions and operations of an aviation organization. All stakeholders in aviation may play a role in the SMS process. Because the actions and operations of an aviation organization may create direct or indirect impacts on stakeholders, it is important for an aviation organization to identify these persons. Also, working to include the stakeholders in the SMS process is a best practice that should be followed to ensure that their input related to safety decisions is considered before decisions impacting safety are made. Both ICAO and the FAA agree with the practice of involving stakeholders in SMS decisions.

The following is a partial list of the primary aviation stakeholders to consider in the SMS process:

- Owners/operators of aircraft (corporate, air carriers, private owners, etc.)
- Regulatory agencies (FAA, OSHA, EPA, etc.)
- International aviation governing bodies and organizations (ICAO, IATA, etc.)
- Aviation manufacturers
- Aviation professionals
- Employees of aviation organizations
- Air traffic control
- Aviation industry trade associations
- Aviation investigation agencies (NTSB, etc.)
- Members of the flying public

Because each of these stakeholders and others have a clear interest in aviation safety, each will likely bring a unique perspective and varied areas of expertise to the SMS process. Stakeholders can assist the decision makers in aviation organizations by ensuring that communication about the safety risks under consideration occur early in and consistently throughout the SMS process in an objective, reasonable, and clear manner. Involving stakeholders should be a fundamental part of the establishment of an SMS in an aviation organization.

Chapter Summary

In this chapter, we have introduced the reader to the basic concept of Safety Management Systems (SMS). We have provided an overview of the SMS, how it functions, and its importance in promoting safety within aviation organizations. We have provided an overview of the eight elements of the safety management process. We have discussed some of the specific ICAO and FAA requirements regarding what should be contained in an SMS, such as identification and mitigation of hazards and areas of risk, continual monitoring and improvement of the SMS, and employee reporting systems. The idea that SMS should be a fundamentally important component of the overall business plan of an aviation organization was stressed.

We have also described how the presence of a functional, properly managed SMS can improve the safety culture of an aviation organization. Also, an overview of the role SMS may play in transforming an organization through moving from being reactive to proactive to predictive was discussed. We have covered the basics of SMS standardization and regulatory issues current at the time of writing. We provided a brief overview of the importance of gap analysis in the SMS process. Also, we touched upon SMS and its role in business aviation, including a short discussion of IS-BAO. Finally, we described the importance of including the various aviation stakeholders in the SMS process.

In summary, SMS is not just a current aviation "hot topic" or fad; it is an extremely important process designed to improve the safety of aviation organizations around the world. The process was designed to be highly adaptable and standardized so that a large number of aviation organizations will follow the same guidelines and procedures for SMS creation, implementation, and management, though with some room for variance to account for size, type, and location of different organizations. Nonetheless, the basic elements and components of SMS remain the same. SMS is a fundamental move away from the reactive safety cultures that have characterized aviation for decades toward a deliberate, intentional, proactive, and predictive safety process that actively seeks to identify and address hazards and areas of risk before problems can occur. SMS is the future of aviation safety, and that future is here and now. SMS may make the aviation industry safer, both domestically and internationally, as more and more organizations establish effective SMS programs and reap the eventual benefits.

Chapter Concept Questions

1. In your own words, what is a Safety Management System (SMS)?
2. What are the primary features and components of an SMS?
3. How can an SMS have a positive impact on the safety culture of an aviation organization? On its overall culture?
4. Describe in detail how an SMS can help an organization transition from reactive response to proactive response to a predictive approach?

5. What similarities exist between ICAO and FAA requirements and guidelines for SMS development and implementation? Explain.

6. Name and discuss in detail the eight components of safety management.

7. Why is it important to include aviation stakeholders in the SMS process for an organization?

8. How might an SMS assist an aviation organization with the process of risk management? Explain in detail; be specific.

9. Why is having an employee reporting process an essential component of SMS?

10. What are the three ICAO SMS features described in this chapter? Discuss each of the three features in detail.

11. Conduct your own research outside this chapter and explain how the ICAO and FAA SMS features are aligned with the International Organization for Standardization (ISO) 9000 and 14000.

12. From a global perspective, why is SMS standardization important from nation to nation?

Chapter References

Alteon, A Boeing Company. n.d. Maintenance Human Factors Program Training for Managers. PowerPoint presentation.

Federal Aviation Administration. 2006. *Introduction to Safety Management Systems*. Advisory Circular 120-92.

Federal Aviation Administration. 2008. *Aviation Safety (AVS) Policy*. FAA Order VS 8000.370 . http://www.faa.gov/documentLibrary/media/Order/VS%208000.370.pdf.

Federal Aviation Administration. 2008. *Safety Management System Guidance*. FAA Order 8000.369. http://www.faa.gov/documentLibrary/media/Order/8000.369.pdf.

Federal Aviation Administration. 2008. *Safety Management System Requirements*. FAA Order 8000.367 Aviation Safety (AVS). http://www.faa.gov/documentLibrary/media/Order/VS%208000.367.pdf.

Federal Aviation Administration. 2010. *Safety Management System (SMS)*. http://www.faa.gov/about/initiatives/sms/.

Federal Aviation Administration. 2010. *Safety Management Systems (SMS)*. Implementation Guide for Safety Management System (SMS) Pilot Project Participants and Voluntary Implementation of Organization SMS Programs.

Federal Aviation Administration. 2010. *Safety Management Systems*. "Read Me First" for Safety Management System (SMS) Pilot Project Participants and Voluntary Implementation of Organization SMS Programs. Advisory Circular 120-92A.

Federal Aviation Administration. 2010. *Safety Management Systems for Aviation Service Providers*. Advisory Circular 120-92A.

Goetsch, D. L. 2008. *Occupational Safety and Health for Technologists, Engineers and Managers*. Upper Saddle River, NJ: Pearson Prentice Hall.

International Civil Aviation Organization. 2008. Implementation of the State Safety Programme (SSP) in States. http://www.icao.int/anb/safetymanagement/070e.pdf.

International Civil Aviation Organization. 2009. *Safety Management Manual. Montreal*, Quebec, Canada: International Civil Aviation Organization.

International Civil Aviation Organization. 2010. SMS Training. http://www.icao.int/anb/safetymanagement/training/training.html.

International Organization for Standardization. 2010. ISO 9000 and ISO 14000. www.iso.org/iso/iso_catalogue/management_standards/iso_9000_iso_14000.htm.

Joint Helicopter Safety Implementation Team of the International Helicopter Safety Team. 2007. Safety Management System Toolkit. http://www.ihst.org/Portals/54/SMS-Toolkit.pdf

National Business Aviation Association. 2010. ICAO Annex 6, Part 2: SMS Provisions. http://www.nbaa.org/admin/sms/icao-annex-6.php.

National Business Aviation Association. 2010. Safety Management System (SMS). http://www.nbaa.org/admin/sms/overview/.

Occupational Safety and Health Administration. 2008. Effective Workplace Safety and Health Management Systems. http://www.osha.gov/Publications/safety-health-management-systems.pdf.

Southern California Safety Institute. n.d. Safety Management Systems – Essentials (SMS-E). http://www.scsi-inc.com/SMS-E.php.

Willemsen, H. 2008. *Safety Management Systems and Safety Culture*, Part 1: Why and What. Contrail Solutions Pty. Ltd. www.contrailsolutions.com.au.

Willemsen, H. 2008. *Safety Management Systems and Safety Culture*, Part 2: How. Contrail Solutions Pty. Ltd. www.contrailsolutions.com.au.

Wood, R. 2003. *Aviation Safety Programs, A Management Handbook* (3rd ed.). Englewood, CO: Jeppesen-Sanderson, Inc.

5 Elements of Effective Aviation Safety Programs

Chapter Learning Objectives

After completing this chapter, the reader should be able to:

- Recognize and discuss some of the basic elements that should be present in an effective aviation safety program.
- Identify the basic concept of Safety Management Systems.
- Describe the importance of leadership support to successful aviation safety programs.
- Evaluate the importance of involving front-line employees in safety programs and initiatives.
- Identify the basic elements of flight safety, ground safety, environmental compliance, and other areas that should be present in a safety program.
- Describe the basics of conducting incident investigation within an aviation organization.
- Discuss the basic features of internal evaluation programs and their applicability within an aviation organization.
- Discuss the practice and importance of internal auditing.
- Understand and describe the basic elements and importance of an aviation internal safety reporting system.
- Have a basic understanding of the integration of ground safety and environmental compliance with an aviation safety program.
- Determine the elements of an effective safety committee and how it promotes employee participation.
- Identify and implement effective strategies for accident prevention.
- Describe the basic elements of emergency response and contingency planning for various types and sizes of aviation organizations.

Key Concepts and Terms

Accident
Active Listening
Auditing
Aviation Safety Action Program (ASAP)
Behavioral Observation Process
Causal Factors
Consequence Management
Demonstrations
Emergency Response Planning
Employee Involvement
Error-Chain
External Decisions
Flight Operations Quality Assurance (FOQA)
Flight Safety Manual
Front-Line Employees
Gap Analysis
Incident
Incident Command
Incident Rate
Internal Decisions
Internal Evaluation Program (IEP)
Job Safety Analysis (JSA)

Loss Type Potential
Lost Workday Rate
Lower to Middle Management
Management Commitment
Near-Miss
No-Fault, No-Jeopardy
Ownership
Postings
Pre-Shift Communications (PSCs)
Preventive Maintenance
Recognition and Rewards
Safety Culture
Safety Management Systems (SMS)
Self-Audits
Signage
Step Analysis
Strategic Leadership
Tactical Leadership
Trend Analysis
Upper Management
Voluntary Protection Program (VPP)
Worksite Analysis

Introduction

A good safety program is at once foundational and paramount in importance to any aviation organization. In many ways, a safety program is analogous to the foundation of a structure. Just as a strong foundation is essential to ensuring the structural integrity and longevity of a building, a safety program is necessary to maintain the viability and long-term success of a company. Any legitimate aviation company should make safety one of its highest priorities. Since at its core safety is a function of actions that are deliberate and intentional, one of the best ways that an organization can demonstrate its commitment to safety is by developing, implementing, and continually managing an appropriate safety program that goes well beyond basic regulatory and legal compliance.

This chapter contains information on some of the primary elements that should be present in an aviation safety program as a part of the larger organizational culture. Due to the wide variety of aviation companies in terms of size, function, location, and other factors, the intention of this chapter is not to propose a "one size fits all" approach to safety program elements, but rather to describe some of the most common and important features of aviation safety programs.

Safety Management Systems

In spite of the economic issues that have characterized some segments of the aviation industry since 9/11 (most notably the airline segment), aviation continues to expand and is becomes increasingly global in nature. A considerable number of airlines around the world have extensive domestic and international route structures, and many corporations that are reliant on commercial and general aviation operate a complex network of locations and contractor supply chains that often expand into other nations. As aviation becomes more global, it follows that safety in aviation must also increase globally. However, there must be an effective, logical approach to integrate safety into the overall operational and business structures of companies. Toward this end, the process of **Safety Management Systems (SMS)** has been developed and implemented by a number of aviation organizations around the world, as discussed in the previous chapter, and this trend will continue.

Following the precepts of a structured SMS, aviation organizations are better able to identify potential hazards, areas of risk, and other weaknesses before problems arise. These issues can then be properly managed by creating specific responses and actions appropriate to each problem area. Thus, SMS is the nucleus of an effective aviation safety program. Ultimately, the SMS *becomes* the safety program throughout a company.

The objective in discussing SMS again here is to stress that SMS should be the overarching structure of an aviation organization's safety system. The SMS concept contains a number of principles, features, and steps that should be followed in creating and implementing an SMS in a company. What follows in this chapter is a discussion of a number of the elements that are essential to an effective aviation safety program. It is important to note that virtually all of the following sections contain information that may be embodied in an SMS as part of the aviation safety process in any organization.

Case Study: A Management System within Aerospace

Introduction

This case study outlines the journey of a major aerospace company to develop, design, and implement a management system for safety, health, and environment. It chronicles 15 to 20 years of history and takes place during times of booming business as well as significant business downturns.

In the Beginning

In business for many years and looked on as a leader in the industry, this aerospace firm had yet to adopt a proactive safety philosophy or any systematic approach to safety. It had grown to an employee population of approximately 8,500 with 12 facilities. Its business mix was both commercial and military. Its reputation for safety in flight and the quality of its products was well known and established throughout the industry, yet the same could not be said for its manufacturing and repair processes. For years its operational processes had resisted safety enhancements, though they had the requisite safety plans. These were in effect, but there was no execution, accountability, or responsibility. The entire staff consisted of a manager, one safety person, and an industrial hygienist. When audited, the company relied on a series of fire and accident prevention standards, which were more related to fire prevention than to accident prevention. There was a significant history of OSHA activity as well as injury/illness litigation.

Due to resignations and terminations, prior to the start of the journey, the firm had only one industrial hygienist for a period of 18 months, who acted as manager and safety engineer in addition to regular industrial hygiene activities. Needless to say, this one individual spent most of the time in a reactionary mode. Not much was being done or could be done to proactively address the safety issues at hand.

Injury rates were in double digits, with lost time rates approaching double digits. Workers' compensation dollars were significantly higher than state and national averages.

The First Steps

The company had hired several managers in hopes of beginning some sort of change process. Yet none of them lasted longer than six months, for various reasons. Finally, a manager was hired who possessed the knowledge, skills, and certifications to begin a turnaround. One of the first acts was to gain approval and hire the necessary resources. To that end, two safety engineers were brought onboard. Soon thereafter the industrial hygienist resigned; burnout was a key factor in the decision. A new industrial hygienist was secured, and the new team started to build not only a program but also a system.

One of the safety engineers had prior experience in drafting and implementing safety policy and standards. A gap analysis was conducted and a project started to build the base of the process. Over a period of 18 months, 23 standards were written, approved, and put in place.

Concurrently, a review of the accident investigation process was begun. Average time from injury to completion of the reporting and investigation was approximately six months. Forms were not appropriate and supervision and management did a poor job of investigation and correction. The mentality was production and getting the product to the customer. Other activities were looked on as unnecessary and something that was the responsibility of the safety department; after all, they had just hired additional resources—let them do the job. Now a formal structure was implemented, and forms revised and put into place to address the issues at hand.

With all these new changes being implemented, it was obvious that in order to drive change, a significant training program was needed that included support from upper management. An eight-hour training program was developed. The CEO was convinced that it was necessary and made the training mandatory. A letter was sent to each member of management from the CEO outlining the training and indicating his support. Over the next six months, in two four-hour sessions, approximately 700 managers and supervisors were put through the training. This initial training was well received and many of the standards started to be implemented, enforced, and produce results. Reductions were beginning to be seen in accidents and recordable injuries. Lost workday cases were declining. The change had begun. Rates were trending downward and some savings were beginning to be seen in terms of compensation dollars.

The Next Phase

After several years of continuing improvement, a plateau was reached and performance leveled off. Though better, things were not progressing as fast as desired both within the company and its corporate parent. Similar issues were being seen at other divisions within the organization. Something was needed to drive the next step change.

One of the safety engineers was selected to participate in a parent corporate initiative to benchmark leading safety processes and develop and recommend a management system for safety, environment, and health that would produce such a step change. A team of six individuals set forth on a 60-day project.

Eight companies were identified as having excellent safety performance and processes. The team visited each, touring facilities, speaking with safety professionals, looking at management systems, and talking with employees to gather the necessary data for their own system. Cutting-edge ideas and technologies were reviewed and best practices documented. The team sought out and listened to plant leaders on how they had driven and implemented the necessary changes to improve safety performance. Each company and person brought something unique to the process. Now, armed with all that data, it was time to get down to design and development.

With the help of a consulting firm, the team conducted a weeklong off-site session, working late into the night on many occasions. Data was reviewed. Best elements of the benchmarks were considered. Current culture was evaluated. At the end of the week, the team had come up with a recommended system containing 14 elements. Each element had been broken down into distinct steps to identify full compliance and implementation. Examples were developed and published for each. It was now time to move forward.

A meeting was held with the consultant, the team, and a cross-section of business unit operations leaders. The system was presented and discussed. Comments from the operations executives were taken and incorporated into the system. A strategy and plan were discussed and developed for the system rollout. Support and a commitment to lead the rollout and implementation were secured from the executives.

The System

Due to proprietary issues, the details of the system cannot be reproduced. However, an overview can be presented to show in broad strokes the system and its contents.

There were 14 elements in the final system. These system elements were similar to those found in the benchmark companies as well as various certifications such as ISO 14000 and OHSAS 18000. The elements also mirrored requirements of OSHA's Voluntary Protection Program (VPP). As one would expect, elements addressed standard safety components such as compliance, hazard identification and correction, training, employee involvement, incident investigation, and follow-up. Other specific elements pointed at targeted aspects to drive injury reduction, ensure awareness, assist in operational integration, and establish processes to deal with emergencies and methodologies for system measurement, evaluation, and improvement. Each element was subdivided into five discrete parts, and many of these had further subdivisions. Scoring was set so that each element could receive a rating of 5 if all activities and subparts were in place, with one point being awarded for each part. Totals from each element were compiled; the scores from all 14 elements were added and then divided by 14 to yield one overall score.

First Pass

With the backing of the operations executives, the system was put in place and each business unit was required to complete a baseline audit to determine the its current system climate. Once that was done, soft goals were set for improvement. Some business units set meaningful targets, while others set goals that were not aggressive at all. One business unit scored a baseline of 3 and then announced that they were satisfied with being a 3. The self-audits were not consistent with third-party findings, and it was determined that credit was being taken for any activity within the organization; the system yielded neither breadth nor depth. There was no correlation with injury/illness reduction. Operations were taking credit for a 5 in elements where they were doing some activities at level 5 but were not doing level 2 or level 3 within the same element. Bottom line: Though some advances were being made, many business units were not utilizing the system as anticipated and were being overly generous in their scoring.

Getting Serious

After several years with no significant progress, a working group was formed to perform a thorough review of the system. The goal of the review was to either fix the system and make it work or eliminate it altogether. Recommendations were to conduct a re-baseline with several new parameters. All parts and subparts had to be in place before taking credit for that element. For example, an organization could be doing all parts and subparts at level 5 within an element, but missing several subparts at level 5. Their score could not be a 5—that organization's score would be a 3. While meeting initial resistance, this concept was ultimately accepted.

Next, the organization got serious with the "what gets measured gets done" philosophy. They were able to integrate a system score into the personal goals of the CEO. The initial re-baseline scores were tabulated and averaged, and a single score was developed for the corporation. The CEO set the initial target at a one-half point improvement from the re-baseline figure. He put this goal in his personal goals reported to the board. Being in the CEO's personal goals meant that responsibility rolled down to each business unit leader. Once at the business unit level, it flowed down to various departments within each of the units. This created traction from the manager's level all the way to the CEO. Corporate-led audits and third-party audits were conducted to ensure that the playing fields were leveled and that each organization was consistent in its application of the system and its elements. If there was a difference of more than one point in any element or in the entire system between the self-assessment and third-party independent assessments, a management-level finding was issued which required correction and follow-up by each organization's senior management.

A training program was implemented so that all could learn and understand what was expected. This training module was included in all future training sessions for new safety professionals entering the organization at either the corporate or plant level. Corporate auditors developed standard protocols and definitions. A single source with prior experience within the corporation was hired as the independent third-party auditor.

Did It Work?

After the CEO put a system score improvement goal into his personal goals, the system started to gain traction. What was important to the boss now became important to all business units from top to bottom. After the first year, an overall improvement was seen that matched the goal. The following year an additional one-half point was set as the target. Once again, the goal was met and improvements were noted. Injuries and illnesses declined significantly. A project was set in place to see if there was any correlation between system scores and improving injury/illness rates. The results showed that there was little to no correlation prior to the re-baseline and setting of the system within management's goals. However, with a new baseline and goal flow-down from the CEO to shop floor, correlations were significantly higher.

After several more years of using the system as a leading metric, the corporation was seeing a management system score in the range of 4½ to 5. Injury rates continued to decline, providing historical best performance levels. It was discovered that some flexibility was needed as, with management changes and reorganizations, new baselines were required. However, due to the attention to the metric and the expectation that business units would continuously improve and sustain top performance levels, rates continued to decline.

Lessons Learned

Systematic approaches to injury/illness reduction can work. What gets measured gets done. Applied in the right way, management systems can be used as a leading metric. System improvements can be correlated to rate improvements. The work is never finished. The system must be designed to be somewhat flexible, yet drive change, especially during periods of organizational change or realignment. The system will work if you work the system.

Gap Analysis

As was discussed briefly in Chapter 4, **gap analysis** is simply a needs assessment of a particular organization or department, sometimes referred to as the sample area or location. In general, a gap analysis uses a set of audit tools to determine the current performance of the sample area that can be compared with its potential performance. For safety and health purposes, the potential performance may mean basic compliance with the standards or a measurement "above and beyond," such as that required for OSHA's **Voluntary Protection Program (VPP)**. VPP is a program for employers who have implemented effective safety and health management systems and maintain injury and illness rates below the national average for their industry. VPP participants work proactively to prevent injuries and illnesses through cooperation with management, labor, and OSHA.

The first step to conducting a compliance gap analysis is to determine the "where we want to be." These decisions must include upper management to determine the depth or comprehensiveness of the analysis that will be supported. The next step is to determine each element that should be included in the analysis as well as the specifics that should be included in each element—in other words, to determine the "need" required to achieve the goal. This can be done through extensive research on each element, benchmarking other organizations that have reached similar goals, or by contracting a professional consultant who specializes in the outcome matrices that are expected.

Examples of elements for safety and health include, but are not limited to, management leadership, employee involvement, hazard prevention and control, and compliance. Within these elements are subelements. An example of a basic subelement of management leadership is safety policy. Within this subelement, certain queries must be answered to meet the subelement's objectives or "needs," such as:

- Does the organization have a written safety policy?
- Is the written safety policy signed by the highest-ranking official of the organization?
- Is the policy clearly posted in a common area?
- Is there evidence that the safety policy is communicated to all employees?
- Does a local written safety policy exist that is tailored to the operation?
- Is the local written policy signed by a member of senior management?
- Is the local safety policy clearly posted?

The answers to these questions reflect the organization's current status. Any "no" answers demonstrate the organization's "gap" in reaching the desired goals. Once the gap analysis is complete, estimates can be made to determine resources that will be required. These include the investment of time, money, and human resources. Priorities should be identified and solutions determined to meet objectives.

Queries are not necessarily requirements. Consequences of not satisfying a query should be determined. Decisions for satisfying the query should be made by management with regard to the recommendations of the safety department or the consultant performing the analysis. Another way of performing a gap analysis to include different levels of compliance or desired goals is by developing a **step analysis**. This type of analysis identifies multiple levels of performance, such as achieving the basic standards, a higher standard such as OSHA VPP, and "world-class" performance.

Gap analysis may also be used as a means for classifying how performance meets a goal or set of requirements. In this case, the "gap" can be classified as a ranking such as "excellent," "good," "average," or "poor." Examples of these include local injury rates versus established goals, behavior analyses (through observations), internal assessments, and employee surveys.

The Support of Leadership

The internal climate within a given company will go a long way in determining how seriously safety is taken by the employees and by the overall organization. With respect to safety, if the organizational climate is lackadaisical or follows a hands-off, laissez-faire approach to safety, it is unlikely that a positive safety culture will be present. A **safety culture** can defined as the internal attitude, level of motivation, and approach that characterize the organization with respect to how seriously safety is regarded. A poor safety culture will generally be synonymous with a weak or insufficient approach to safety. At this point, it is extremely important to point out that, left to its own devices, a safety culture will develop on its own in any organization. But, in the absence of strong, consistent leadership, the safety culture that forms by default will most likely not be desirable or truly safe. This is because a default safety culture will usually not be deliberate, intentional, or proactive, but rather will most likely be reactionary and event-driven instead of being out in front of problems and issues. Effective, visionary leadership is absolutely essential to establishing and maintaining a strong safety culture that does not leave the process to chance.

One of the most fundamental elements that must be present in any organization that endeavors to create, implement, and effectively manage a successful aviation safety program is firm, consistent, and unwavering support from all levels of company leadership. One will virtually never find a truly successful safety program or a strong safety culture in an organization that is lacking in leadership support of safety initiatives. There are numerous reasons why this is true. To give further insight and clarification, we will briefly consider this issue from both strategic and tactical leadership perspectives. Although there are numerous leadership theories and models in existence, the following section will focus on the general application of basic strategic and tactical leadership principles to safety.

Strategic and Tactical Leadership

Strategic leadership in an organization is primarily concerned with guiding the overall direction of the organization as a whole. Persons in these positions create short-, mid-, and long-term strategies, conduct planning activities, and make many of the decisions that help to determine where the company is going, how it will get there, and other pertinent considerations. Strategic leadership positions are generally found in the **upper management** or senior leadership ranks of an organization. Some examples of these types of positions would include directors, vice presidents, senior vice presidents, presidents, CEOs, boards of directors, and so on, depending on the hierarchical structure of a given organization. By way of comparison, **tactical leadership** in an organization is generally responsible for supervising and managing the day-to-day activities and functions of departments, divisions, and operations. Although at times these individuals may be engaged in making mid- or long-term decisions about their segment of the organization, persons in these positions tend to make many short-term tactical decisions that address issues in the here and now, based on the immediacy of situations that demand attention promptly. It is important to understand that tactical leadership decisions should be made in support of the larger strategic leadership decisions concerning the overall direction of the organization. Tactical leadership positions are those generally found in the **lower to middle management** ranks of an organization. Some examples of these types of positions would be shift leads, supervisors, shift managers, managers, regional managers, and so on, depending on the hierarchical structure.

With respect to both strategic and tactical leadership, anyone in an organizational leadership position should play an active and consistent role in support of safety programs and initiatives. While the roles played by leaders at different levels will vary in scope and areas of responsibility and will vary between organizations, it is crucial that all persons in leadership positions stand in firm support of safety throughout the organization. The following subsections will examine some of the ways that leaders can support safety programs.

Vocal Support of Safety by Leadership

Real support of safety by leadership in any organization needs to be expressed in several ways: vocal, written, financial, consistency of actions, and accountability. To begin, leaders must be vocal in their support of safety. They must consistently voice their support of safety initiatives to all of the people in their respective areas of responsibility, and the employees need to hear their leaders expressing support for safety on a regular basis. The safety message from senior leadership should be heard throughout the entire organization. Lower and middle management should ensure that their support of safety is heard by all employees in their departments, shifts, and/or divisions. Vocal support requires that leaders develop effective, proactive means of communicating with all the employees of the organization. This imperative of establishing effective communication is also directly related to human factors, as proper communication between all levels in an organization is an essential component of any safety program.

Another thing that leaders can do to be vocal in their support of safety is by recognition of employees who demonstrate positive safety behaviors. Quite commonly, employees hear about the negative things they have done, but do not always receive praise when they demonstrate desirable safety actions and behaviors. The overall safety culture within an organization can be greatly enhanced by leaders who are vocal in giving positive affirmation to employees who demonstrate good safety behaviors. This simple action can help foster trust between employees and their leaders, and may also encourage others to demonstrate proper safety characteristics.

Written Support of Safety

Members of leadership need to ensure that they do not limit their support of safety to expressing themselves verbally; they need to back up their safety proclamations by putting them in writing. This can be accomplished by email communications or by memos that are posted and distributed throughout their departments, or even throughout the company, as appropriate. Leaders should be active in the process of helping to create specific written safety policies and procedures to be utilized within the organization. This is perhaps one of best ways to demonstrate leadership support of safety. Senior leadership should start by writing and distributing a specific safety policy statement for the entire organization. In a fashion similar to the OHSA General Duty Clause, the organizational safety statement should clearly indicate that safety is truly the highest priority of the company, and that the organization stands in firm support of creating a working environment that is as safe as possible for all employees, customers, contractors, the general public, and other stakeholders. In addition, the safety statement should include specific language that encourages people to report any potential safety problems, issues, areas of risk or hazard, and any other unsafe situations or behaviors without fear of retribution against them or their jobs.

Resource Allocation

Members of leadership are usually the ones who control the allocation of available resources within a company. They have the say-so over the how, why, where, and amounts of financial resources that can be provided for specific departments and programs. Therefore, leaders should demonstrate their support of safety by ensuring that legitimate safety initiatives and programs have the necessary funding to create, implement, and manage them in the short and long term. Short-term financial allocation is needed to ensure that funding is available to finance a safety program or initiative. Long-term resource allocation is necessary to ensure the sustainability over time of a valid safety process. Further, the potential benefits of a new safety enhancement are often not going to be seen immediately. In many cases it may take some time before a negative safety trend will level off and take a downward turn in terms of number of occurrences. For example, if an airline decides to implement an observation process on the ramp to try to reduce the number of aircraft ground damages attributable to the operation of ground support equipment, the results will most likely not be seen until the process has been in place for a certain period of time. Long-term resource allocation will help to ensure that valid, potentially effective safety protocols are adequately funded, particularly if the protocols prove to be successful in preventing accidents and incidents.

Consistency of Actions

Leaders should also strive to do their best in showing support for safety by setting a proper safety example, demonstrating consistency in their actions. There are a few different ways that consistency of actions can be viewed. First, leaders should consistently remind all employees in their respective departments/divisions/groups about the importance of safety through specific references and examples, and ensure that procedures, policies, and guidelines are being followed by all employees under their responsibility. Next, leaders need to ensure that they themselves are following all of the safety rules and policies as well, and not just expecting these types of behaviors from their employees. Arguably, few things will erode the trust and credibility of a leader in the eyes of employees quicker than holding others accountable to standards and rules which they themselves are not following consistently. Finally, leaders must uniformly apply safety rules among all employees, and not follow double standards. If a given leader permits favored employees to get by without following the rules for safety but holds others accountable for the same basic actions, or simply fails to uniformly apply safety rules across employee groups, this will send the message that certain individuals or groups can get away with not consistently abiding by established safety protocols, while others cannot. Such a situation will harm the safety culture and will erode employee trust in leadership. Each of these issues may prove to be highly detrimental to safety in an organization.

Accountability

Leaders must also support safety in their organizations by establishing a process of accountability to be applied to all employees, as necessary. When a violation of policy or procedure occurs, it is important that leadership step up and hold the person(s) involved accountable for their actions. This is not to imply that every safety violation should result in harsh consequences, although at times this may be necessary, depending on the issue and situation. However, it is important to establish and maintain standards for holding organization members accountable for their actions. True accountability can only be effective if leaders actively let their employees know that they are responsible for abiding by all applicable safety standards, and for their individual actions; as such they will be held accountable for violations of established policies and procedures. Importantly, leaders need to communicate that they themselves will be held accountable for any violation of safety policies and procedures.

Balanced Safety Leadership and Decision Making

An important factor to consider is the necessity for balance in safety decision making. A balanced leadership approach to safety will consider the impact of decisions from both internal and external perspectives. **Internal decisions** are those decisions made within the organization that are likely to create one or more impacts within the organization. These are decisions that may affect some or all of the frontline employees, management, and other stakeholders. Internal decisions can also play a major role in affecting operations and, ultimately, profitability. **External decisions** are those made within the organization that are likely to create one or more impacts outside the organization. Some examples of general external decisions are those that may impact the community in which a company is located, those that may impact other organizations from the standpoint of competition or safety, and decisions that could impact the environment. It is important to note that both internal and external decisions can create impacts that may be positive, negative, or something in between, depending on the situation and the type of decision made. If the overall direction of the organization does not include the creation and support of initiatives designed to improve or enhance safety, it follows that the overall company will not have safety as a primary factor from a strategic planning perspective. Senior leadership must be supportive of safety initiatives in order to have a successful safety program and a consistently strong safety culture. This factor is absolutely essential and cannot be overstated. Front-line employees cannot be expected to take safety seriously if the leaders of the organization do not stand in direct support of relevant safety initiatives.

Involvement of Employees in the Safety Process

Beyond question, the support and involvement of organizational leaders is an essential component of any successful aviation safety program. However, effective and successful organizational safety systems will take the time and effort to make absolutely certain that front-line employees are actively involved in the safety process. Employee involvement in safety program is essential for a number of reasons, which are discussed in the following subsections.

Front-Line Employees: The Voice of Practical Knowledge and Experience

Front-line employees perform specific jobs on a day-to-day basis, often over time periods that span a number of years. These positions are often non-leadership roles that serve to support the operation of a department or division. Front-line positions are generally numerous, constituting the majority of the employee base of an organization. Although the types of front-line positions are far too numerous and varied to list here, a few examples of front-line aviation employees would be aircraft maintenance technicians, customer service agents, flight attendants, and line service technicians working for a fixed-base operator.

During the time front-line employees spend conducting the specific job responsibilities and duties associated with their positions, they have often learned a tremendous amount about the jobs they perform and the physical areas in which they work. As a result, these individuals may have determined what types of changes might be beneficial to make from a safety perspective, based on their knowledge and experience. Therefore, including the people employed in front-line positions in safety program development and management is not just a good idea, it is an essential part of a successful safety system. Who better to go to for ideas about improving safety than the very ones who work in these positions? This is an example of a practical approach to aviation safety. A skilled, visionary leader will understand the benefits of tapping into the vast knowledge base of proven front-line employees in order to enhance organizational safety.

The potential benefits of involving employees in safety reach far beyond tapping into their knowledge and experience, however. When organizational leaders take steps to involve front-line employees, they are also helping to create a safety culture that is inclusive and in which trust can be generated between management and the various employee groups. An open safety process that includes representatives from both management and the front line may help to bridge the gaps that sometimes exist between leaders and employees. From a practical perspective, if only organizational leaders and job-specified "safety" personnel are involved in the process, there is a risk that the employees may feel excluded from the program. If this is the case, some employees may stand a greater chance of not supporting the safety program. Non-support, in turn, could contribute to higher incident or injury rates by virtue of increased front-line non-participation in safety.

Perhaps one of the strongest reasons for the inclusion of front-line employees is found in this fact: People tend to support that which they helped to create. When employees are actively involved in creating, revising, and managing a safety program, they will be much more supportive of the safety initiatives in the organization. In other words, employee involvement creates a sense of "ownership" in the program because they were directly involved with development. As a result, the safety program in effect becomes "theirs"; it is personal and tangible to them. These employees will sometimes be enthusiastic and perhaps even more outspoken when it comes to safety in their respective workgroups and in the company as a whole. When this type of positive response occurs among front-line employees, an added benefit may sometimes be realized: Front-line employees may become "safety leaders" in their own right, even if they are not in official leadership capacities. They will tend to become "opinion leaders" among their peers with respect to safety and will often begin to hold themselves and their fellow employees accountable for their actions and behaviors, reminding them regularly about the need to be safe in everything they do on the job. Harnessing this type of employee peer influence can be extremely powerful within an organization. Involving front-line employees in the entire safety process also helps to demonstrate the commitment of leadership to safety.

However, as important as the involvement of front-line employees is to the overall safety process, it is important that the right types of employees be chosen for involvement in the safety program development process. Although the goal is to have everyone in an organization involved in safety on the job, it is important to note that the direct involvement of front-line employees in the development process is representative in nature; that is, certain front-line employees will represent the larger working groups in which their jobs are located. Ideally, front-line representatives chosen to be a part of the safety development process should be proven, reliable employees with an overall good reputation within the organization. They should have a good working record that contains little in the way of disciplinary problems or job-related actions taken against them. These employees should have the respect of their direct leaders and peers, because it is more likely that others will listen to them and their opinions. Also, they should be interested in safety and motivated to make the organization a safer place. In short, these individuals should have at least a reasonably strong dedication to safety.

Safety and Health Committees

As discussed previously, two keys to a successful safety program are management commitment and employee involvement. **Management commitment** can be demonstrated within the organization and in safety and health committees by participation in committee activities, providing the time for committee members to plan and conduct activities, and providing authority in key positions to make decisions that effect change. The simplest and most effective way to establish and maintain **employee involvement** is through the development of employee safety and health committees, or "safety committees." Providing the time to prepare and conduct safety meetings and activities surrounding them is a demonstration of management's commitment to the safety process. Safety committees are critical for identifying and communicating safety-related issues between the workforce and management or those in a position of authority to mitigate hazards. Other important contributions of a safety committee include incident/accident investigation, training and education, pre-shift safety message communication, behavioral observations, assistance with facility safety inspections, and more.

Note: The discussion in this section relates to typical safety committees. Some organizations may also utilize supervisor, union, divisional, and even inter-company committees for the same purposes as well as to identify and address industry trends, best practices, or cooperative problem solving.

Key Questions and Answers Used for Committee Development

How should a safety committee be established? This determination is based on the size of the facility and its individual operations. A small facility may establish a safety committee for each work shift, with representatives from each work area and/or job type. A large facility may need separate committees for each operation. The larger facilities (mid-size to large airline hub facilities, for instance) often will have committees within each operation plus a hub safety committee to communicate interdepartmental issues. The decision should be based on the number of employees represented by each committee member and his/her ability to effectively communicate to that workgroup (preferably in person).

The committee should also be represented by a member of management with the authority to make decisions on behalf of the committee and provide direction (likely the operation's manager, division manager, etc.). If possible, the safety director/manager should participate to provide expertise. Inclusion of other support management positions such as environmental, security, and so on may also be beneficial. The safety of the organization requires a joint commitment between management and non-management employees. Non-management personnel on the committee should participate on a voluntary basis. The non-management members should select a responsible individual to serve as a chairperson for the committee. This person should be organized and capable of delegating activities among committee members. The committee should also have a person responsible for a roster of members present, taking the minutes of the meeting, and posting and distributing the minutes appropriately to communicate with the workforce. To support a wellness program (where applicable), the inclusion of a wellness champion, who gathers materials and conducts related activities, would be necessary as well.

The committee membership should be rotated periodically to involve more employees with a fresh perspective. This will improve the safety education of the workforce through committee training in areas such as hazard recognition and incident investigation. It should also help to reinvigorate the membership through the mentoring of new participants.

How often should the committee meet? Meetings should be held at least once a month on a consistent basis. Some operations may choose to meet more frequently if injury rates or numbers of hazards are high. Special or extra meetings may be called to conduct investigations or handle issues of high concern.

How long should a meeting last? Whenever possible, meetings should last no more than an hour, sometimes as little as 30 minutes, depending on the frequency and topics for discussion. A well-organized committee chairperson can develop an agenda to cover all necessary topics within the allotted time frame. It may take a few meetings to determine how long is appropriate for certain topics and to arrange the agenda accordingly, but being organized and timely is important to a successful meeting. Try not to schedule meetings for more than an hour. An extended meeting tends to lose focus, momentum, and sometimes the acceptance or support of management. Stay on task!

What should be covered during the meeting? There are a few key things to cover during the meeting. First, consider starting each meeting with a "safety first" topic. This topic may come from the monthly safety theme, or perhaps a safety policy. Review items from the previous safety meeting minutes and provide an update to any closed or outstanding topics.

A safety meeting should always cover any injuries and accidents that have occurred since the last meeting. The discussion should focus on the facts of the incident only and not include the names of those involved, even though sometimes those present will know those involved. The key is to discuss the root cause and other causal factors in the incident and the prevention activities that were implemented. Committee members should be aware of these prevention activities to share with the work area they represent.

The meeting should discuss injury and accident goals and how the operation compares with others. If possible, goals and actual numbers or frequencies should be established within each department of the operation. It is important to celebrate successes and to determine areas that are struggling. At this point, any successes (areas significantly ahead of goal) should be recognized.

The committee should discuss planning for the next month or other period between meetings. Planning should consist of delegating activities to committee members such as area hazard walks or safety assessments, employee observations, surveys, and pre-shift meeting safety communications. This delegation of activities is critical to promote employee involvement in the safety process. If the operation conducts employee observations, discuss any high at-risk behaviors and solicit new ideas for changing them. Track activities assigned to committee members and the specific date(s) by which those activities are to be achieved.

There may be some other topics in addition to these that can certainly be included. Just be certain that the nature of the topic is appropriate for the committee. Once all topics are complete, the remainder of the meeting time should be devoted to addressing employee concerns. These might be concerns that representatives have brought from their area employees, questions that may arise regarding safety or operating policy, or issues of awareness that should be communicated with other workgroups. To be more efficient in mitigating hazards, get into the habit of asking committee members for solutions to the issues they have identified.

What are the chances that a safety committee might fail? The chances can be great that a committee might not work well if it falls into some of the common traps such as:

- *Falling off task:* Allowing the committee members and not the chairperson to run the committee. If the meeting is not well controlled, members may tend to discuss repairs that have not been completed or specific employees who break safety rules. This will often turn the committee into a "gripe session" and will accomplish little toward improving safety.
- *Too much planning:* Spending a lot of time planning and not so much on activities is a poor way to show the value of the efforts made. This is why it is so important for the chairperson to be organized. Planned goals and objectives should be realistic (attainable) and measurable.
- *Poor communication:* An effective communication link should be established between each member of the committee and the workgroup they represent. The workgroup should be familiar with their representative, be able to approach with any safety concern, and be provided feedback from the committee based on all concerns that affect the workgroup. A follow-up posting of this feedback may increase communication for those who were unable to receive the initial message.
- *No money for safety:* The committee may not have a budget. Without adequate financial resources, the committee will be unable to pay for solutions, training, safety equipment, or assistance through consulting.
- *No buy-in by the boss:* The committee may not have full support of management. The management member must have a good understanding of the safety process and buy into it. This person should also be in a position to make key decisions on behalf of the company for the committee.
- *No follow-though:* As with "poor communication," when an employee brings up a concern, it is very important to keep the employee informed of the status of the concern, even if the outcome is not what they wanted, in which case a valid explanation should be provided for why such a decision was made.
- *Little to no participation:* An organized and effective chairperson will provide activities for each member between meetings. A good member will provide evidence once a task has been completed. However, the chairperson should follow up with the member to remind him/her of this expectation. The management member should be made aware of any committee members who regularly do not complete assignments. The management person should verify the member's commitment to the committee and determine if a change in representation is necessary.

Regardless of how safety committees are formed and used, the concept is the same. Safety within the organization should be viewed as a process that is ever-changing rather than a program. The process depends on people from all levels to actively participate and work together to recognize and reduce workplace hazards that affect employees' health and safety.

Accident Prevention Strategies

In this section, the term "accident" will be used to describe all unwanted events that occur within an aviation organization, such as those involving work-related injuries and illnesses, aircraft damages, and damage to ground service equipment (GSE). Most references will contend that about 80–90 percent of accidents can be attributed to human error. The FAA has indicated that human error accounts for approximately 80 percent of aviation accidents. However, most of these references do not stress the fact that the root causes of many of these accidents can be attributed to a poorly designed system, job setup, product, or the environment. Oftentimes, these types of issues are considered human error as well.

Figure 5–1 provides a good illustration of Heinrich's Law and how it is commonly used in various industries. Although neither Heinrich's Law nor von Thaden's theory, nor the many other similar theories have been proven—or disproven, for that matter—the generalization that human behavior is a factor in a large percentage of accidents and injuries is commonly accepted. Because human error has been identified as a major factor, many industries and manufacturers have made adjustments to take the "human" out of the equation as much as possible. For aircraft, technology and automation improvements reduce the amount of human interaction necessary. Similarly for other areas and industries, machines and equipment have been designed and guarded by various means to reduce and oftentimes eliminate the human element that has historically led to accidents.

Worksite Analysis

In order to establish an accident prevention strategy, a study of the "history" of accidents must be conducted. This "history" research is known as a **worksite analysis**. In aviation, worksite analyses are commonly conducted on injuries or workers' compensation claims and ground damage events. However, they may also be conducted for other events such as damage involving ground support equipment, hazardous material or spill events, and so on. In any case, the information that is extracted from the analysis is only as good as the information submitted, which in turn depends on the system parameters or requests for details, as well as complete details submitted by the user or interface. In other words, the more quality details accepted

Figure 5–1. An example of a safety risk pyramid based on Heinrich's Law.

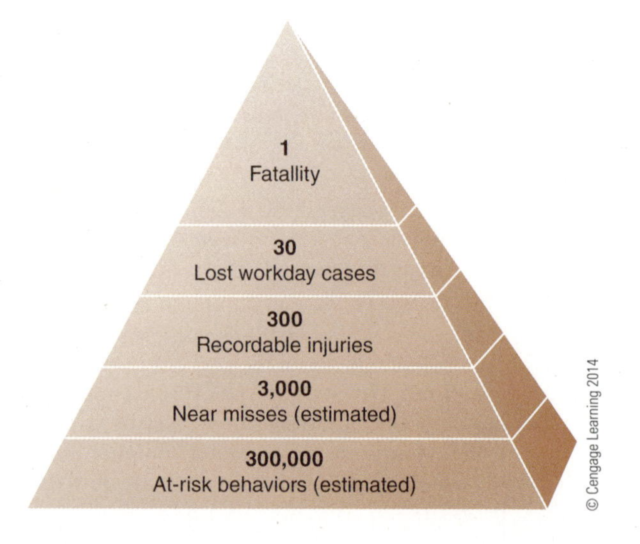

by the analysis system and provided by the end user, the more ways the information can be analyzed and used for prevention purposes. If data submitted to the system is too often categorized as "other," the system parameters should be changed to incorporate some of the details that are most common, or training around and demand for better information to be provided by the submitter may be necessary.

A medley of information can be extracted from the worksite analysis and exhibited in a number of formats that can communicate the information to multiple levels within the organization. A basic worksite analysis that can be posted for communication to the workforce might include information that can be sorted to the operation level's most frequent and severe (costly) injury sources. Beneficial information to include at this level is the injury type (lift/lower, struck by/against, etc.) and a history that includes similar periods of time for comparison or trend analysis (previous three years or seasons validated at the same date for each year). This data can be used by safety committees to discover basic trends for which preventive activities can be implemented.

A more complex analysis, usually used at a higher level within management to drive change, may include information on employee tenure, specific location of the injury, time/day of occurrence, job type, equipment involved, and so on. This information helps to determine trends related to these factors. **Figures 5–2** and **5–3** are examples of worksite analyses.

From the basic example in Figure 5–2, the most frequent and severe injury source is related to lifting/lowering, followed by pushing/pulling. A different analysis (see Figure 5–3) may reveal what

Figure 5–2. Sample Worksite Analysis of injuries by type.

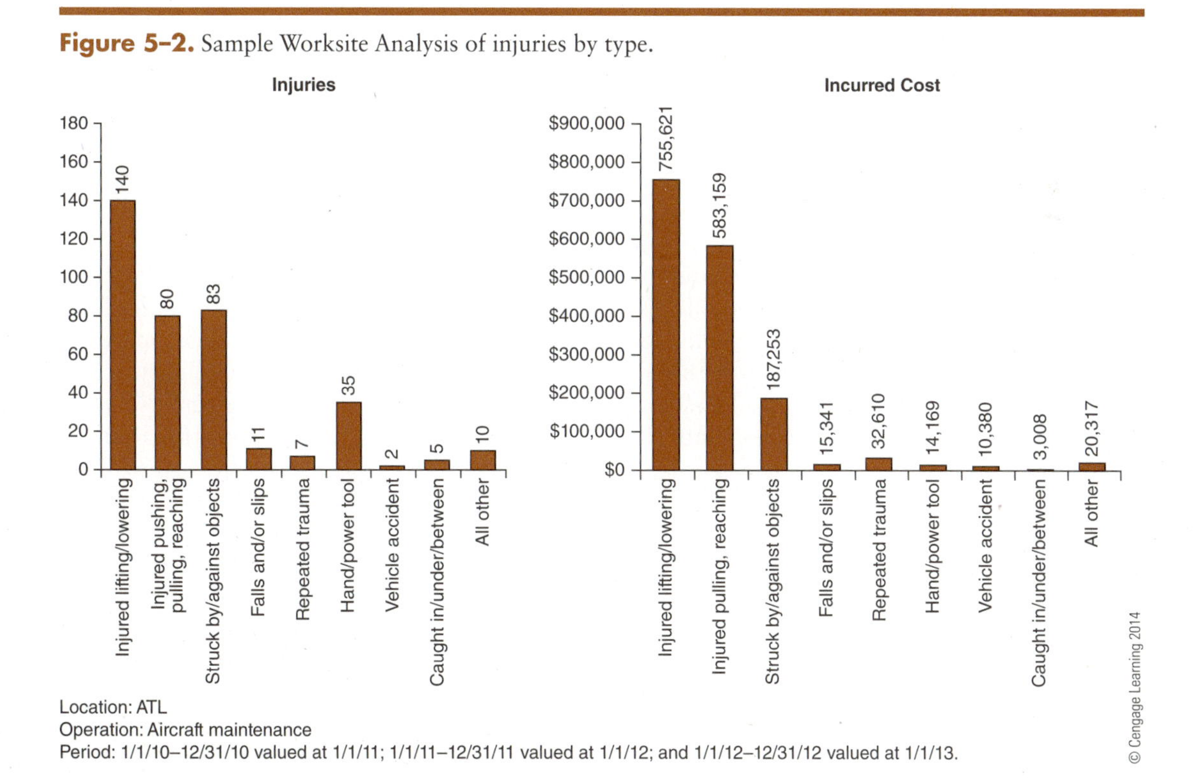

Figure 5–3. Worksite Analysis of injuries by job.

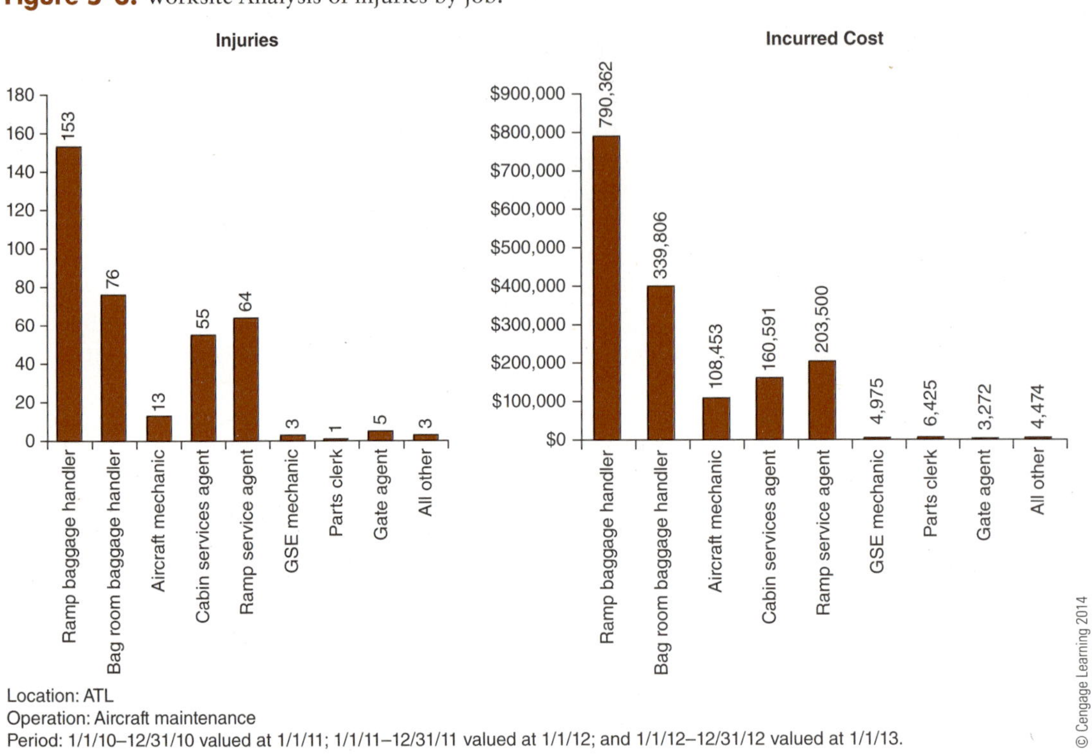

job types are experiencing these injuries. And a more in-depth analysis could reveal that a certain percentage of the lift/lower injuries are attributed to baggage handling. Data can be arranged in a number of ways to identify areas of challenge.

Without studying the descriptions and investigations of each of these injuries, certain factors could be considered as possible causes. Does the company have appropriate work standards or methods in place? Most companies spend a lot of time and energy putting good work standards in place, but typically don't do enough with regard to enforcement. In this case, disciplinary action does not imply "enforcement," but addresses at-risk behaviors as they occur in an effort to change unwanted behaviors. (See Chapter 6, "Introduction to Human Factors," for more information on behavioral science.)

After studying the descriptions and investigations of the lift/lower injuries, some assumptions could be made as to the actions that led to the injuries, but likely wouldn't identify the specific behaviors that were involved unless the company utilizes a **behavioral observation process.** Observations that are made in this effort focus specifically on the job function being performed against the appropriate methods which may or may not exist until the study, or **job safety analysis (JSA)**, has been completed for each task. A JSA is a method that breaks down job tasks, identifies each risk element (lifting, chemical, etc.), and details measures that should be taken (specific proper tasks or methods, proper personal protective equipment that should be used, etc.) to complete the task in a safe manner. JSAs are suitable for jobs such as baggage handling, machine operation, changing tires, and so on, where the job assignment is not too broad or narrow. Jobs selected for JSA should be prioritized according to those with the worst injury experience. For each job, the successive steps or activities should be broken down to determine how they are performed or should be performed to eliminate injury. The hazards or potential hazards should then be identified. Finally, safe job methods and/or personal protection measures should be identified and developed to prevent injuries.

A good JSA will be specific to each element of the job task. For example, a JSA for handling a bag from the bag cart to the belt loader (one task) may include illustrations and be set up with the following type of instructions or recommended actions:

1. Position baggage cart so that the angle of movement between the cart and belt loader is approximately 45 degrees.
2. Grasp the top bag firmly, using the handle (where available) to slowly pull the bag from the cart while testing the weight.
3. Grasp the rear corner of the bag for support and bring close to the body.
4. Pivot feet (don't twist) to move the bag over the belt loader.
5. While bending the knees to the appropriate level to deliver the bag, place one end on the belt and allow the movement to assist to lower the other end. Keep the back straight and the bag as close as possible during this step of the process.
6. Pivot feet (don't twist) back to the baggage cart and repeat until completed, using a top-down method for removal. Never pull from the center of the pile.

Figure 5–4 is one example of how a JSA may be developed. Notice how it shows the risk element or potential for injury as well as the proper methods for completing the job in a safe manner.

With these types of instructions or methods in place for each job task, a supervisor or other observer has the ability to understand how to perform the job using "safe" behaviors versus potentially risky ones and can effectively demonstrate appropriate methods in an effort to reinforce the "wanted" behaviors. Illustrations aren't always used with JSAs, but they can benefit an observer in better identifying safe and at-risk behaviors. When paired with effective feedback using **consequence management** to explain the effects certain at-risk behaviors may pose on the body and how certain safe behaviors will positively impact the employee's future, a decline of these types of injuries is sure to follow.

Hazard Recognition/Identification

The intent of a safety inspection, whether for a general facility, weekly operations, or daily pre-shift checklist inspection, is to identify hazards in the specific targeted inspection area. However, this is a quick "snapshot" in time of the area. As the day progresses, many factors can introduce new hazards into the work area. For example, a tug may overheat, a chemical spill may occur in the composite shop, or contact with a piece of equipment or machine may cause a sharp protrusion. If the tug operator does not recognize the water gauge is rising or steam is coming from the engine area, the radiator hose could burst and spray the operator or a nearby employee with hot water, not to mention the damage that could be caused to the engine. If the composite shop employee does not recognize the spill or correctly identify the chemical product, the potential effect could be as serious as death if handled improperly. Some chemicals can violently react when mixed with other chemicals, or even with water. And finally, if the employee who made contact with the piece

Figure 5–4. Job Safety Analysis example.

Job Safety Analysis (JSA)

Job: Baggage Handling | **Department:** Ramp Operations | **Job No:** R38 | **Page** 1 **of** 1

Date Developed: 30-Jul-11 | Revised Date: None | Reviewed Date:

PPE Requirements: *Hearing protection (muffs, ear plugs), high visibility vest, gloves, and non-slip sturdy work boot.*

General Requirements: *Baggage Handling–Load/unload baggage from plane/transfer cart.*

Steps of Job	***Potential Hazards***	***Safe Work Practices Required***
Lift without jerking the load. Secure baggage close to body with bent knees and back straight. Extended reach/lift while twisted places more strain on the lower back. Getting close to lift maximizes employee strength. Extended reach/lift–steps or bags in the way can prevent employee from getting up close.	Potential injury if the load is jerked. Potential back strain if knees are straight and back is bent while lifting. Potential injury to shoulder/back if "one arm swinging" bag into bin or up on high shelf of transfer cart. Potential back strain if lifting and a twist motion is performed during the lift without moving the feet. Potential for extended reach/lift if steps are in the way. Bag should be lifted facing bin and as close as possible so an extended lift or twist is not necessary. Potential shoulder/back strain for extended reach/lifts.	Make sure proper PPE is being worn. For loading to/from cart, position at 45 degree angle to the belt loader so limited body turning is required. Position body so that an extended reach/lift is not required to reach the bin or cart. Grasp the top bag or object firmly, slowly pulling the object close to test the weight. Do not jerk the load. Keep lifted object close to the body to reduce pressure on the lower back. Make sure the knees are bent and back is straight when lifting. Do not twist while lifting. Pivot feet to intended destination. Ensure firm grasp of two handles on any large bag. Do not "swing" bags into cart or up on belt loader. Get up close to loader before setting bag down.
When loading/unloading conveyor, use stronger leg muscles to lift rather than weaker muscles of the back. Make sure you are close to conveyor so an extended reach/lift is not required.	Good use of bending legs/straight back with secure hold (two hands) on bag.	Do not twist with the load. Turn your feet, keeping your nose between your toes. Keep lifting height of bag between the knees and shoulders for optimum lifting range. Keep the back straight and bend your knees. Keep area around feet clear of any dropped bags, etc. Do not jerk bag or get in a hurry. Wait on bag, do not reach out of bin to lift bag with an extended reach/lift.

of equipment or machine decides to ignore it, the potential could be as obvious as the struck-by potential, but could also lead to a more critical incident because of the behavior that led to the incident in the first place.

In the last part of the example above, what if the supervisor didn't address the hit or a situation in which the operator nearly hit the equipment or machine? For an established program backed by upper-level leadership, one of two things is possible in this situation: Either the supervisor is not committed to the safety policy established or the supervisor does not recognize this type of behavior as at-risk. For the latter, training should be provided to supervisors to recognize such situations. Training should include visual aids and/or one-on-one interaction to help develop the skills necessary for identification of the hazards and risks associated.

Preventive Maintenance

Preventive maintenance is the practice of replacing critical parts or components of a piece of equipment or machinery before failure occurs. The most commonly known is the act of changing the oil and filter in a vehicle to keep the engine running smoothly. There are many other areas where preventive maintenance helps to protect not only the equipment itself, but also the operator. Cables and belts may be pertinent to the safe operation of certain equipment. They may require frequent inspection and a replacement schedule of a year or per *x* number of hours or revolutions.

In some situations, preventive maintenance plays a key role in safety because a breakdown means the operator may continue to use the equipment in an unsafe manner, or not having the piece of equipment forces manual manipulation of the product. Sometimes the breakdown of a piece of equipment contributes to overuse of other pieces of equipment, or perhaps use of another piece of equipment for reasons other than its intended purpose (e.g., using a belt loaded as a lift for engine repair).

Communication

Communication is the most effective component of accident prevention. There are several different strategies or elements that can be used to increase awareness, hazard recognition skills, and safe behaviors. One of the most effective is through training. The frequency and content of training will affect the product or expected outcome of the level of safety. Many of the other elements of communication discussed in this section could also be considered forms of training as well, and are typically used much more frequently than "traditional" training.

During the initial training of an employee, an overwhelming amount of information may be provided to the employee in addition to all the safety rules that apply to the position. Because of this, retention may be quite low unless the training is spread out over a period of time and safety elements are included frequently and repeatedly for particular high hazard elements. Periodic training is usually provided during specific times of the year or around the employee's employment anniversary. Again, the effectiveness of this training depends on the delivery method.

Pre-shift communications (PSCs) occur every day before the shift and often include production information in addition to a safety message. However, the safety message should always come first to emphasize the operation's commitment to the safety process by communicating safety first. The PSC should include information about any incidents that occurred the previous day or shift, hazards that may have been introduced to the operation for the day (adverse weather, equipment issues, etc.), and a safety or wellness tip for injury/illness prevention.

Demonstrations are effective during PSCs as visual aids to employees. Demonstrations can be used for proper lifting technique, effects of behaviors or techniques on the body, and much more. They should involve employees where possible to engage them in the process.

Signage serves as a reminder to follow certain rules or behaviors. These may be as typical as stop and other traffic signs or they may include information on required personal protective equipment. In addition, **postings** can be used to further communication to the workgroup. Posting information about the number of safe workdays can serve as a motivator to employees. Posting information from the investigation of a recent injury or accident can help to deliver a message to prevent recurrence. Posting other information on the activities conducted by safety committees demonstrates the company's commitment to safety through employee involvement.

Finally, **recognition and rewards** are important to a company's accident prevention strategy by improving morale. This is done by recognizing individuals, workgroups, or the facility for specific safety achievements. Even though it's important to provide recognition on the facility and workgroup level, more frequent recognition is required for the individual to make a better impact. Examples of individual recognition and rewards include immediate one-on-one feedback for safe behaviors, recognition for safe milestones (e.g., five years with no injuries or accidents, etc.), and tangible rewards for demonstrating safety knowledge or submission of a safety best practice. Workgroup and facility rewards might, for example, be given for achieving/exceeding an established goal for the operation or exceptional improvement on a specific injury type.

For workgroup and facility (and some individual) recognition and rewards, a celebration should be planned and well communicated to the target participants. When doing so, the operation should also demonstrate how the achievement has made the operation more efficient for production in addition to protecting the health and well-being of the employees. Some of the ways facilities or operations have been known to celebrate these successes are through cookouts, tangible rewards, or group gatherings/team meetings hosted by a member of upper-level management.

Emergency Response

With regard to emergency response in the aviation industry, the first thing that may come to mind is an aircraft accident. However, there are many other areas of focus for which preparation is necessary to minimize injury or loss.

Emergency Response Planning

Emergency response planning incorporates the organization of emergency response efforts on the federal, state, and local levels. It defines public health priorities for effective consequence management and explains the concepts of communication and the incident command structure. The emergency response plan should take into consideration many factors important to successful planning. These include, but are not limited to:

- The size of the organization and each of its independent sites
- Resources of the organization
- Location of each site
- Loss type potential
- Local emergency resources (fire, chemical response, etc.)
- Drills

Organization Size

The size of an organization will factor into determining how expansive the emergency response plan should be. By way of example, an FBO (fixed-base operator) would only need to plan for emergencies on a local basis, perhaps only with local and/or state agencies. But a major or regional air carrier would need to incorporate much broader demographics and multi-departmental disciplines as well as federal and, in some cases, international agencies in addition to those at the local and state level. Independent sites or field cities often have far fewer resources at their disposal. For these operations, planning with nearby agencies and/or quick mobility of resources may be necessary.

Organization Resources

Resources of the organization are important for obvious reasons. If the organization cannot provide adequate resources, the response effort could be compromised. A poor response effort is usually evident in poor communication, which could lead to negative media attention. However, resources are also important for performing a proper investigation, responding to families' needs, and a fast restoration to operational status. Larger and most mid-size organizations typically have an Emergency Response Manager or Coordinator and a budget for response equipment, supplies, and drills.

Site Location

Location of each site is important for many reasons. Several factors must be considered when developing each individual element of the plan. These include asking the following questions:

- Where is the location geographically?
  - What are the typical weather challenges for the area (tornado, hurricane, ice/snow, etc.)?
  - Are there large bodies of water nearby?
- What is the distance from the organization's headquarters or **Incident Command**?
  - Is the time to get to this location (and points in between) acceptable?
  - If not, is there another organization that can be contracted to assist in the event of an emergency?
- What agencies have jurisdiction over the location (local, state, federal, and/or international)?
- Is the area of operation predominantly industrial, residential, or rural?

Loss type potential is simply the type of loss exposures that exist. For aviation, what is considered the "worst case scenario" is usually an aircraft crash. However, there are other types of losses including ground damage, serious injury, chemical spills, fuel spills (usually listed separately from chemical spills because of size and frequency potential), weather-related, bomb threat, sabotage or terrorism, and biological exposures (blood-borne pathogens, illness epidemics, etc.). Each of these requires individual response planning and preparation.

Local Emergency Resources

Local emergency resources must be determined to make alternate resourcing plans where necessary. Is there a fire response on-site or nearby? Does the fire department or a local FBO have the capability to respond to the maximum potential fuel spill? Maintenance facilities often have hazardous chemicals as part of certain production requirements. These typically have flammable or corrosive properties. If the on-site or nearby fire department does not have the capability of responding to a spill, a local contract must be acquired or the facility must determine and train its own response personnel and provide the appropriate equipment for the maximum potential response.

Emergency Drills

Finally, the whole program is just written material and costly supplies if drills are not conducted to prepare for the "real thing." Drills help each responder better understand their job tasks and how their actions affect other areas of the response (communication between departments and outside agencies, etc.). It also helps to make necessary alterations to the plan in areas that need to be improved. It gives the organization the opportunity to see where and how they can fail, and to make adjustments accordingly, reducing the potential for failure during a real emergency.

Chapter 10, "Emergency Response," will cover each of these elements in more depth. It will include all facets of emergency response from aircraft accidents to fire response, tornadoes and other weather-related contingencies, spill response, and so on.

Safety Program Reporting System

An effective, successful aviation safety program must contain an internal reporting process as a fundamental part of the system. As it pertains to this chapter, a safety reporting process should be used to report any accidents, incidents, near-miss events (as defined in the incident investigation section in this chapter), and any hazards or areas of risk that have been identified. The absence of a safety reporting system in an aviation organization represents a significant gap and is thus an area of potential threat. Safety-related changes cannot be made in the absence of specific information about what is going on in a company. The success of an aviation safety program is largely dependent on obtaining current information that can be analyzed so that corrective actions can be put into place as needed. These are just a few of the many reasons that having a safety reporting system is so vital. Whether a company is small or extremely large in terms of employees, assets, and number of locations does not matter; any aviation operator should have a formalized reporting system as a part of its safety program and culture.

Note: Although there are a significant number of different types of aviation safety reporting systems in existence (mandatory, voluntary, FAA, NASA, etc.), the type that is going to be the focus in the following subsections represents a voluntary reporting system that could be used in almost any aviation company to report common issues that may pose a threat as well as actual safety events.

Basic Elements of a Voluntary Safety Reporting System

In order for a reporting system to be effective, some fundamental elements are needed to define how the process works within a given organization. The following are of some of the basic elements that should characterize an effective safety reporting system.

- *Accessible, simple, and easy to use:* Employees do not want to be bothered with using a system or process where the forms and other required information are not readily available, nor do they want to be burdened by a process that is time-consuming and non–user friendly. Therefore, it is imperative that the necessary reporting system forms and other documentation sources be readily available and easily accessible throughout the company. Some suggestions for accomplishing this would be to place the reporting form(s) on a company website that can be accessed by all employees at any time. Online forms can be set up so that they are also submitted online, which is an excellent way to increase efficiency and have less paperwork to file and monitor. As a backup plan, it is a good idea to have printed copies of the reporting forms available in each location. Another good idea is to have the safety reporting system integrated with the company human resources system. In this way, any reports and situations that affect specific employees will already be present in the human resources department.
- *Standardized:* The reporting system should, for the most part, be standardized in nature; that is, all employees should be using the same basic forms (with some exceptions which are discussed in the following sections) and processes for submitting information.
- *De-Identication of Reporting:* Although some may disagree with this, a voluntary reporting system may be best served by de-identifying the reports so that the privacy of the reporter is

protected. However, the person who submitted the report can be contacted for follow-up, as needed. This option may also encourage more people to report issues and situations without fear of consequences.

- *No-Fault, No-Jeopardy:* In order to encourage maximum participation in the safety reporting process, it is a good idea to have the program advertised as **no-fault, no-jeopardy**. Employees will be able to submit reports without fear of actions being taken against them and their jobs. This will help to establish trust in the safety program and in the persons responsible for the program. However, a word of caution is needed at this point: If the reporting process is no-fault, then those responsible for the program need to ensure that they stay true to their word and not take action against participating employees. Such actions will destroy trust very swiftly, and harm the credibility of the program. Some notable exceptions would be if employees committed a criminal act or demonstrated gross negligence or disregard for company standard procedures and caused an accident or other undesired event.
- *Acknowledgment and Feedback:* All reports need to be acknowledged soon after they are submitted. Written acknowledgment would be considered the preferred method. A simple message thanking the employee for their submission and a brief acknowledgment that their report will be examined and they will be apprised of the progress of the situation may prove helpful. In general, employees will not be inclined to participate in a process in which no one acknowledges their contributions and efforts. Also, feedback should be provided to the employee and other appropriate persons about the results of any investigation based on their report. Feedback helps to complete the process by getting back to the person who submitted the report and letting them know what is going on in the process. In general, it is a good idea for initial feedback to take place within seven days after the report was submitted if possible.
- *Multiple Means of Reporting Information:* To make the process easier, a company should consider providing more than one way of submitting information. Although standardization is the goal, providing more than one way to submit is a good idea. This can be accomplished through fax, email, or the mail system, if need be. Another good idea is to consider establishing an employee safety "hotline," which would be set up as an 800 number that employees can call to report safety information. Employees should also verbally tell their immediate supervisor about safety events and issues as soon as possible after they occur. However, a verbal report or hotline call should always be followed up immediately with the completion and submission of a reporting form. All employees share the responsibility of reporting safety concerns and issues immediately.

What Should Be Reported

For the purposes of the voluntary internal safety reporting process being discussed here, any accident (aircraft-related or otherwise), incident, near-miss, potential or actual hazard, or area of risk should be reported as soon as possible after the person(s) making the report become aware of the issue. This will ensure that reports and follow-ups are made in an expeditious manner.

Safety Reporting Forms

Since many aviation organizations consist of a number of different departments and divisions, a good best practice would be to design safety reporting forms that are tailored to fit the needs of specific departments or divisions. For example, flight departments are going to have different types of safety issues than ramp operations, which are going to be different from a maintenance department's, and so on. Of course, since these departments will interact on a regular basis, there will be some common issues. However, it is still a good idea to create specific reporting forms designed for each department. Since the number and types of departments in many organizations are far too extensive to cover here, the following sections provide a basic idea of some of the many issues that could be included on a safety reporting form for flight operations and for ground operations, respectively. A brief discussion of common elements that may be included on all voluntary reporting forms is also included.

Reporting Form Commonalities All safety reporting forms should contain the following elements:

- Basic instructions for completing and submitting the form
- Name of the person submitting the report
- Employee identification number
- Name of other crew members/involved employees/witnesses
- Date, exact time, and exact location of the event
- Did the event result in injury or death?
- Aircraft number and type (as applicable)
- Other equipment involved in the event
- Weather/environmental conditions (as applicable)
- A checkbox area containing a number of possible types of events or other issues
- Space for a written narrative of the event
- Signature line
- A box for entering the date the report was received in order to follow up and track
- A simple thank-you and acknowledgment printed conspicuously on the form

Flight Operations Safety reporting forms for flight operations might contain the following elements:

- Flight-related information (route, etc.)
- Emergency declared (a yes/no checkbox)
- ATC communication frequency and facility
- Phase of operation (climb, cruise, taxi, landing, etc.)
- A checkbox of numerous types of possible flight-related events (aircraft damage, air return, altitude/course deviation, diversion, runway incursion, FOD, unstable approach, terrain separation, aborted takeoff, unsafe behaviors, identified potential or actual hazard or risk, etc.)

Ground Operations Safety reporting forms for ground-based operations might contain the following elements:

- Flight/event-related information (as applicable)
- Phase of operation (hangar, towing, taxi, ramp, loading/unloading, etc.)
- A checkbox of numerous types of possible ground-related events (aircraft ground damage, employee injury, aircraft misloading, aircraft misfueling, ground support equipment accident, runway incursion, unsafe behaviors, FOD, identified potential or actual hazard or risk, etc.)

Potential Benefits of a Safety Reporting System

The establishment of an effective safety reporting system as a portion of an aviation safety program is a necessity that carries numerous potential benefits. Among the possible benefits are:

- *A proactive safety culture:* When created, implemented, and managed properly, a safety reporting system may help to establish a safety culture actively identifying and reporting problems and potential areas of concern consistently.
- *Evaluation of the overall safety status of company:* Again, if used properly and consistently, a safety reporting system enables employees to know where the company stands with respect to safety and compliance.
- *The opportunity to make positive corrections:* Possessing knowledge about problems and events gives decision makers the capacity to make corrections in the organization or in individual departments as issues are identified and reported. These actions may help to make the company safer by being better able to protect employees, customers, and assets.
- *Trend analysis:* Safety reports are very useful in conducting the all-important safety practice of **trend analysis**. As discussed in greater detail in Chapter 2, trend analysis is the practice of analyzing data submitted over a period of time and identifying common issues and problems or "trends" that have occurred over a specified time frame. Trend analysis enables a company to identify its most common and perhaps most hazardous safety issues. This practice helps to establish a system whereby events and issues can be rank-ordered by the level of risk. Issues can be addressed swiftly, with sufficient resources allocated in support of the actions.
- *Provide employees with the opportunity to participate in the safety process:* A safety reporting system provides employees the chance to be involved in the safety program and to make a positive difference by reporting safety-related problems, concerns, and possible solutions.

Accident and Incident Investigation

Another key element that must be present in any successful aviation safety program is a standardized internal process that is used to investigate the accidents, incidents, and near-misses that occur. *It is important to make a distinction between the terms 'accident', 'incident', and 'near-miss' for the purposes of this section.* For many people, the first thing that often comes to mind when accident investigation in aviation is mentioned is the occurrence of an aircraft accident. However, the most common types of accidents that occur in aviation are not traditional aircraft accidents as defined by the National Transportation Safety Board (NTSB; see Chapter 4), but other types of events that directly and negatively impact the organization. While it is extremely important that actual aircraft accidents be thoroughly investigated, it is also important that other types of accidents within an organization be investigated, along with incidents and near-misses. These types of events must be investigated in order to prevent similar events from occurring in the future. For the purpose of further distinction and clarification, the following definitions will be used to describe the types of event investigation that will be covered in this section.

Definitions for This Section

- **Accident:** An actual unexpected and undesired event that causes injury or death to employees or others, or causes damage to aircraft, equipment, facilities, or other assets.

- **Incident:** An actual undesired occurrence that is likely to cause injury, death, or damage to assets, if left unreported and uninvestigated.
- **Near-miss:** An actual undesired "close-call" type of event that does not cause injury or damage, but could have.

Why a Standardized Investigation Process for Accidents, Incidents, and Near-Misses?

Having a standardized process for investigation is very important to any aviation organization. A standardized process will help to ensure that all persons within the organization are using the same investigative strategies, methods, and techniques, no matter where the event occurred or what actually happened. Herein lies another advantage of standardization. Since many aviation organizations experience such a variety of accidents and incidents, a standardized investigation process can be applied across a wide spectrum of events with success, provided that quality training is provided and that qualified individuals are actually conducting the investigation of events.

Also, having a standardized investigation process that is part of an organization's safety program will help to ensure that consistent training will be given to all persons who will be analyzing events. As a result, these individuals will be much more likely to consistently follow the same basic procedures when conducting investigations of events that have occurred in their respective working areas, departments, or divisions.

Why Accidents and Incidents Must Be Investigated

Accidents and incidents are extremely expensive to the organizations in which they occur. In addition, accidents and incidents sometimes result in injuries, death, and damage to aircraft and/or other assets. Accidents and incidents will often lead to interruptions, cancellations, and delays, all of which are undesirable and generally impose a significant impact on aviation operations and overall efficiency. Although it may seem obvious, accidents, incidents, and near-misses must be investigated by qualified individuals who have successfully completed investigational training. It is very important to determine all of the **causal factors** that led to the occurrence of the event. As implied by the term itself, causal factors are the individual issues or events that contributed directly or indirectly to the occurrence of the event under investigation. Determining and understanding all of the factors that led to the occurrence of a given event is necessary in order to learn what happened, what went wrong, and what can be done to prevent a similar event from happening in the future. Ultimately, the rationale for investigations of accidents, incidents, and near-misses really comes down to one word: prevention. Accidents and incidents are extremely costly in a number of ways, and it is in the best interest of an aviation company to try to prevent undesired events from occurring. When the causes of events are found, effective strategies can be developed to address each contributing factor individually in order to prevent other occurrences. It is difficult if not outright impossible to prevent undesired events from happening if accidents, incidents, and near misses are left uninvestigated. Therefore, a good aviation safety program should include a standardized investigation process as part of the organizational requirements.

What Should Be Investigated?

If a person were to ask an average employee of virtually any aviation organization what types of events should be investigated, most responses would probably come down to what is considered to be the big stuff: aircraft accidents, a fatality or serious injury, a major fuel spill, and so on. Beyond

Figure 5–5. Event pyramid.

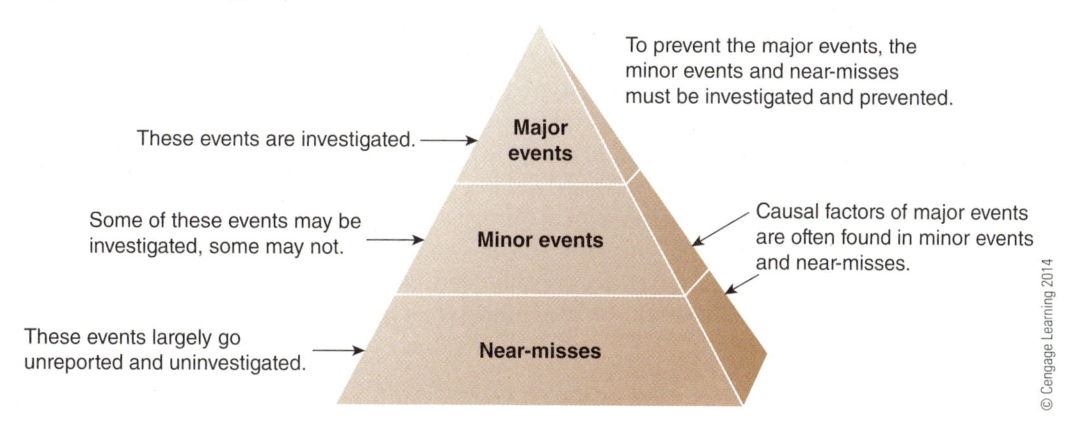

question, events of this magnitude are very serious and must be investigated. However, although major events are almost always higher cost than those of a less serious nature, they also occur at a much lower frequency. Here is an important point to consider: Minor events and near-misses, even those that are seemingly unimportant or insignificant on the surface, should, as much as practical and possible, be investigated because of their potential to have become more serious. In most organizations, there are a large number of minor events and near-misses that occur with a high degree of frequency (refer to **Figure 5–5**). For every major event that occurs, there are many more minor events that have also happened. The majority of these events are known to many front-line employees, while leadership may not know about the occurrences.

For example, in the world of commercial aviation maintenance, it is estimated that errors made in aircraft maintenance are a contributing factor in between 15 and 40 percent of all serious accidents (as defined by the NTSB). These types of events are often dramatic in their consequences, and are investigated in tremendous detail. However, it is further estimated that for every maintenance error that results in an aircraft accident, there are over 600 other errors that have occurred, but produced less significant results. Unfortunately, most of these issues tend go unreported and uninvestigated, while the big events receive a high degree of attention. This statement is applicable not just to maintenance, but to organizations in general. This is a very serious problem because the root causes of many major events may be found in the smaller issues that were never reported. An event that is not reported cannot be investigated. Typically, an uninvestigated event cannot be prevented from happening again. Therefore, all accidents, incidents, and near-misses need to be reported, documented, and investigated. Investigating all minor events and near-misses may seem excessive and unnecessary until one considers the importance of preventing events from occurring. Taking the time to investigate events may help to reduce the number of at-risk behaviors and minor events, which may in turn help prevent major events.

The Basics of Incident Investigation

One of the key elements that should be present in incident investigation is the establishment of the assurance that it is designed to be a no-fault, no-jeopardy process. Because the primary goal of investigation is to prevent future occurrences, persons who are involved in events need to understand that the purpose is not to place them or their jobs in jeopardy, but to find out exactly what happened in order to prevent similar events in the future. If the involved parties believe that what they report or say will be used against them, they will be much less likely to participate in the process out of fear for the well-being of their jobs. Therefore, incident investigation should ultimately be established as no-fault, no-jeopardy. However, actions involving gross negligence or criminal acts that led to the occurrence may result in consequences for the employee, even when the process is no-fault in nature.

As with most processes, the investigation of events involves several basic steps. These steps will be discussed briefly in the following subsections. The process is sequential in nature; that is, each step will be completed before moving on to the next. Time is of the essence in an incident investigation, so it is important that the process begin as soon as possible after the event has occurred so that information is as factual, fresh, and accurate as possible. The sooner the investigation commences, the better, as those who were involved will be more likely to remember important details of the event closer to time of occurrence.

Note: The intent of covering this information is to provide the reader with a basic familiarity with the investigation process, not to provide actual "how-to" training.

Collection of Initial Information After the occurrence of the event, the first step in the investigation process is to gather all available preliminary information pertaining to the event. Examples of this type of information and initial actions would be getting medical attention for persons involved in the event (as necessary); identifying all involved parties to the event and any witnesses; obtaining written statements from the involved person(s) and any witnesses; taking photographs and drawing diagrams of the event and the area in which it happened; and contacting applicable leadership to inform them about the event. It is very important that incident investigation be synonymous with fact management. All information must be factual to ensure an accurate investigation and a greater likelihood of preventing similar events in the future. As the preliminary information is obtained, written documentation of all data and information must be completed in a progressive (ongoing) manner as the investigation continues. This is to further ensure accuracy and timeliness, and to have a written record of the investigation steps.

Conduct a Risk Assessment After all factual initial information pertaining to the event has been gathered, a risk assessment of the event should be performed right away. Time is of the essence: A potentially hazardous situation or event needs to have the degree of risk determined as soon as possible after the occurrence of the event, as a rapid response may be needed if it is possible that the event could happen again or if the event could have a possible negative impact throughout the organization's entire system. Here is a hypothetical example based on actual events. An airline has a particular type and model of belt loader that is used by ramp agents in the loading and unloading of baggage and cargo transported in aircraft. Identified as part of the investigation was the fact that the airline has many of these belt loaders located in a number of stations throughout the country. In a relatively short period of time, the airline experienced some significant ground damage events involving some of their aircraft as the result of a design flaw in the belt loader. The airline conducted a risk assessment as part of the investigation and determined that the degree of risk was rather high. Because of this factor, representatives of the airline contacted the belt loader manufacturer and reported the problems. The airline worked with the manufacturer to develop a design solution that was installed on all of the belt loaders relatively quickly. A swiftly expedited risk assessment enabled the airline to understand the potential hazard posed by the belt loaders, and to eventually develop an effective, specific solution to the problem. Thus, a risk assessment should be conducted as soon as possible after each accident, incident, and near-miss.

To briefly reiterate the process previously covered in Chapter 3, "Risk and Risk Management," a basic risk assessment as it pertains to incident investigation involves analyzing each individual event by the application of two questions:

1. Determining the severity of an event. In other words, how significant in terms of severity was the event?
2. What is the probability (likelihood) of this event or similar events recurring?

Figure 5–6. Risk Assessment Matrix for airline belt loader issue.

1 = High risk 2 = Medium risk 3 = Low risk

Hazard	Likely	Probable	Improbable
Catastrophic	(1)	1	2
Critical	1	1	2
Marginal	2	2	3
Negligible	3	3	3

Probability

It is important that investigators do their best to apply and answer each of these questions as accurately as possible to ensure an appropriate response to the occurrence. Once these questions have been answered, the next step is to employ a risk assessment matrix in order to determine the level of risk. Using the example of the belt loader issue that was discussed previously, **Figure 5–6** shows the application of the risk assessment matrix to the belt loader issue. The sample matrix is a simple *X* and *Y* axis, where the intersection where the two points meet is the level of risk. In answering the first question as applied to this case, the worst-case scenario could be a serious injury or death of an employee as a result of the problem, or at least a significant aircraft ground damage event. This places the potential severity of the event in the "critical to catastrophic" range, which equates to a risk level of "1" for severity. For the second question, regarding probability, the airline owns and operates a large number of belt loaders in a variety of different locations around the country. Since the issue with the belt loaders is related to their design and they are used on a consistent basis every day, it is likely that a similar event may occur again in the future. Thus, this places the probability level also at "1," a high level of frequency. The intersection where severity and probability meet (depicted in Figure 5–6 by the circled number) defines the level of risk. Note that the lower the number, the higher the level of potential risk present in a given situation. The risk code number for the belt loader is "1," placing it at a high level of potential risk.

Determining the potential level of risk is very important in order to understand the seriousness of the event. Knowing this information will enable the leadership of an organization to determine the level of involvement that needs to take place with specific events. A basic example: Can the event be handled (investigated) directly by the leadership in the location where it happened, or does the corporate safety department or an outside entity need to be called in to assist with the investigation? Also, assessing the potential of severity and probability helps to ensure that the right priority and an appropriate degree of attention and response are given to high-risk events. In general, issues that pose a high level of risk should be given the most immediate and highest degree of attention. Other events can then be rank-ordered for attention and response based on their corresponding risk codes.

Determining Causal Factors of the Event One of the most obvious yet important parts of conducting an incident investigation is determining all the specific factors that caused the event to occur. The reason for this is simple: In order to prevent something from occurring, we must know what caused it to happen in the first place. In most cases, there is usually not just one single causal factor; rather, there are usually several factors that were present and that caused the event to occur.

This is sometimes referred to as an **error-chain**, in which one issue is linked to the next, and so on, causing an undesired event to occur. It is imperative that each causal factor be determined as quickly as possible during the course of the investigation.

Basic Strategies for Determining Causal Factors Because the types of issues that cause events to occur are extremely varied in nature, there is no single method used to determine what caused an event to occur. Rather, there are several basic methods that can be followed to assist the investigator. Some of these methods are discussed briefly in the following subsections.

Interviews Insofar as possible, all employees or others who were directly involved in the event or who witnessed the event should be interviewed by the investigator. The interview(s) should take place as soon as practical after the event to ensure that the memory of those to be interviewed is fresh and that details can be more easily remembered. It is imperative that the interviewer be thoroughly prepared ahead of time for the interview so that it is productive and accurate. The investigator conducting the interview should have all of the initial factual information (written statements, pictures, etc.) at hand during the interview, and should have read through everything beforehand in preparation. Steps should be taken to ensure that employees and witnesses are kept separate from each other to eliminate the possibility of corroboration.

The investigator should prepare a list of specific questions pertinent to the event to ask during the interview. Rather than asking closed-ended questions that can be answered with "yes" or "no," open-ended questions should be asked during the interview. A simple example of an open-ended question or query would be as follows: "What are the exact procedures that you were training to follow?" The advantage of asking open-ended questions is that they cannot be answered with one-word answers; they require the person to expound in a greater degree of detail. Asking the right types of open-ended questions gives the investigator an increased chance of gaining valuable insights into what occurred that may have contributed to the event.

While there will be some exceptions, interviews should take place in a one-on-one format, that is, between the investigator and the person being interviewed. This is to make the person more comfortable and relaxed. In general, when the person being interviewed is relaxed, they will be more likely to bring forth important information pertaining to the case. A notable exception to this rule of thumb would be if an investigator desires to interview an employee who is a member of an organized labor union. In these types of situations, the employee possesses the right to have a union representative present during the interview, should they choose.

Interviews should take place in a quiet location that is free from interruptions. During the interview, the investigator needs to engage in the practice of **active listening**. Active listening consists of listening carefully to what the person is saying, taking accurate notes, and repeating back to the person what was said in order to ensure clarity and accurate information. The investigator needs to take measures to ensure full understanding of what the interviewee is saying, and the practice of active listening is a valuable and necessary tool in order to conduct a successful, factual interview.

Review of All Available Information and Documentation As important as interviews are during the course of conducting an investigation, it is equally important that other available sources of factual information be examined as well. Some examples of other items that could potentially be examined by an investigator are aircraft maintenance records, employee and or company training records, audit or inspection information, ground support equipment records, injury and aircraft damage rates at a given location or company-wide, employee involvement with previous events, operating manuals, and so on. The list of possible resources that may be related to a specific event is rather extensive, and the investigator needs to take the time to examine all pertinent information.

Review of Factual Information Pertaining to the Event In order to determine the causal factors of an event, the investigator needs to take some time and reflect on everything that has been done during the course of the investigation. All of the facts of the event need to be scrutinized carefully. What does the factual information suggest about the likely causal factors of the event? It is important that the investigator take the time and effort to look beneath the surface of an event to the deeper issues that lie underneath and that may have caused the occurrence. Proper incident investigation is synonymous with fact management. The investigator should concern him/herself only with factual information, not with conjecture, blame, finger-pointing, or arguments. The focus should be on that which is pertinent and factual.

Developing Corrective Actions The incident investigation process is incomplete unless specific, realistic corrective actions are developed in response to the findings of the investigation. As the phrase implies, corrective actions are just that; they are designed to make effective, specific changes in the work environment in order to prevent future occurrences based on the event that has occurred. Corrective action development is one of the most important elements of the investigative process, as it provides the opportunity to identify hazards and areas of risk and to take the appropriate steps to address these types of issues.

Like many issues that may have an impact on an organization, there are a number of factors that will serve to influence corrective actions. Among these factors are:

- *Economics (cost):* How much will the corrective action cost? How likely is it that the potential benefits will outweigh the costs?
- *Operational impacts:* What will be the most likely impacts to the operation?
- *Effectiveness:* How likely is it that it will produce the desired results?
- *Feasibility:* Is it realistic?
- *Employee involvement:* Having members of the safety committee involved in developing corrective actions may enhance employee buy-in.
- *Acceptance by leadership and employees:* Will the organization accept and support this action?

It is important that each of these factors (and other related issues) be considered as corrective actions are under development.

In order to be effective in the prevention of future events and in addressing identified areas of risk, corrective actions need to be measured against several important criteria. To begin with, corrective actions need to be very specific in nature. A proposed corrective action that is too broad in scope or not closely related to the factors of the event is most likely not going to be effective. Also, corrective actions need to be tangible, realistic, and achievable. If a proposed corrective action is not realistic and beyond the realm of what can feasibly be accomplished, then appropriate modifications to the proposed action should be made. Next, corrective actions should be designed in such a way that they are measurable. In other words, can the organization see tangible improvements that can be assessed as a result of the corrective action? Here is a simple, generalized example: Has the number of runway incursions by flight crews decreased over a period of time as a result of a corrective action? Being able to measure the effectiveness of corrective actions is important in order to determine if it is actually working and producing the desired results. If the measurable results are positive, it is likely that the action is working. If the results do not demonstrate any change over time, then the corrective action may need to be modified or replaced with another that is more specific and appropriate. Finally, corrective actions should produce long-term benefits to an organization by helping to mitigate or even eliminate hazardous events, behaviors, or weak policies. Corrective actions need to be designed in such a way that immediate needs are met, but also with an eye toward producing a safer working environment in the future.

Other Elements of Corrective Actions As a portion of the overall incident investigation process, each corrective action that is developed must contain two basic elements: ownership and a deadline. **Ownership** means assigning a specific person or group the full responsibility for implementing the corrective action in the organization. These persons are charged with putting the action into place, helping to communicate with other employees about the action, and assisting in monitoring the success of the action. In other words, a specific corrective action becomes theirs; in this sense they *own* it and are thus responsible for the action. Ownership ensures that accountability is established, and takes the guesswork out of who is responsible for each corrective action. If the corrective action is not completed and implemented, leadership will know who is responsible and can take the appropriate steps to address the issue with those responsible. Next, each corrective action must have a deadline for completion and implementation. An estimated date and time should be established for when each corrective action will be put into place. Similarly, having a deadline helps to ensure that accountability for the action is firmly in place, and will ensure that each action is brought online by a specific date and/or time.

Feedback The last basic element of the incident investigation process is feedback. Feedback about the investigation and corrective actions needs to be provided to all persons in the organization with responsibility for the department or area in which the event occurred, and to other persons who have a vested interest in the event. Feedback should be given to anyone who was involved in the event; to all employees affected by the event or by the corrective actions; to the leadership of other departments if the information may be pertinent or useful to them; and to other persons within the organization as necessary. Feedback will ensure that proper communication about the results of the investigation and the corrective action has taken place within the organization. Also, providing feedback demonstrates to leadership and employees that a proactive series of steps is being taken to make the overall organization a safer place to work.

Internal Evaluation Programs

An **internal evaluation program (IEP)** represents a proactive approach to safety by providing the means to continually assess the current state of the various departments of an aviation organization with respect to compliance or noncompliance with any corresponding regulatory agencies and company policies, safety, areas of risk and hazards, areas of deficiency, and other related issues. An IEP should be set up to function as a continuous process of organizational assessment, evaluation, and corrective action featuring a scheduled series of internal audits and inspections designed to ascertain exactly where the company stands with respect to the areas mentioned above. The effective use and management of an IEP also provide leadership and employees with a useful tool for helping them to understand the exact state of their respective work areas, as well as their individual responsibilities in promoting safety and regulatory compliance, and in working to ensure these things.

Many aviation organizations benefit from the presence of an internal evaluation program. Noncompliance with local, state, and/or federal regulations can and sometimes does result in monetary fines or operating sanctions being levied against certain aviation operators. Some fines can be rather significant in terms of total dollar amount, and some sanctions can limit or restrict the ability to operate as normal. An IEP can enable an organization to avoid costly fines, operating sanctions, and potential bad publicity by identifying issues that may be problematic, and to develop effective corrective actions to address these issues ahead of time. However, an IEP is only as effective as the people who are responsible for its creation and management. For this reason, an IEP must be properly managed to be effective and to produce positive benefits. Ideally, an IEP should be integrated throughout the company until it becomes a part of the organizational culture and the majority of employees are in step with the program.

Although there is at present no regulatory mandate that requires aviation organizations to have an IEP, the FAA has produced Advisory Circulars (AC) that encourage both aviation maintenance repair stations and air carriers to develop and implement an IEP within their respective organizations. AC 120-59A, dated April 17, 2006, provides guidelines and pertinent information for air carriers desiring to implement an IEP. AC 145-5, dated September 27, 1995, provides the same basic guidelines and framework for the establishment of an IEP to aviation maintenance repair organizations. As mentioned previously, these programs are not mandatory. They are strictly voluntary in nature and can be developed at the discretion of the individual organization. However, even though they are voluntary, an IEP does represent what may be considered a best practice within a company, and having this type of process in place carries the potential for providing numerous benefits. AC 120-59A (2006) states that "an IEP is a fundamental element to ensure compliance with external regulatory requirements, identify nonconformance to internal company policies and procedures, and identify opportunities to improve organizational policies, procedures, and processes." An IEP will also enable an aviation organization to be in line with FAA policy that encourages certain aviation operators to identify and voluntarily report issues that are noncompliant. AC 120-59A (2006) states that "the FAA encourages certificate holders to identify, correct, and disclose instances of regulatory noncompliance. Therefore, the development and implementation of an IEP will benefit both the certificate holder and the flying public." Thus, it is evident that the FAA strongly supports the development of an IEP and views this as a favorable and even desirable action.

IEP Fundamental Elements

AC 120-59A (2006) states that an IEP should contain specific fundamental elements in order to be effective in an organization. Some of the primary elements are listed below, derived directly from AC 120-59A. Some of the actual words have been changed or modified, but the content remains generally consistent with the AC in each topical area.

- *Systems-Oriented Process:* Internal evaluation should be a continuous process utilizing inspections, audits, and evaluations to analyze the effectiveness of managerial controls and processes in critical systems within an organization. IEP should function to constantly improve these various systems

Source: U.S. Department of Transportation/FAA

based upon the results of ongoing audits and evaluations. IEP should also include processes that provide for the continual monitoring and feedback of information on critical processes, and frequent trend analysis of any data that was derived from the program.

- *Beyond Regulatory Compliance:* IEP should extend beyond minimal regulatory compliance to determine the causal factors of deficiencies. IEP should also work to detect and implement needed enhancements to the operating practices of a company before deficiencies have the chance to occur.
- *Independent:* IEP should be designed as an independent process that has direct reporting responsibility to senior management in the company or organization.
- *Defined Responsibility and Authority:* IEP should establish a specific person or group within the company who will be granted the responsibility and authority to develop, implement, and modify the IEP; conduct audits, inspections, evaluations, and data analysis on a continuous basis; initiate, recommend, or provide corrective actions, including prevention strategies; track and confirm that corrective actions have been implemented within predetermined time frames; and communicate and coordinate activities pertaining to the IEP with FAA representatives regularly.
- *Senior Management Review:* The support and involvement of senior management in an IEP is critical to ensure the success of the program. An IEP should include the regular review by senior management of the results of IEP audits and evaluations. This should be done at least annually, if not more frequently. For IEP program purposes, senior management includes the chief executive officer, president, chief operating officer, or an equivalent position that has the authority to take action toward the resolution of issues. IEP reviews by senior management need to be documented and should include the content and any resulting management-directed action items. A basic precept of IEP is that senior management is accountable for acting on the information it receives from persons responsible for the program. As such, the IEP should include an assessment of the effectiveness of the management review process and opportunities for improvement of the process as they are identified.
- *Feedback Loop:* The quality control of an IEP is accomplished through the use of a feedback loop. Feedback is a process in which the output of a system is passed to the input to continually achieve the desired results. An IEP should be designed to utilize a closed-loop feedback, which is the continuous monitoring of the differences between IEP objectives and outcomes while also implementing appropriate changes to the system. An effective IEP should provide quality-related information to all affected employees and members of leadership. Feedback to employees may include best practices, system/process weaknesses, common management errors, and other items. This information may be passed along by more informal means such as "Read and Sign" memos, bulletins, newsletters, or web-based information. Conversely, information may be more formally incorporated as program/policy/procedural changes, training enhancements, manual revisions, or reorganization of employee groups.
- *Continual Process:* An IEP should be a continual, ongoing process in order to effectively anticipate problems and correct them before actual findings can occur. A continual IEP process is needed to verify whether findings are isolated events or actual symptoms of systemic policy, process, or procedural problems. An IEP should involve more than planned evaluations, tracking corrective actions, follow-up audits, and special evaluations based on trend analysis. IEP should also include continuous data collection and analysis to initiate actions designed to prevent undesired events before they can occur. A well-structured IEP ensures that all areas of operations are assessed at proper time intervals. A continuous IEP process is also important to aiding in the identification of problems and issues that may otherwise be missed by regular audits and evaluations.

Source: U.S. Department of Transportation/FAA

Safety Auditing

An effective aviation safety program needs to include the process of conducting ongoing auditing and interval evaluation. From the perspective of aviation safety, **auditing** is generally synonymous with conducting a comprehensive inspection of the various areas within an organization. The practice of auditing seeks to ascertain the status of every department and division within an aviation organization in order to identify any areas of regulatory noncompliance, areas of risk and hazard, noncompliance with any company rules/policies/procedures, issues of organizational management, evaluation of programs, and areas of strength and weakness. Audits may be conducted by direct employees, by an outside consulting firm, by an individual, by an auditing team, or by a combination of one or more of these. The auditing practice is also a key component of IEP, which was discussed in the preceding section. Auditing is an extremely important piece of a successful aviation safety program and needs to be performed on an ongoing basis. There are a few common ways in which audits are conducted in many aviation organizations, and some of the more common approaches are briefly discussed below.

The Practice of Self-Audits

As is implied by the name, **self-audits** are conducted by the persons who work in a certain location or department, auditing the area(s) in which they work. Self-audits are very important because they help the employees to be aware of the status of their respective work areas at all times, and give them the ability to identify problems swiftly and take rapid measures to address areas of concern. Informal self-audits can be done daily or weekly; an example would be a daily foreign object debris (FOD) inspection on an aircraft parking ramp. However, self-inspections can be much more detailed in nature, depending on the organization. For example, a large regional airline in the United States developed a series of comprehensive self-audit checklists. Numerous versions of these checklists were created and tailored to fit the specifics of a number of different operating divisions within the airline. Members of departmental leadership were provided with training on the basics of performing a self-audit of their respective locations using the checklist. These detailed self-audits were completed once a year, and the completed checklists were then submitted electronically to the auditing program manager. This action provided written documentation of the completion of the audit, and also established accountability. After the self-audits were submitted, members of the airline's auditing team selected a sample of departments and locations to visit. The members of the team then compared the completed self-audit checklist with what they themselves found at the location. If any discrepancies were noted, local leadership had to give a plausible explanation for each item that proved to be different from what was noted on the checklist. Steps were then taken to ensure that corrections were made.

Comprehensive Safety Audits

In terms of time to complete and degree of work involved, a comprehensive safety audit is often an arduous, yet necessary undertaking. These audits are usually highly detailed, extremely thorough, and sometimes require significant amounts of time to complete. These types of audits can be applied to a specific facility, area, or an entire organization, depending on the nature of the inspection. Comprehensive safety audits are generally planned well in advance of the scheduled inspection date so that the auditor(s) have adequate time in which to prepare for their task.

Comprehensive safety audits are often conducted by specific individuals within the company, or by members of an auditing team. These persons have usually received and completed training on how to conduct an audit/inspection. Often these types of safety audits make use of detailed checklists to guide the auditor as the inspection is taking place and to serve as a reminder of the areas and items that need to be examined. Depending on the company and the areas being audited, the team may consist of several direct employees of the company, and in some cases qualified persons or firms from outside the company may be brought in to participate.

For example, a corporate safety manager from the same regional airline discussed above in relation to self-audits completed an audit of the carrier's primary hub. The hub comprised a very large area, and the entire facility (inside and outside), including numerous documentation elements (employee training records, ground support equipment maintenance records, etc.), was inspected. The corporate safety manager conducted the inspection, accompanied by the hub safety manager and an outside safety consultant. The consultant provided an expert, objective safety perspective and was highly instrumental in the completion of the hub audit.

A properly conducted comprehensive safety audit will examine the facility, operation, or organization thoroughly, and will provide a written report on the findings to members of leadership in their respective areas of responsibility. In addition to completing a written report about the audit, a good best practice is to schedule a face-to-face audit debriefing between the primary auditors and the leadership of a given area. A copy of the written report should be provided to the leaders well in advance of the meeting. Holding a post-audit meeting will give the leadership an opportunity to hear firsthand about the findings in detail, provide them with the ability to ask specific questions, and give the auditors and the leaders the chance to work together to develop and implement potential solutions to problem areas that were discovered during the audit.

Flight Safety

Although it may seem self-evident because we are dealing with the subject of aviation, virtually any safety program for a company that operates aircraft needs to include specific information addressing flight-related issues. Depending on the size and type of aviation organization, a flight safety program may be relatively small and simple, or it may be large and extremely comprehensive. The latter is especially true for companies such as major and regional airlines, which have significant flight safety programs for the purposes of regulatory compliance and to ensure that they are operating as safely as possible. However, regardless of the size of the company, it is imperative that a solid flight safety program be put into place, revised as necessary, and followed very closely by all flight crew members and leadership. The following sections briefly discuss a few of the basic elements that should be present in a flight safety program.

Note: A more expository, detailed analysis of flight safety programs and related issues is covered in Chapter 8, "Flight Safety Programs."

A Written Program

Having a written program for flight safety is a fundamental necessity. A written program helps to ensure regulatory compliance, standardization, and accountability, and provides a means of organizing the information, making it readily available for flight crew members and others who are involved in flight operations. A written program also provides single-source information for flight crews, which makes information easier to obtain and more efficient to disseminate. Some organizations make their flight safety manual available by posting it on a company website accessible by flight crews and others within the organization needing access to this information.

A written program will often consist of some form of **flight safety manual**. This type of manual will generally include all of the company policies, procedures, guidelines, and rules pertaining to flight-related activities that take place within the company and/or are required by the FAA and other regulatory agencies. The manual will also include policies, procedures, and rules for the operation of aircraft in accordance with manufacturer specifications. For example, an aircraft manufacturer may create a recommendation or procedure that calls for no more than one engine to be started on a certain type of jet aircraft during pushback from the gate. This type of information would likely be included in the flight safety manual, and would be covered during flight crew classroom training.

A Responsible Program Leader

It is very important that a specifically designated individual be in charge of the flight safety program. For air carriers operating under FAR 121, a Director of Safety is a mandatory position that must be present in the organization. The Director of Safety is ultimately responsible for the oversight of all safety-related functions of the airline. This includes maintenance, ground operations, and flight safety–related activities. Some air carriers may have several managers in a flight safety department, each of whom is responsible for a different aspect of flight safety. These managers often report to the Director of Safety. This is the case in some large U.S.-based major and regional airlines. For FAR 135 operations, a Director of Safety is not required, but is still recommended by the FAA.

Even in smaller aviation organizations, it is still important to have a specific individual who is responsible for safety, even if not required by regulations. For example, in some corporate flight departments, one flight crew member is given specific responsibility for focusing on safety-related issues, and this becomes a regular part of their job. These individuals may revise flight safety programs and manuals, create new additions to the flight safety program in accordance with others in the departments, communicate with aircraft and power plant manufacturer representatives on safety issues, and so on.

Investigation of Events

Any flight safety program is incomplete unless it includes a process for investigating aircraft-related accidents and incidents. Depending on the size and personnel arrangement of the organization, responsibility for investigations may be given to a specific air or flight safety investigator, a manager of flight safety, or another specific person or group. Information on these types of actions can be found in the incident investigation section of this chapter, and also in Chapter 8.

Safety Reporting System

As discussed later in this chapter, an effective flight safety program should include a safety reporting system to convey events related to aircraft operations. Flight crews should be trained in the use of the program and encouraged to do so in order to discover and correct issues before they become a problem. Irrespective of size, any aviation organization engaged in aircraft operations should have and use a good safety reporting system. Larger operators such as airlines generally make use of programs such as the **Aviation Safety Action Program (ASAP)** and **Flight Operations Quality Assurance (FOQA)**. ASAP is a voluntary reporting program designed to function as a partnership between the FAA and air carriers and maintenance repair stations in order to prevent accidents through the reporting of safety issues and incidents. Air carriers typically have a manager or other person in charge of their ASAP program. Advisory Circular (AC) 120-66B covers ASAP in detail. FOQA is a program that incorporates digital recording devices installed on aircraft. These systems record a significant number of flight parameters as the aircraft is in operation. This provides the carrier with a significant amount of data on their aircraft that are FOQA-equipped. Not all domestic carriers have an active FOQA program; however, these systems are becoming more common in the United States. This is because both ASAP and FOQA are considered to be fundamental parts of the Safety Management System (SMS) of an air carrier. Many international carriers already have active FOQA programs in place.

Note: An interview with the FOQA manager of a large U.S.-based major airline revealed that a couple of other items should also be part of a flight safety program. The FOQA manager, speaking from experience, stated that all of the various managers and other key persons comprising an air carrier's flight safety department must openly communicate on a continuous basis to keep each other abreast of problems, issues, changes, successes, failures, and other pertinent information. In addition, the FOQA manager stated that any successful flight safety program will also work to achieve the support and buy-in of flight crew members. This is extremely important because pilots are a primary source of flight-related safety information, and their support is crucial to program success.

Ground Safety

Prior to this century, accident prevention programs focused primarily on flight operations and flight safety concerns, rather than on ground-related issues. Typically when attention was given to ground operations, it was reactive rather than preventive. Ironically, the majority of injuries and costs associated with these injuries belong to workers in ground and maintenance operations.

For 2007, the Bureau of Labor Statistics (BLS) reports that the airline industry **lost workday rate** was 5.2. That is, the industry experienced 5.2 lost workday cases per 100 full-time employees. By way of comparison, the lost workday rate over the same time period for combined private industry was 1.2 cases per 100 full-time employees. For the same year, BLS also reported an overall **incident rate** for the air transportation industry of 9.9 per 100 full-time employees. Most injuries that occur in the air transportation industry involve persons working in ground-based positions as opposed to flight operations positions. The above statistics and factors may indicate that not only does the air transportation industry have a high rate of injuries, but that a sufficient amount of attention and resources may not be allocated to ground safety activities.

The most common types of injuries that occur in air transportation are musculoskeletal disorders, which can be basically defined as strains, sprains, and similar injuries. Back, shoulder, knee, and ankle injuries are some of the most common musculoskeletal disorders in air transportation. Although the causes of these types of injuries are extremely numerous, the very nature of the air transportation industry is conducive to injury, especially in the absence of an effective ground safety program. Aircraft baggage compartments, for instance, are not designed with the employee in mind. Range of motion in many cases is very limited, particularly with regional aircraft, which make up about 50 percent of the aircraft flying today.

Another factor is that air transportation is highly reliant on being on time, and employees are often in job situations where there is time pressure, which can sometimes contribute to injuries when employees rush to complete tasks. Certain ground-based positions are characterized by excessive repetitive motion and the repeated lifting of heavy objects. Because ground-based operations for air carriers (passenger and cargo) require extensive material movement, equipment is necessary to reduce the amount of manual handling. Some airlines are confined by their financial limitations and rely on older, less reliable assist equipment. Others may not even have the appropriate equipment to reduce manual handling, drastically increasing the opportunity for strain-related injuries. Bob Vandel, Executive Vice President of the Flight Safety Foundation, estimates that ground-based accidents cost airlines around the world at least $4 billion a year, with an additional $1 billion in costs for the corporate and general aviation sector. According to Vandel, these figures do not include an additional $5.8 billion a year in the cost of personal injuries. Clearly, these factors indicate that an effective ground safety program is absolutely essential for virtually any aviation organization.

In Chapter 7, "Ground Safety," the reader will be provided with specifics in ground safety applicable to various areas throughout an aviation organization. Topics will include ramp operations, cabin services, fueling, aircraft maintenance, warehousing, dangerous goods transportation, hazardous materials handling, aircraft ground handling, and other support functions.

Environmental Compliance in Aviation

Environmental compliance in the aviation industry has become an ever-growing subject over the last two decades. It's an area some smaller airlines were once reluctant to face because of the program costs. However, with the liabilities involved and the irreversible damage that can be caused, these organizations now realize that it is actually cheaper to develop and manage a system now than to pay for a spill or hazardous exposure, both literally and through bad press exposure.

The Environmental Protection Agency (EPA) is the federal governing body, but many states also have a state-sponsored agency whose regulations exceed the federal standards. There are several facets of environmental compliance within the aviation industry. Governing agencies oftentimes will require environmental studies for new airports as well as new or expanded service to airports. These may include noise estimates or effects on surrounding neighborhoods, air emissions, and/or surface (ground and water) impact studies. Other facets governed by federal, state, and perhaps local agencies include management of:

- air emissions—Clean Air Act
- storm water pollution—National Pollution Discharge Elimination System (NPDES)
- hazardous, non-hazardous, and universal waste generation
- storage tanks

Because of the complexities of the environmental regulations on the aviation industry, this book will not cover them in detail. However, a high-level overview of programs applicable to the aviation industry was covered in Chapter 2, "Regulatory Oversight."

Chapter Concept Questions

1. Describe in detail the importance of leadership support for safety programs and initiatives. What are some of the ways in which leaders can demonstrate support for leadership within their organizations?
2. Why is it important for front-line employees to be included in the development and implementation of safety programs? Describe some of the assets that front-line employees may bring into the safety planning process.
3. Describe the basics of an internal evaluation program (IEP). What are the basic elements of the IEP process? Discuss some of the ways that an IEP may benefit an aviation company.
4. Why is the practice of internal auditing so important for aviation companies? Compare and contrast self-auditing with comprehensive auditing.
5. As described in this chapter, what are the basic steps of conducting an incident investigation? Why is it important to identify all of the causal factors of an event? How does the development of corrective actions factor into the investigation process?
6. Describe in detail why a safety reporting system is important for an aviation company to employ. What are some of the possible benefits of a reporting process?
7. Why should a comprehensive and effective aviation safety program include specific information and training on ground safety for employees?
8. An organization has experienced an increase in the number of injury claims over the last three consecutive years. Describe the process by which the organization would determine which issues are driving the losses and any methodologies for generating the biggest gains.
9. The same organization above experienced 205 injuries in 2007, 233 injuries in 2008, and 256 injuries in 2009 over the same time frame during the year. The total actual hours worked were 11,205,505, 10,605,876, and 11,342,110 for each year, respectively. Calculate the injury rates for each year and the percentage of increase for each year.

Chapter References

Alteon, A. Boeing Company. n.d. Maintenance Human Factors Program Training for Managers. PowerPoint presentation.

Armstrong, K. n.d. Incident Analysis: Getting Below the Surface of Events to Correct Breakdowns. Unpublished presentation.

Bureau of Labor Statistics. 2009. Occupational Injuries and Illnesses: Industry Data. http://www.bls.gov.

Department of Transportation (DOT). 2010. *Title 49 Code of Federal Regulations (CFR)*, Part 171.8: *Definitions and Abbreviations*. http://www.phmsa.dot.gov/regulations.

Environmental Protection Agency. 2007. The Plain English Guide to the Clean Air Act. http://www.epa.gov/air/peg/peg.pdf.

Environmental Protection Agency. 2010. Clean Water Act. http://www.epa.gov/oecaagct/lcwa.html#Summary.

Environmental Protection Agency. 2010. Resource Conservation and Recovery Act (RCRA), Appendix C: Glossary. http://www.epa.gov/epawaste/inforesources/pubs/orientat/romapc.pdf.

Environmental Protection Agency (EPA). 2010. Title 40, Code of Federal Regulations (CFR), Subpart A: General, Part 261: Identification and Listing of Hazardous Waste.

Environmental Protection Agency (EPA). 2010. Wastes: Information Resources. http://www.epa.gov/epawaste/inforesources/index.htm.

Federal Aviation Administration. 1995. Advisory Circular 145-5. http://www.airweb.faa.gov/Regulatory_and_Guidance_Library/rgAdvisoryCircular.nsf/0/6d35dfced414e703862569e00074d653/$FILE/AC145-5.pdf.

Federal Aviation Administration. 2002. Advisory Circular 120-66B. http://www.airweb.faa.gov/Regulatory_and_Guidance_Library/rgAdvisoryCircular.nsf/0/61c319d7a04907a886256c7900648358/$FILE/AC120-66B.pdf.

Federal Aviation Administration. 2006. Advisory Circular 120-59A. http://www.airweb.faa.gov/Regulatory_and_Guidance_Library/rgAdvisoryCircular.nsf/0/fd8e4c96f2eca30886257156006b3d07/$FILE/AC%20120-59a.pdf.

Federal Aviation Administration. 2006. Advisory Circular 120-92. http://www.airweb.faa.gov/Regulatory_and_Guidance_Library/rgAdvisoryCircular.nsf/0/6485143d5ec81aae8625719b0055c9e5/$FILE/AC%20120-92.pdf.

Fritzsche, D. 2005. *Business Ethics: A Global and Managerial Perspective* (2nd ed.). New York: McGraw-Hill.

Hagan, P. E., Montgomery, J. F., O'Reilly, J. T. 2001. *Accident Prevention Manual for Business and Industry, Administration and Programs* (12th ed). Washington, DC: National Safety Council.

International Civil Aviation Organization. n.d. Welcome to the Safety Management Website. http://www.icao.int/anb/safetymanagement/index.html.

Jones, C. 2006. Airline Focuses on Ramp Safety After Rash of Mishaps. *Knight-Ridder Business Tribune*, July. http://goliath.ecnext.com/coms2/gi_0198-336792/Airport-focuses-on-ramp-safety.html.

National Business Aviation Association. 2009. An SMS Can Raise the Safety Bar. *Business Aviation Insider*, March/April.

National Business Aviation Association. 2009. Safety Management System. http://www.nbaa.org/admin/sms/overview/.

National Fire Protection Association. 2008. NFPA 30: Standard for Flammable and Combustible Liquids. NFPA.

Occupational Safety and Health Administration (OSHA). 2005. Title 29, Code of Federal Regulations (CFR), Part 1910, Section 106: Flammable and Combustible Liquids.

Visual Experts Human Factors. 2003. An Attorney's Guide to Perception and Human Factors. http://www.visualexpert.com.

Wild, B. 2009. Personal interview on flight safety program elements.

Wood, R. 2003. *Aviation Safety Programs: A Management Handbook* (3rd ed.). Englewood, CO: Jeppesen Sanderson, Inc.

Introduction to Human Factors

Chapter Learning Objectives

After completing this chapter, the reader should be able to:

- Define human factors, identify the various disciplines involved in human factors, and discuss the importance of human factors in aviation.
- Describe the fundamentals of some common human factors models used in aviation.
- Discuss the background of the human factors movement.
- Discuss some of the basics of human physiology.
- Describe the importance of proper judgment and decision making in aviation.
- Explain the basics of NREM and REM sleep, the four stages of the sleep cycle, and understand the importance of sleep to physical and mental well-being.
- Identify and describe fatigue, and discuss some of its many causes and its symptoms and effects on the human body and on human performance.
- Discuss some of the origins of human factors as a discipline.
- Discuss the basics of situational awareness and its importance in safety.
- Discuss and describe the significance of stress on the human body and on judgment and decision making.

Key Concepts and Terms

5 M Model
Active Risk
Aeronautical Decision Making (ADM)
Aircraft Ground Damage
Conventional Decision Making (CDM)
Crew Resource Management (CRM)
DECIDE Model
Decision Making (DM)
Fatigue
Fatigue Risk Management
Five Hazardous Attitudes
Hardware
Human Factors (HF)
IMSAFE Checklist
Insomnia
- Acute
- Transient
- Chronic

Judgment
Latent Risk
Non-Rapid Eye Movement (NREM) Sleep
Rapid Eye Movement (REM) Sleep
Reason's Model
Self-Assessment
SHEL Model
Situational Awareness (SA)
Sleep Inertia
Software
Stress
Stressor
Stress Tolerance Level
Systems Approach

Introduction

The topic of human factors (HF) in aviation is very important for a number of reasons. First and most obvious, the aviation industry is run by and composed of a wide variety of people who make decisions, fly and repair aircraft, and work in a number of ground-based positions in support of the end product of aviation: flight. The direct actions, inactions, judgments, and decisions as well as the individual physical and mental status of these persons cause both direct and indirect impacts on aviation operational efficiency and safety.

However, perhaps the most significant reason HF is so important in aviation is found in the records of aviation safety data. A number of different organizations including the Federal Aviation Administration (FAA), the National Transportation Safety Board (NTSB), the International Civil Aviation Organization (ICAO), and a number of individual researchers have all determined that somewhere between 60 and 80 percent of all aviation accidents, incidents, and other undesired events have their root causes in human factors. This statistic remains at least reasonably accurate at present, in spite of the rapid advances in knowledge and technology that are occurring at a seemingly exponential rate. Technology, as important as it is, is not sufficient in and of itself to overcome HF-related events in aviation. Human factors, arguably, remain the most statistically and operationally significant safety issue in aviation.

The purpose of this chapter is to introduce the reader to some of the most common and important basic aspects of human factors and their applicability to the aviation industry. Human factors models, judgment and decision making, sleep, fatigue, and other HF-related topics will be covered in this chapter. While our purpose here is not to be rigorously scientific and comprehensive on the topic, we believe that the information contained in this chapter will be useful for understanding basic concepts and information relating to human factors.

Human factors is a very large, complex subject, and a comprehensive coverage of the topic is well beyond the scope of this chapter. Our objective is to provide a basic, general explanation of the background and application of some primary concepts in the field as related to aviation safety. Other, more comprehensive works are available for those who desire to conduct additional research, and some of these were consulted in the writing of this chapter.

Case Study: United Airlines Flight 232

On July 19, 1989, United Airlines (UA) flight 232, operated with a McDonnell-Douglas DC-10-10 series aircraft (registration number N1819U), crashed while attempting an emergency landing at Sioux Gateway Airport (SUX) in Sioux City, Iowa (**Figure 6–1**). The aircraft was relatively full, with 285 passengers and 11 crew members onboard. Of these passengers and crew members, 111 were killed in the crash and/or the post-accident fire, a significant single-event loss of life by any measure.

A video of the crash, shot by a local television station in Sioux City, showed a high-speed, hard ground impact, with the aircraft breaking apart as it slid along the runway. Even a cursory viewing of the video makes it even more remarkable that there were such a large number of survivors, considering the impact forces and fire. Despite the nature of the crash, it is quite remarkable that anyone survived, let alone 185 persons.

UA 232 departed Stapleton International Airport (DEN) in Denver, CO, at 1409 (2:09 P.M.)

Figure 6–1. Debris from United Airlines flight 232 after crash-landing in Sioux City, Iowa on July 19, 1989

CDT bound for Philadelphia, PA (PHL), with a scheduled stop at Chicago's O'Hare International Airport (ORD). The flight crew consisted of Captain Al Haynes, First Officer (FO) William Records, and Flight Engineer Dudley Dvorak. The aircraft was being flown by FO Records. Serendipitously, Dennis Fitch, an off-duty United Airlines DC-10 training check airman, happened to be onboard as well. This factor would prove to be significant as the event progressed.

Takeoff and climb to the assigned cruise altitude of 37,000 feet were normal, with nothing out of the ordinary occurring. However, at 1516 CDT (3:16 P.M.), approximately one hour and seven minutes after departure from DEN, the flight crew heard an explosion, followed by a shuddering of the aircraft. After scanning the aircraft instrument panel, the crew quickly determined that engine No. 2 (mounted in the tail assembly) had failed. What had happened was actually a catastrophic failure of No. 2 during cruise flight. The fan disk assembly of the engine structurally failed, separating and severing the aircraft's three hydraulic systems. In essence, the hydraulic systems "bled out" as hydraulic fluid was lost. The loss of hydraulics also resulted in the immediate loss of all flight control systems of the DC-10. This created a situation that made control of the aircraft extremely difficult for the flight crew.

Eventually, the crew was able to establish some directional control by jockeying the throttle levers of engines 1 and 3 back and forth to steer the aircraft, but this was a very difficult task. Fitch, the off-duty UA DC-10 check airman, volunteered to assist, and joined the crew in the flightdeck. Captain Haynes tasked Fitch with control of the throttle levers to permit himself (Haynes) and the other crew members to devote time and attention to other decisions they needed to make. Fitch began to manipulate the thrust of engines 1 and 3 using the throttles to establish and maintain some control and steering of the aircraft. During this time, the aircraft completed two gradual right turns while descending toward the landing site.

The crew was given vectors to SUX by Minneapolis air traffic control. Given the extreme control limitations, turning the aircraft toward SUX, descending from altitude, maintaining control, slowing the aircraft's speed, and lining up with the runway at SUX were monumentally difficult tasks for the flight crew. However, they managed to perform most of these tasks in such a manner that they were able to minimize the loss of life that occurred in the accident. If not for the appropriate response of the crew in the execution of judgment, task performance, crew resource management, and decision making, it is likely that many more lives would have been lost. **Crew resource management (CRM)** is the proper utilization of all available resources by a person or crew, including people, information, and available equipment. CRM is a fundamental part of human factors training for flight crews and involves a direct application of human factors in a specific aviation process. CRM training involves communication of basic human factors knowledge relating to aviation and providing the tools to apply HF concepts operationally.

In the case of UA 232, it is evident that both positive and negative human factors responses played a significant role in both the occurrence and the outcome of this accident. On the negative side, the NTSB revealed that the probable cause of the accident was insufficient consideration given to human factors limitations in the inspection and quality control procedures and processes of United Airlines' engine overhaul facility. This resulted in maintenance inspectors failing to detect a fatigue crack in the fan disk assembly of engine 2. This crack caused the disk to disintegrate in flight, resulting in the loss of all three of the aircraft's hydraulics systems.

The positive human factors in the accident, discussed above, included the excellent CRM coordination, task performance, and decision making of the flight crew, which resulted in the saving of many lives.

United Flight 232 Case Outcome

The following is a brief overview of the National Transportation Safety Board's (NTSB) findings and some of the safety recommendations issued as a result of the investigation of United 232.

The events, causal factors, and response of the crew in the case of UA 232 strongly demonstrate a dramatic example of the impact and importance of human factors in aviation. Significantly, both the causal factors and crew actions impacting survival outcome of the accident are firmly rooted in human factors. With respect to the statement of probable cause of the accident, the NTSB in their 1990 accident report concluded:

> The National Transportation Safety Board determines that the probable cause of this accident was the inadequate consideration given to human factors limitations in the inspection and quality control procedures used by United Airlines' engine overhaul facility which resulted in the failure to detect a fatigue crack originating from a previously undetected metallurgical defect located in a critical area of the stage 1 fan disk that was manufactured by General Electric Aircraft Engines. The subsequent catastrophic disintegration of the disk resulted in the liberation of debris in a pattern of distribution and with energy levels that exceeded the level of protection provided by design features of the hydraulic systems that operate the DC-10's flight controls.

In short, there was a failure in maintenance procedures at UA attributable to human factors that led to this occurrence. Given that HF-related maintenance inspection issues were identified as a primary causal factor, it can be asserted that the accident was preventable.

Courtesy National Transportation Safety Board

The actions and decision making of the flight crew demonstrated a positive human factors response and outcome, by-products of proper training and coordination. The crew of UA 232 worked with each other in a manner characterized by efficient communication while performing multiple tasks, employed a fourth crew member (Fitch), and performed proper task prioritization, all while under extreme pressure and stress. The NTSB concluded that proper crew resource management was a key factor in the survivability for some passengers and crew members in the accident.

As a result of the accident and subsequent detailed investigation by the NTSB and its associated investigation parties, a number of safety recommendations were put forth by the NTSB. Included among the many recommendations were the following (derived verbatim from the NTSB report):

- The Safety Board recommends that the FAA establish a system to monitor the engine rotary parts failure history of turbine engines and to support a database sufficient for design assessment, comparative safety analysis among manufacturers, and more importantly, to establish a verifiable background for the FAA to research during certification review. This system should collect worldwide data by means of the reporting requirements for manufacturers contained in 14 CFR Part 21.3.
- Intensify research in the nondestructive inspection field to identify emerging technologies that can serve to simplify, automate, or otherwise improve the reliability of the inspection process. Such research should encourage the development and implementation of redundant ("second set of eyes") inspection oversight for critical part inspections, such as for engine rotating components.
- Conduct system safety reviews of currently certificated aircraft as a result of the lessons learned from the July 19, 1989, Sioux City, Iowa, DC-10 accident to give all possible consideration to the redundancy of, and protection for, power sources for flight and engine controls.
- Conduct a comprehensive evaluation of aircraft and engine manufacturers' recordkeeping and internal audit procedures to evaluate the need to keep long-term records and to ensure that quality assurance verification and traceability of critical airplane parts can be accomplished when necessary at all manufacturing facilities.
- Issue an Air Carrier Operations Bulletin for all air carrier flight crew training departments to review this accident scenario and reiterate the importance of time management in the preparation of the cabin for an impending emergency landing.

Courtesy National Transportation Safety Board

What Is Human Factors?

In order to ensure readers' understanding of the basic concepts, it will be helpful to consider the definition of human factors. This section contains a brief discussion of basic human factors definitions and concepts. Consideration of human factors has become one of the most dynamic forces in aviation because it is continually developing based on new research, ideas, and findings, and it directly impacts virtually any operation, decision, policy, procedure, or action taken throughout the aviation industry. In its simplest form, **human factors (HF)** is about people in their working and living environments and their relationships with technology, equipment, machines, policies, procedures, regulations, other people, and the environment in which all are located. One definition found while researching this chapter stated that HF is the study of the interaction between people and their environment (equipment, tools, vehicles, systems, processes, etc.). Despite significant advances in technology, humans are still responsible for ensuring the success and safety of the aviation industry and must be knowledgeable, adaptable, properly trained, and efficient in task performance, and they must exercise good judgment and decision making. Human factors is about all aspects of people interacting with everything around them and is concerned with the physical, emotional, and overall status of each individual person.

It is important to understand that HF is a multifaceted science incorporating principles and concepts from a number of various disciplines. Contemporary HF has developed into a multidisciplinary field that involves the social and behavioral sciences, various forms of engineering, and human physiology (including ergonomics). The goal is to determine ways to optimize human performance and to reduce the commission of human error. HF also involves the use of appropriate (and increasingly necessary) technology used to improve interactions between people, activities, and the environment by the systematic application of various human sciences. According to its website, the Human Factors and Ergonomics Society describes HF as the systematic use of specific knowledge to obtain compatibility in design of interactive systems of machines, people, and environments to ensure safety, effectiveness, and ease of performance. Thus, it can be said that in aviation, HF functions to integrate these various disciplines together as a cohesive whole to increase efficiency, safety, and understanding of how humans function within a given system.

The Importance of Human Factors in Aviation

In aviation, HF plays an extremely crucial role. A number of studies—and the clear statistical reality of the causes of accidents—demonstrate that human error is a direct contributing factor in the majority of accidents and incidents occurring in aviation. Dating back to 1940, the data shows that three of every four aircraft accidents result from inadequate performance by the human being in the man–machine interaction. Research demonstrates that 80 percent of accidents occurring in general aviation (GA) are the result of human error, and over 70 percent of all accidents and incidents in aviation as a whole are attributable in some manner to human factors issues. As mentioned, the FAA has indicated that human error is a causal factor in approximately 60–80 percent of aircraft accidents; the numbers are between 60 percent and 90 percent for accidents involving major airlines. Although there is some disparity in these statistics, it is clear that human error is the single greatest causal factor in the vast majority of accidents and incidents in aviation. Thus, it follows naturally that human factors is of paramount importance in aviation. Frankly, one could assert that HF is a contributing factor in well over 80% of aviation accidents, given that humans are so interwoven in the system.

Typically, most of the attention to human factors in aviation has been focused on flight crews and aircraft operations, and for good reason: Flight is the end product of aviation activities. Human factors started with the development of aviation and was initially concerned only with the design of equipment and the training of pilots. However, as aviation has progressed over the decades, the scope of HF has moved from focusing predominantly on the pilot and the aircraft to the broader consideration of all the human activities of the whole aviation system, including the ground crew, cabin crew, maintenance personnel, air traffic controllers, dispatch, regulatory agencies, and management/leadership of the organization. Thus, HF is now focused on a much broader range of human activities, having expanded from a focus on individuals to entire aviation organizations and systems. This is not to downplay the importance of HF in flight operations, but rather is in recognition of the expanding nature of human factors throughout virtually all aviation activities and operations.

This broader perspective is not simply a good idea—it is absolutely essential. Although HF is a primary concept with respect to the realm of flight, it is equally important to understand that HF is significant to safety for all ground-based operations in the aviation industry. Without minimizing the importance of pilots and other flight operations employees, by far the largest numbers of jobs in aviation are those located on the ground.

For example, a major airline such as Southwest, Delta, or United employs tens of thousands of people, and the total number of pilots, while significant, is typically relatively small as a percentage of the entire employee base. Customer service agents, ramp workers, maintenance technicians, dispatchers, line service workers, administrative employees, management, and others constitute a significant number of aviation industry employees. These numbers indicate that HF must also be given an appropriate amount of attention aimed at improving overall safety for ground-based employees through the creation of organizational policies and procedures commensurate with their required responsibilities, ensuring that proper training is provided for all employees, and providing the tools necessary to perform their various tasks safely and efficiently.

Consider, for example, aircraft ground damage events, which are highly problematic in the aviation industry. An **aircraft ground damage** event, as implied by the term, occurs when an aircraft is damaged while on the ground in a non–flight-related activity, such as maintenance or ground handling while under the control of ground personnel. Aircraft ground damage can occur for a number of reasons, including mechanical failure of equipment and other issues. However, most aircraft ground damages are attributable to the actions or inactions of ground-based personnel. In the airline industry, ground damages are typically caused by ramp workers, maintenance technicians, and other personnel working in and around the aircraft. Thus, most aircraft ground damage events result from human error, are directly related to HF, and are largely preventable events. Most aircraft ground damages are the results of events that should not occur, but do because of human error.

We can personally attest to the importance of HF in aviation ground safety. As a former and current airline corporate safety manager, we observed repeated incidents of human error that contributed to a variety of accidents and incidents. The overwhelming majority of aircraft damages that occurred during our employment tenures with a certain airline were the direct result of human error. Among these events were some very serious occurrences that caused significant amounts of damage to the aircraft involved in the events. In fact, all of the most serious ground damage events in which one of us was involved in the investigation and reporting were the direct result of mistakes made by one or more individuals working on or around the aircraft. Some of these occurrences resulted in damage not only to aircraft, but to ground support equipment, facilities, and other company or airport assets. Human error was overwhelmingly the most common causal factor in the majority of these damage events.

Human factors also figure significantly in aviation maintenance. While the role of HF in the flightdeck has received considerable emphasis, in recent years much more attention has been directed toward identifying and reducing human error in aviation maintenance. Efficient and properly performed maintenance is absolutely essential to safety and to an aviation organization's financial health. Flight crews, passengers, organizations, the general public, and others rely on maintenance activities to be as safe and error-free as possible; much depends upon proper aircraft maintenance. The safety requirements of regulatory agencies (FAA, etc.) dictate that aviation maintenance operations remain largely error-free and that aircraft inspectors and maintenance technicians work in environments using procedures and equipment designed to minimize the potential for error. As with the issue of aircraft ground damage, HF is very significant in aviation maintenance. Aircraft inspectors and maintenance technicians perform work that is crucial to safety and to ensuring an efficient air transportation system. Errors in the maintenance system can be extremely costly in terms of loss of human life and financial impact, as demonstrated in the accident involving United Airlines 232.

In recognition of the importance of human factors in maintenance, some companies have implemented a human factors program for their maintenance divisions. For example, in 2002, Delta Airlines (DL) initiated a human factors component in their Technical Operations (TechOps) division and achieved some significant, measurable positive results attributable to the HF program. An interview with a Delta HF specialist published in the October 2004 edition of *Overhaul and Maintenance* magazine stated that the DL maintenance HF program had resulted in a reduction of 52 percent in cost per accident, a 28 percent reduction in employee injuries, and a reduction in overall absentee rates for all maintenance personnel. In addition, DL TechOps also experienced a 79 percent reduction in aircraft ground damages as a result of their program. These numbers would seem to indicate that the HF program at DL achieved a significant degree of success.

A Brief History of Human Factors in Aviation

An essential element of the study of HF is found in its background and origin. As discussed previously, HF is an interdisciplinary science of human–system interaction. One of the primary functions of HF in aviation is to improve overall operational safety, efficiency, and quality of the products and/or services produced by an organization. This is because most aviation accidents and incidents do not result from problems with the aircraft (mechanical issue, etc.), but from the actions and decision making of the people responsible for the handling and/or operation of the aircraft. With these factors in mind, this section looks at the how the HF field originated.

Significantly, the phrase "human factors" has its origins in aviation and its related activities. World War I proved to be a catalyst for the acceleration of the development of HF as a science. A number of human factors and ergonomic advances had their origins in support of military interests. In 1917–18, 2 million new recruits to the U.S. military were given intelligence tests to assign them effectively to job responsibilities in the service. World War I also brought about the need to quickly select and train pilots, which precipitated the development of aviation psychology and early aeromedical research.

However, it was World War II (WWII) when some of the most significant strides were made in HF. WWII accelerated what would eventually become known as HF because more technologically advanced aircraft were outpacing human capability to operate them with maximum efficiency and safety. In the United States, the discipline of human factors and ergonomics is generally considered to have originated during WWII. At Cambridge University in England, an aircraft flight simulator was developed. The device was known as "the Cambridge Cockpit" and was used to conduct early research on human factors in flight. Through the work conducted with the Cockpit, it was found that skilled pilot behavior was largely dependent on the design, layout, and interpretation of the controls and flight instruments. The tests conducted with the Cockpit were important, as they demonstrated that aircraft needed to be matched to human characteristics rather than the reverse. Prior to WWII, the normal condition was for the person to fit the existing aircraft instead of designing aircraft to fit the human. The change is important, as it represents a move forward in the science of ergonomics as applied to aviation, designing the workspace (aircraft) to fit the pilot.

A similar innovation which was also utilized during this time period was the Link Trainer, which is credited as being the forerunner to modern flight simulators, which are used in training throughout the aviation industry. The Link Trainer was a small wooden machine roughly shaped

Figure 6–2. Air Force officer at panel training with flight students in Link Trainers.

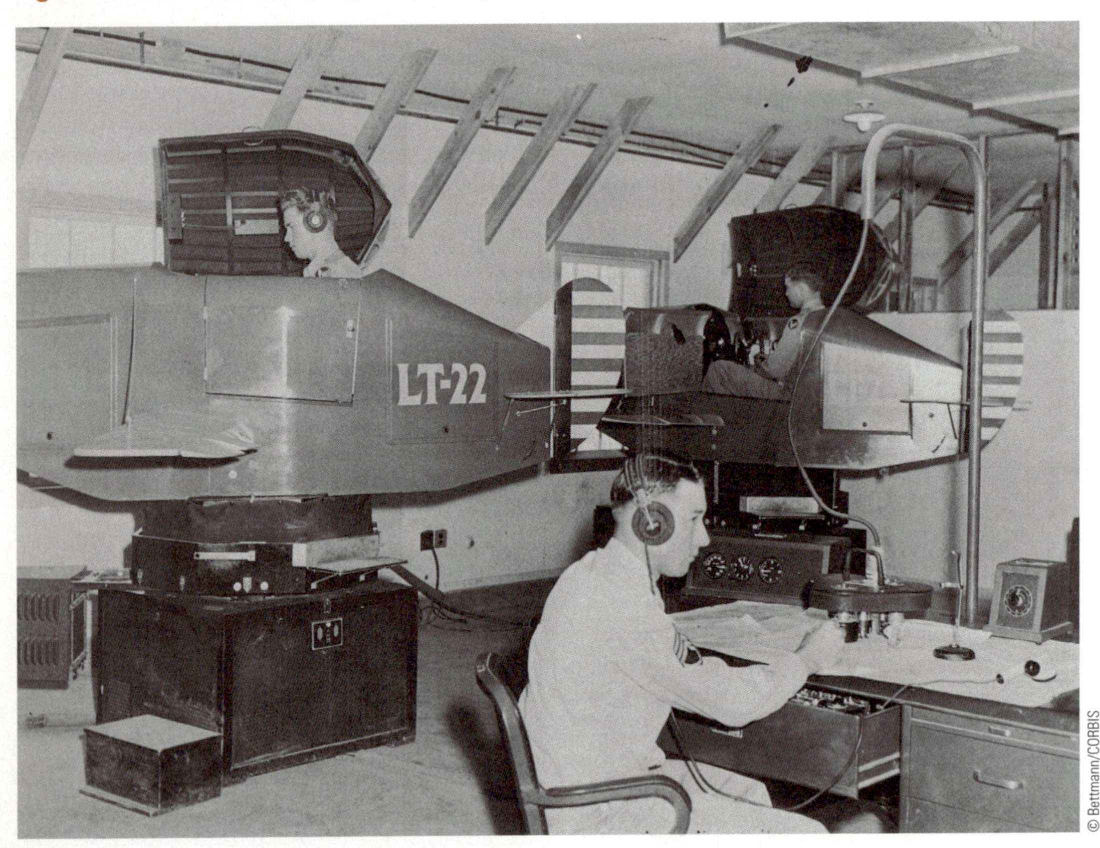

like a fixed-wing aircraft (See **Figure 6–2**). This device was designed to provide flight training for pilots by displaying only the instruments of the aircraft without any outside visual references.

The trainer was equipped only with the necessary instruments for non-visual flight and basic aircraft controls. It was considered an important innovation, and by the start of WWII, some 35 nations were using the trainer for military pilot selection and training. The instruments and controls were connected to an external instructor's console. This connection of the machine to the instructor's console is an example of human–machine interaction, a primary component of HF. This interactive feature provided instructors with the ability to assess a pilot's flight skills while allowing the pilot to learn instrument flight without the risks associated with actually flying an aircraft. However, while this new information was beneficial, it also served to promote the idea that "pilot error" was the only real explanation for aircraft accidents not involving a mechanical failure of the aircraft.

After WWII, the scope of HF grew quickly as new knowledge, techniques, and more stringent medical standards for pilot selection and qualification were introduced. New research in areas such as spatial disorientation, information processing, and fatigue were conducted, providing a large amount of new and insightful information on HF in flight. The two decades after WWII saw the continuation of military-sponsored HF research, driven in part by the pressures of the Cold War as new technologies were developed rapidly and aircraft were flying much faster, higher, and farther than previously. Military research facilities established during WWII were expanded and new facilities such as the Army Human Engineering Laboratory, the Air Force Personnel and Training Research Center, and the Naval Electronics Laboratory were all created to continue the work of researching HF in aviation.

In addition to human research in the military, it was recognized that basic education in HF was needed throughout aviation, which led to the creation of HF programs at higher education institutions and in some private companies. In 1971, Loughborough University in England established a two-week course called "Human Factors in Transport Aircraft Operation." The University of Southern California also implemented a short HF training course, and the Airline Pilots Association (ALPA) began to conduct an HF accident investigation training program for selected members. The Ohio State University and the University of Illinois both established laboratories to further HF studies. In the private sector, aviation companies such as Boeing, Grumman, and McDonnell-Douglas established HF and ergonomics groups within their organizations. In 1957, the term "human factors" was first used to describe what would become the modern practice. Also in 1957, the Human Factors Society was formed. In 1992, the organization changed its name to the Human Factors and Ergonomics Society; it has more than 4,500 members at present.

In the mid-1960s, the discipline of HF continued to develop, expanding into other pertinent areas including computer hardware and software; weapons systems; adaptive technologies; and other fields.

At present, many aviation organizations have incorporated human factors into their overall scope of business. Even though HF maintenance training is not required by regulation in the U.S., many aviation maintenance organizations have adopted HF training because they see the potential benefits of integrating such a program into their business plan; in addition, because the HF training may eventually be required by regulation, they may desire to be in compliance with regulations before these come into force. The European Aviation Safety Agency and Transport Canada have regulations in place requiring human factors initiatives in aviation maintenance organizations; these programs include coverage of areas such as initial and recurrent training; investigation of events, reporting and tracking of data, and fostering a safety culture that focuses on identification of potential hazards and risk management.

Human factors continues to be a multifaceted science today. Persons from a number of disciplines including psychology, engineering, physiology, and flight and ground operations work to understand how people interact with systems in order to improve the overall level of safety in the aviation industry.

Systems Approach

Another area to consider in the development of HF is found in the **systems approach**. By definition, the systems approach is the study of understanding the connections and interactions between the various components of a given system (e.g., social, economic and environmental).

The systems approach is a discipline that has proven useful over many years in a number of fields including electronics, engineering, and economics. In aviation, the systems approach has been used in the air transportation industry to model, analyze, design, evaluate, and manage. In essence, the systems approach involves analyzing a specific portion of the real world in its identifiable pieces and examining how the pieces interact with each other. The systems approach has been defined as a set of techniques which enables an analyst to identify a problem and divide it into specific elements (observable/measurable inputs and outputs and the quantitative relationships between them), and then determine the interaction of the elements. Prediction and analysis of system performance is the main objective, but identifying and quantifying the elements may result in understanding the nature of problems in the system.

The systems approach provides some specific advantages: It enables the analyst to identify and formulate the problem, from which a computer model can be designed to simulate the behavior of the system under hypothetical inputs and parameter changes. The systems approach thus has the ability to allow for quantitative prediction by the analyst, researcher, designer, pilot, or computer to predict the behavior of the system from the interaction of its components.

Also, the systems approach may have a positive impact on financial planning in an organization by providing an accounting framework to see which variables should be considered when performing budget planning and cost accounting, designing, training, operating, repairing, and so on. The systems approach is designed to provide a clear and orderly manner in which data on the design and operation of an aviation system can be codified so that others may access and understand the information.

Significantly, the FAA has stated that the aviation industry as a whole must establish a systems approach to the application of HF, asserting that doing so would establish systematic, disciplined procedures applicable to acquisition programs, regulatory functions, and internal aviation company operations. The FAA has also mandated that the aviation industry must consistently apply explicit consideration of HF in a continual manner, and the scope must be comprehensive, including all aspects of human performance, staffing, training, and safety and health—in short, all phases of the system. HF should be an integral part of all development and implementation efforts in an aviation organization, and be strongly considered in all analyses and decisions; safety depends on HF being a proactive component for improving total system performance.

Human Factors Models

Over the years, a number of different models have been developed that demonstrate how humans interact with their environments, systems, processes, and other people. Although somewhat theoretical, human factors models are very useful in helping to understand human interaction with and impact on a system and can actually be applied to real-world situations. This section discusses some of the more prominent and widely used HF-related models.

Reason's Model (The Swiss Cheese Model)

Reason's Model, or the "Swiss Cheese" Model of accident causation (**Figure 6–3**), is a well-known HF model that has garnered significant attention in the realm of HF research. Developed in 1990 by Dr. James Reason, this model is commonly used in aviation in the identification and management of human error occurring in organizations. The use of Reason's Model has been advocated by the International Civil Aviation Organization (ICAO) and the FAA. Since 1993, the Australian Bureau of Air Safety Investigation (BASI) has employed Reason's Model.

Reason's Model describes various organizational "barriers" in place designed to prevent errors from occurring. Among these barriers are decisions made by various levels of management, engineering controls (aircraft configuration, tools and equipment, etc.), system controls

Figure 6–3. An example of Reason's Model.

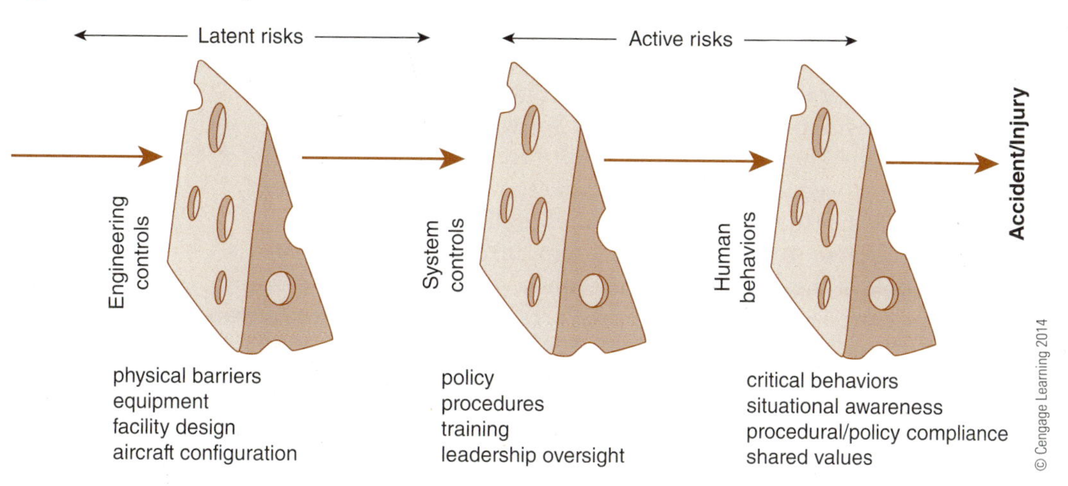

(training, procedures, oversight of leadership, etc.), and human behaviors. Reason's Model likens organizational system defenses or "barriers" to a series of randomly holed Swiss cheese slices arranged vertically and parallel to each other with gaps between pieces. The holes in each slice represent weaknesses in the individual barriers of the system. System failures occur when an individual error or series of errors, hazard, poor decision, or human action or other event passes through the weaknesses in each of the barriers, leading to an accident or other undesired occurrence. Reason's Model demonstrates that most aviation accidents and incidents can be traced to one or more levels of failure, including organizational influences, management practices, unsafe or inadequate supervision, preexisting system conditions, and the actual unsafe acts or behaviors committed by individuals or groups. When the "holes in the Swiss cheese" line up with each other, an accident may occur.

Reason's Model also describes the presence and role of latent risks and active risks in accident causation. **Latent risks** are areas within the system that pose a hazard (contributory factors) but are not always obvious, tending to lurk beneath the surface. Latent risks are the result of an action, policy, or decision made long before an accident occurs; their accompanying areas of hazard and consequences may be dormant within the system for a long time before the occurrence. **Active risks** are those that pose a more obvious hazard and are often the direct causes of an accident. An active risk is an action, decision, error, or hazard that results in an immediate undesired event.

A simple illustration of latent and active risks can be found in the example of a jet engine. For workers on the ramp or a pilot performing a preflight aircraft inspection, a non-operating jet engine poses a latent risk. However, the hazard posed by the engine becomes an active risk when it is running. If an employee has developed the habitual undesirable behavior of walking into the ingestion zone of the engine when it is not running (latent risk), there may be a greater probability that this same employee may walk into the ingestion danger zone when it is running (active risk). This is one illustration to contrast latent and active risk in an applied situation as a function of behavioral characteristics. Often latent risks are present in the overall management system of an organization and, left unchecked, may result in safety being compromised.

The SHEL Model

Another widely known HF model is the **SHEL Model** (**Figure 6–4**), sometimes spelled SHELL. This variation is attributable to Frank Hawkins (1975), who added an additional "L" component in his iteration of the model. However, we will focus on the "SHEL" variation in this section.

Figure 6–4.
The SHEL Model.

SHEL is an acronym for the components of Software, Hardware, Environment, and Liveware. In the SHEL Model, **software** represents the training, policies, rules, procedures, operating manuals, and other documents that constitute the standard operating procedures (SOP) of a given organization. **Hardware** represents the tools, equipment, machines, and components used by people in the execution of their job responsibilities.

Environment represents the entirety of the area, location, or system in which individuals are performing tasks, that is, in the flightdeck of an aircraft, on the ramp, in the maintenance hangar, and so on. Finally, **liveware** represents all the humans within the system.

The primary point made by the SHEL Model is that humans are at the center of a complex aviation system designed to ensure a safe, efficient operation and mission. A closely related formulation is as follows: Given that people are at the center of the system, humans have the ability to impact and be impacted by each of the other components by virtue of interactions with them. Humans are the most important component in the SHEL Model, as they are on the front line of aviation operations, perform actions, make decisions, create and implement policy, and interact with each component of the system. The SHEL Model is very useful in describing the importance of human interactions with the various system components. The SHEL Model places emphasis on human interactions with the other components and features within the aviation system; it can be used to assist in visualizing the relationships between the various components.

The initial framework for what would later become the SHEL Model was developed by E. Edwards in 1972. Putting the concepts into the "building block" form of the model is the result of the work of Frank Hawkins, the results of which were first published as the SHELL model. Like Reason's Swiss Cheese Model, the SHEL Model has gained recognition and is used in the aviation industry to describe the interaction of humans with machines, equipment, procures, rules, and other humans. As an example, the International Civil Aviation Organization (ICAO) has adopted the SHEL Model to define the role of HF in aviation and show the importance of SHEL interactions. The SHEL Model is described by ICAO in Circular 216-AN31 and in their Safety Management System (SMS) manual.

There is a rationale for the model being constructed in the form of building blocks. Referring to Figure 6–4, the reader will note that the edges of the SHEL blocks are not smooth, straight lines; they are curved and somewhat uneven. This is because the SHEL Model attempts to visually depict that while people are very adaptable, they are not standardized in any way, due to a number of variables. People have varying abilities, tendencies, limitations, physical characteristics, and so on, so the block edges in the model are not drawn as straight lines. According to ICAO, humans do not interface perfectly with other components, and in order to avoid problems that may hinder human performance, the problems occurring at the interaction points between the SHEL blocks need to be understood. The components of the system must be carefully matched as closely as possible to humans if stresses in the system are to be avoided; a mismatch in the blocks may result in human error occurring within the system.

ICAO has identified several important human considerations that factor into SHEL interactions and tend impact human performance. These factors are as follows:

- *Physical factors:* Human physical capabilities to perform the required tasks, such as dexterity, strength, height, vision, and so on.
- *Physiological factors:* Variables affecting human internal physical processes that may impact physical and cognitive performance, such as aerobic capacity, fitness, overall health, use of alcohol and/or drugs, and so on.
- *Psychological factors:* Factors affecting the psychological state of the human in meeting all potential circumstances that may be encountered, such as adequacy of training, recency of experience, memory, task knowledge, stress, workload, and so on.
- *Psycho-social factors:* External factors in the social interaction of humans that bring pressure in their working and non-working environments, such as family/personal problems, relationship with co-workers and/or supervisors, personal financial problems, and so on.

Source: ICAO

The SHEL Model is very useful in providing a means to visualize the interfaces between the various components within a system. A brief discussion of these interactions follows. Most of the following information was derived from the ICAO Safety Management Manual.

- *Liveware–Hardware (L-H):* This interface between the human and technology is the most commonly considered with respect to man–machine systems. This determines how humans interface with the physical work environment and all components and equipment utilized.
- *Liveware–Software (L-S):* Interaction between the human and non-physical aspects of the system, such as regulations, policies, training, manuals, checklists, SOP, computer software, and so on.
- *Liveware–Liveware (L-L):* The interaction between human beings in the workplace; it is thus largely relational. Flight crews, air traffic controllers, ground crews, maintenance technicians, management, and other operational personnel play a role in determining human performance. Increasing focus on crew resource management (CRM) has put significant emphasis on this component. This interaction is also concerned with leadership, crew coordination personality differences and interactions, and so on.
- *Liveware–Environment (L-E):* This interaction is concerned with the human and both internal and external environments. Internal environment includes heat and cold, degree and type of light, noise, air quality, and so on. External environment includes weather, visibility, turbulence, topography of an area, and so on.

Source: ICAO

The 5 M Model

The **5 M Model** is another HF-related model widely employed in aviation safety. The 5 M Model is a commonly used approach for investigating systemic issues that contribute to aviation accidents and incidents. The 5 "M's" of the model are as follows: man, machine, media, management, mission. Each of these components is intertwined in a given organization or system and thus they are shown as overlapping in written depictions of the model, as shown in **Figure 6–5**.

Since most accidents and incidents are multifaceted, it is important that investigators analyze the system as a whole to identify all causal factors so that specific, effective corrective actions can be developed and implemented. The 5 M Model provides a basic yet effective tool for examining the various interactions of the components of the system and how each may have contributed to the event. The individual components are as follows:

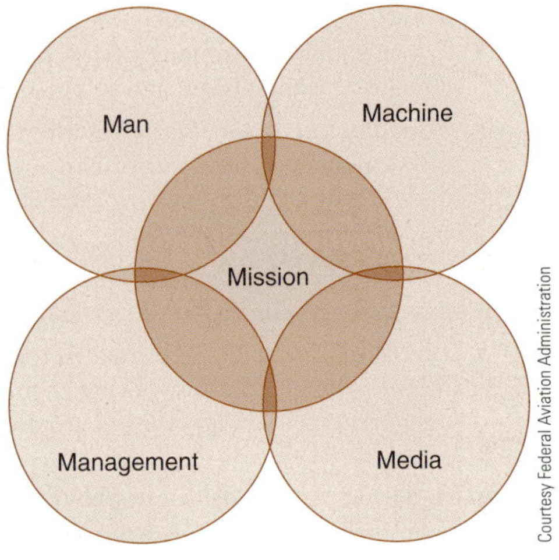

Figure 6–5. The 5 M Model.

- *Man:* The human component of the system. Areas of focus include task performance, training, physical and psychological factors, fatigue, knowledge, and many other factors pertaining to the human condition. Example: What was the physical condition of the pilot at the time of the event?
- *Machine:* The tools, equipment, and machines used in job and task performance. Example: When did the most recent inspection of the nose gear take place? What discrepancies, if any, were noted?
- *Media:* The environment in which the work is taking place or in which the event occurred. Example: What were the wind conditions at the arrival airport at the time of the event? What were the reported runway surface conditions?
- *Management:* The overall organizational and its culture, including procedures, policies, leadership oversight of operations, and so on. Example: When presented with the findings of previous related safety events, what did management of the affected departments do in response? Are all work rules being actively enforced? Were all corrective actions put into place?
- *Mission:* Ultimately, the goal or "mission" of all aviation activities is a safe, efficient operation, flight, and so on. Example: What factors may have impacted the overall mission or operation?

The Dirty Dozen

The FAA, Transport Canada, and a large number of aviation companies recognize that there are a number of common causes of human factors errors present in many aviation systems. This group of factors is commonly referred to as the "Dirty Dozen" (see **Figure 6–6**). Although the Dirty Dozen is used to describe the most common errors in aviation maintenance, these same factors are also common causes or errors in other aviation operations such as flight operations, air traffic control, and ground operations. In examining the evidence from some aviation accidents, it is rather easy to identify a number of these issues as having contributed to the occurrence. Therefore, it is important for all aviation operations employees to be alert to the Dirty Dozen and their potential impacts on safety.

The Dirty Dozen

1. Lack of Communication
2. Complacency

Figure 6–6. The Dirty Dozen.

3. Lack of Knowledge
4. Distraction
5. Lack of Teamwork
6. Fatigue
7. Lack of Resources
8. Pressure
9. Lack of Assertiveness
10. Stress
11. Lack of Awareness
12. Norms

Currently, many aviation companies provide training and supplemental materials on the Dirty Dozen to employees in their operating departments to raise awareness about the real and potential problems caused by these 12 factors.

Decision Making in Aviation

In practical application, a significant portion of safety and all activities in aviation comes down to judgment and decision making by pilots, maintenance personnel, air traffic controllers, ground-based personnel, management, and anyone else involved in an aviation system. Intuitively, many people understand that the ability to make decisions and render judgments is an important aspect of human factors, as it can be said that the decision a person or group makes is the end product of a process that is generally transformed into a course of action. It is widely recognized that proper decision making and judgment are the foundation of aviation safety. This section will briefly consider some factors related to decision making and judgment in aviation.

Definitions

In order to ensure understanding, it will be helpful to briefly define judgment and decision making. **Judgment** can be likened to the assessment of a situation, a cognitive process that enables us to make conclusions. One definition found while researching this chapter stated that judgment is the mental process that humans use in making decisions; it includes gathering information, processing information, and choosing a course of action. The FAA defines judgment as the mental process of recognizing and analyzing all pertinent information in a specific situation, conducting a rational evaluation of alternative actions, and making a timely decision as to which action to take. Judgments can be made immediately or may be carefully thought out over a period of time, depending on the situation and other factors. An in-flight emergency often requires swift judgment and action. However, it is important that proper judgments be made. After all, swift reaction may not be good if the wrong judgment is made, resulting in an improper course of action. While human beings possess the capacity to make judgments at least at a rudimentary level, this factor does not always translate into individuals in making the best possible judgments in aviation. However, it is important to note that the ability to make good judgments can actually be taught. Most people learn how to make judgments mostly from experience (their own and that of others), and with appropriate training people can learn to recognize limitations in their information processing and improve their judgment-making abilities. However, even the most effective training does not guarantee that people will make the best decisions, as there are many variables that impact human judgment.

Decision Making (DM) is probably easier to define than judgment, as in its most basic form it is simply the process of making decisions. This is rather simple, but it is not always easy to accomplish in practical and correct application to real-life situations in aviation. Research shows that there are different forms of decision making.

The FAA describes a form of DM called **conventional decision making (CDM)**, which can be likened to a norm—that is, the typical way in which a person makes decisions. CDM involves making a decision as a result of a recognition that something has changed or that an expected change did not occur. Recognition of the change or non-change is an important step in any decision-making process, and failure to take note of the change may lead to an undesired event. Thus, situational awareness is necessary to make a proper decision. Situational awareness will be discussed later in this chapter. The change or non-change demonstrates that an appropriate response and/or action is necessary in order to affect the situation and create a desired outcome. At this point, the person is faced with a need to evaluate the entire range of alternatives and determine the best course of action.

Aeronautical Decision Making

Over the last two decades, the FAA has adopted and promoted a well-known DM concept called **aeronautical decision making (ADM)**. ADM is defined by the FAA as a systematic approach to the mental process used by pilots to consistently determine the best course of action in response to a given set of circumstances. ADM is described in FAA Advisory Circular (AC) 60-22. According to AC 60-22, ADM builds on the foundation of CDM but enhances the process to decrease the probability of human error. The FAA (1991) wrote that ADM enhances CDM by providing a focus on the importance of attitudes in decision making, learning to seek and establish the relevance of available information, and establishing motivation to choose and execute actions ensuring safety within the time allotted in the situation. ADM provides a structured, systematic approach to analyzing changes during flight and how the changes may affect the safe outcome of the flight. ADM addresses all aspects of decision making in the flightdeck of an aircraft and identifies the steps involved in proper decision making.

According to AC 60-22, there are several steps involved in proper decision making:

- Identification of personal hazardous attitudes
- Learning techniques of behavioral modification
- Learning to recognize and deal with stress and pressure
- Development and application of risk assessment skills
- Using all available resources in a multi-crew environment (CRM)
- Evaluation of the effectiveness of personal ADM skills

The Five Hazardous Attitudes A key element of ADM is recognition of hazardous attitudes within an individual person. ADM discusses five specific attitudes that have become widely known as the **five hazardous attitudes**. Even a cursory glance at these attitudes reveals that these are very common in humans. These attitudes are as follows:

- *Machismo:* This attitude is often displayed as the "I'm tough and can do it even if you can't" mentality. This may lead a pilot or other person to make poor decisions, leading to taking unnecessary chances in order to prove themselves. An aircraft captain who will not listen to the input of the first officer is creating risk.
- *Invulnerability:* This is an attitude of invincibility; the worst will not happen to me. Overconfidence in one's own abilities, training, experience, and education may be some of the factors leading to this attitude. Note that this is an attitude of complacency and perhaps even self-delusion.
- *Resignation:* This is an attitude of powerlessness and basically giving up. A person who feels powerless believes they cannot exert any control or offer any valuable input to a given situation. An example: "What's the point, the captain won't listen to me anyway."
- *Anti-authority:* This person does not like to listen to others, may not like to be told what to do, may not like following the rules and/or procedures, or may have a problem with coming under the authority of another person. A first officer who is reluctant to heed the legitimate input of their captain is a hazard.
- *Impulsiveness:* This is the "fire-ready-aim" approach. Do something, anything, NOW!!! This person will not think or plan before acting, and this could result in a wrong action taken, or perhaps doing the right steps out of sequence, which can be very dangerous, depending on the situation.

Because these attitudes are present in people, it follows that means can be devised for effectively dealing with these attitudes. The first step is **self-assessment**, which is nothing more than the practice of examining oneself, being honest about real characteristics, traits, tendencies, and so on, recognizing one's tendencies, and acting in a manner that that is non-hazardous.

Stress and Stress Management

Recognition and management of stress in a person's life is another key element in decision making. Like other concepts discussed in this chapter, **stress** can be defined in different ways. The FAA has described stress as a response to circumstances inducing a change in a person's physiological and/or psychological patterns, forcing the person to adapt to the change. However, in this context stress will be defined as those things in life that may place a person under pressure exceeding what is normal for them and may have negative impacts on the body, mind, judgment, decision making, and performance. Anything in the life of a person that causes stress is called a **stressor**. Some of the more common categories of stressors are emotional and physical. Emotional stressors are those things that bring stress to one's mind and emotions, and that may create a negative impact on the body. Physical stressors are things having an impact on the human body. The aviation industry is characterized by innumerable stressors, more than can be exhaustively listed here. Some examples of common aviation industry stressors are tight time schedules, on-time performance (in all its myriad forms), weather, repetitive physical activity, variable flight schedules, and maintenance problems; the list could go on indefinitely.

Ironically, a degree of stress is healthy and even necessary for humans to flourish. In fact, some stress is necessary in order to avoid complacency in task performance. Stress is an inevitable and even necessary part of life that often creates motivation and heightens a person's response to meet and successfully handle various challenges. Interestingly, task performance will typically improve with the onset of stress but will peak and degrade rapidly as stress levels exceed the individual's ability to adapt to the situation. Often, when stress is nonexistent or low, an individual's motivation and attention to tasks are minimal, and performance is generally poor. However, when stress becomes excessive or continual over a period of time, the results of prolonged exposure can be harmful.

Stress Effects

A characteristic of stress in humans is that it has a gradual, cumulative effect that develops slowly and can be present in a person long before it becomes evident. A person may believe they are handling everything in their life well, when in fact they may have exceeded their personal coping level abilities. An undesired stress reaction may occur as a result of cumulative stress exposure. **Stress tolerance level** is a person's individual threshold for handling stress. If the number or intensity of stressors becomes too great, the person becomes susceptible to overload. When this happens, mental pressure builds, performance declines, and judgment deteriorates, hindering proper decision making. This type of situation can be hazardous.

Stress can cause a number of effects in a person. These effects may be mental, physical, or a combination of the two. While different people may experience different impacts of stress, some of the more common impacts experienced may be anxiety, unhappiness, panic, depression, sleep disruption, fixation, reduction in verbal communication leading to isolation, upset stomach, headaches, diarrhea, and fatigue. Psychologically speaking, stress causes people to fixate on how badly things could turn out (focusing on worst-case scenarios or on "what-ifs"), continually dwell on their situation, and even obsess over what is happening. A person in this type of mental state is not in a position to exercise good judgment and decision making. Thus, stress can affect decision making.

Managing Stress

While the intent of this section is not to research all aspects of the topic, it will be helpful to briefly touch on some ideas for managing stress. First, recognition of stress is crucial. A person who is not aware of their individual physical and mental state is less likely to be properly equipped to deal with stress, as they are not fully aware of their individual symptoms and condition. The only real way for a person to effectively deal with stress is to recognize that it is present in their life and may be affecting them adversely. People needs to be honest with themselves about stress in their lives. One helpful practice is to identify and write down all stressors and then to prioritize them by virtue of severity and/or importance. This may help with stress recognition and self-assessment.

Next, an individual should talk to someone about what is going on in their life. What is stressing them out? How are they feeling and being affected by stress? Having and maintaining a strong support network is a key component of effectively managing stress. Sometimes just simply talking to a willing (and hopefully trusted) listener can help to unburden a person by releasing what has been bottled up inside. Also, the insights of another person can bring a rational, objective perspective on reality to the stressed-out person that may be helpful. In some cases a person may need to seek professional counseling to assist in dealing with stress.

Another good tactic for handling stress is this: Take a break! During stressful times and situations, a person should take a few minutes' break whenever possible and step away from the tasks and responsibilities in order to regroup, refocus, and relax. In the case of flying, it is obviously not practical or safe for a pilot to step away from the flightdeck in some situations, but even in this type of scenario a pilot can do some mental preparation before entering into anticipated high-stress situations. According to the FAA, good flightdeck stress management begins with good personal stress management, and pilots need to condition themselves to think rationally and to relax. This also holds true for virtually anyone in moments of stress, regardless of the situation.

Another stress management practice is found in basic relaxation techniques. Simple deep breathing exercises are a very powerful means for reducing tension and helping a person relax. Taking a few moments, focusing, and taking a few slow, deep breaths may actually help a person to relax. This can be done almost any time in any place.

Regular physical exercise is an excellent means of managing stress. Not only does exercise produce physical benefit, it may also help a person reduce or manage the stress in their life. Physical exertion helps to release tension and anxiety and may increase a person's health and overall quality of life. Exercise can be likened to a life management technique that also is a stress management strategy.

Finally, any discussion of stress management should include nutrition. Much has been written on the importance of proper nutrition. In short, a proper diet is conducive to a healthy lifestyle, and a healthy lifestyle is a significant part of stress management.

In AC 60-22, the FAA included a brief list of stress management techniques, some of which have already been discussed. These include the following:

- Learn about stress and its effects.
- Conduct a self-assessment.

- Use a systematic approach to problem solving.
- Develop a lifestyle that will protect against the effects of stress.
- Practice behavioral management techniques.
- Establish and maintain a strong support network.

Situational Awareness

Situational awareness (SA) is another widely known component of safety in the aviation industry. However, despite its commonness in aviation, it is not necessarily easy to define, as SA may mean somewhat different things to different people in different situations. To bring some clarifity in defining SA, we will look at a few variations of definitions.

Mica Endsley has conducted extensive research on SA and has defined SA as the perception of the elements in the environment within a volume of space and time, comprehending their meaning, and projecting their status in the near future. Endsley's definition of SA is one of the most widely accepted in the research community. Endsley's work further shows that SA involves perceiving critical factors in the immediate environment (Level 1 SA); understanding the meaning of the factors (Level 2 SA); and, at the highest level, comprehension of what will occur in the system in the near future (Level 3 SA).

Another definition found states that SA in the human/machine system is the pilot's (or flight crew's) continuous perception of self and the aircraft in relation to the changing environment of flight, threats, and mission, and the ability to forecast and perform tasks based on that perception.

Each of these definitions supports the notion that defining SA varies to an extent between different people and organizations. Still, the basic idea is that SA is about being alert in the present environment, knowing when a change has occurred, and responding appropriately to the change. Therefore, for the purposes of this chapter, SA will be considered to have four basic elements, which will be defined as follows:

- Continuous awareness of the immediate environment (flight, ground, etc.)
- Immediate awareness of when a change occurs in the environment
- Reacting in an appropriate manner to the change, including how the change may impact the operation in the near future (relative to current tasks being performed)
- Continuing in and completing the activity with a heightened level of alertness

Technology can be (and is) used as a significant SA enhancement in certain types of aircraft. For example, technology such as onboard weather radar and (in some advanced aircraft) a heads-up display (HUD) are examples of devices used to help significantly enhance flightcrew SA. However, it is extremely important to understand that technology alone will not always serve to increase SA in a person.

Although SA is commonly (and correctly) applied to flight-related activities, it is important to note that SA is crucial in both flight and ground-based environments. Although very important to flight safety, consideration of SA should not be solely restricted to the realm of flight.

Situational Awareness and Decision Making

Even with the difficulty in defining SA, one unambiguous factor is that SA is a primary function of proper decision making. Irrespective of the situation, humans must be able to correctly interpret information and respond in a manner commensurate with properly addressing the issues at hand.

One HF resource found while researching this chapter stated that decision making and proper resolution of an issue require an accurate assessment of situations and that strong CRM principles underlying good leadership and management skills are connected to achieving situational awareness and proper workload distribution among flight crew members. Another HF resource asserted that SA is a component in characterizing crew decision making, and effective crews are vigilant in monitoring the environment for threats requiring a response and in monitoring the overall progress of the flight in accordance with the operational flight plan. SA researcher Endsley said that SA can be viewed as an internal mental picture forming the central component from which all decision making and action occurs, and a lack of SA may result in poor decisions creating human error and that without SA even the most well-trained crews may make poor decisions. Thus, it is evident that SA is crucial to proper decision making in aviation.

The Importance of Situational Awareness in Aviation

In addition to the importance of SA in decision making, research has shown that a lack of SA has been a persistent causal factor in a number of aviation accidents and incidents over the years. In fact, an increasing number of accident and incident reports list SA as a contributing factor in accidents. Endsley studied the causal factors of a number of major airline accidents based on National Transportation Safety Board (NTSB) reports and found that flightcrew SA issues were present in 88 percent of these events. SA is also a primary causal factor in military aviation accidents and incidents.

Although there are many issues and problems that can compromise SA, here is a partial list of some of the most common factors:

- Stress (physical and/or social/psychological)
- Complacency (low mental stimulation; overconfidence; over-familiarity with process, tasks, etc.)
- Overload or underload (associated with degree of mental stimulation in situations)
- Distractions
- Fatigue, sleep deprivation
- Pressures leading to hurrying, rushing
- System design and system complexity; automation
- Confusion, ambiguity, misinterpretation of information
- Failure to properly follow established procedures (SOP, etc.)

The DECIDE Model

Another well-known model in aviation related to HF is the **DECIDE Model.** This model was developed by Brenner in 1975 and is featured in AC 60-22 on aeronautical decision making. The DECIDE Model was designed as a tool to aid pilots in their decision making related to flight operations. The model consists of six components designed to be a continuous loop to assist pilots in the decision-making process when they are faced with a change in a situation that requires a judgment. According to AC 60-22, the DECIDE Model is focused mostly on the intellectual component but may have an impact on the motivational element of judgment; if a pilot practices the DECIDE Model in all decision making, the result may be better decisions made in all types of situations. This assertion implies that the conscious use of this model could result in the development of positive habitual behavioral attributes through practice that could make a person more likely to exercise sound judgment in the decision-making process. Similar to the SHEL Model, DECIDE is an acronym:

- Detect: Detection of the fact that change has occurred
- Estimate: Estimate the need to counter or react appropriately to the change
- Choose: Choose a desirable/safe outcome for the flight or other operation
- Identify: Identify actions to control the change
- Do: Decision maker takes necessary action(s); "do" the best option
- Evaluate: Evaluate the effect of action(s) countering the change

Depending on the level of the reader's knowledge, it may have been noted or the association made that the DECIDE Model is very similar to how some define the concept of situational awareness, an extremely important element of human factors and safety in aviation.

Basic Physiology of Sleep and Fatigue

It is a rather obvious point that we as humans are directly impacted in a number of ways by the physical condition of our bodies and the mental condition of our minds. Although these factors can be considered common knowledge, many people do not always do what is best for their bodies and minds with respect to care, conditioning, and fostering overall good emotional and physical health. We sometimes have a tendency to ignore or even neglect our physical and mental status, which constitutes yet another aspect of human factors that is important to consider.

The aviation industry is filled with varying work schedules and requirements, job responsibilities, duties, and adherence to tight time schedules which are often very physically and mentally demanding and, in some cases, irregular and even unpredictable. For example, depending on the company and its scope of operation, many pilots operate several flight segments between various airports in the course of a typical working day, often crossing multiple times zones. Some pilots' working schedules may not be predictable or consistent from week to week or even day to day. Factors such as these may result in a negative impact on a pilot's sleep cycle, nutritional habits, daily life patterns, and overall well-being, physical and mental. Over the course of time, such situations can be detrimental to a pilot's health. An unhealthy pilot may pose a safety risk to themselves and to others (flight safety, passengers, other company personnel, etc.). This is not to imply that pilots are the only ones whose actions or status may impact others. On the contrary, it has already been well established in this chapter that HF issues affect everyone.

This section is intended to introduce the reader to two important areas of basic human physiology: sleep and fatigue. While this section is not intended to be extremely detailed or comprehensive on the subject of sleep physiology, it does contain important information directly related to the human factors of sleep and fatigue.

Sleep and Fatigue

The topics of sleep and fatigue are the focal point of this section because they are some of the most important aspects of human physiology and thus potentially impact aviation safety. Most people know intuitively (and through some education) that sleep is important and that fatigue can be hazardous. Fatigue is probably the most frequently discussed issue in HF due to its almost universal applicability to the human condition in aviation. Fatigue may be defined as a state that results in the decreased ability of a person to maintain normal mental and physical functions because of stress, lack of proper sleep, poor nutrition, and so on.

However, sleep and fatigue are topics that some do not always take seriously or truly understand the importance of in assessing their individual condition. Fatigue poses a very real and direct threat to safety in aviation because it impairs alertness and hinders performance. As individuals, most of us know how we feel when we are well rested, and also when we have not gotten adequate sleep. However, this self-knowledge does not always translate into good judgment and decision making when conducting aviation-related tasks. It is imperative that all pilots and other persons in aviation be familiar with the causes of fatigue and accompanying symptoms that they experience personally. This statement could be easily applied to nearly any operationally significant position in the aviation industry, including flight, maintenance, dispatching, air traffic control, and all manner of other ground operations positions.

The Essential Nature of Sleep

Much has been written on the importance of sleep, so not much space will be devoted to this aspect of the topic. However, sleep is extremely important to a person's overall health and well-being and thus to ensuring safety. Along with proper hydration and nourishment, sleep is something the human body must have in order to function normally. Sleep is a vital physiological function, and it is necessary for human survival. However, in spite of the fact that sleep is both physically and mentally important, many people experience some form of sleep problem and as a result are not adequately rested.

Specifically, research conducted by the American Psychological Association (APA), citing a study by the National Sleep Foundation (NSF), found that at least 40 million persons in the United States experience at least one of more than 70 known sleep disorders, and 60 percent of adults report having sleep problems a few nights per week, and in some cases even more. In addition, the study found that 69 percent of children experience sleep problems a few nights per week or more. These numbers are very alarming, demonstrating as they do that the majority of Americans suffer from some form of sleep problem, some chronically. With the general population percentages as high as they are, it is rather easy to draw the conclusion that many people in the aviation industry are among those suffering from a sleep problem.

Insomnia

While there are a number of different sleep disorders, the term that generally is used to describe sleeplessness is **insomnia.** Insomnia, by basic definition, is a disruption of a person's normal sleep cycle and/or the experiencing of inadequate sleep. Three levels of insomnia have been identified: transient, acute, and chronic. **Transient** insomnia is the most common form and is usually rather short-term in nature, perhaps lasting only a few days. **Acute** insomnia is a more serious form than transient and may be associated with the occurrence of an immediate life issue (sickness, stress, etc.) that impacts a person's ability to sleep adequately. The effects of transient insomnia are generally minimal to a person and may be overcome after a few nights of good sleep, depending on variables such as overall health, age, and so on. **Chronic** insomnia is the most serious form; it occurs when a person experiences a more continual state of sleep deprivation that may last days, weeks, months, or even years. One of the most common causes of chronic insomnia in members of the general public is unresolved stress and/or illness. Chronic insomnia can be very serious; left undiagnosed and untreated, it may lead to physical and/or mental problems. An important point to understand about sleep deprivation is that is has a cumulative impact on the human body; that is, it builds in a person over time.

Effects of Sleeplessness

When the human body is sleep-deprived, the brain may transition over time from alertness to sleepiness in an uncontrolled, unintended manner in response to the need for sleep. Obviously, this poses a danger to aviation safety if sleep-deprived people are at the controls of an aircraft or are working on or around aircraft (maintenance, ground crews, etc.). According to the APA, irritability, moodiness, and disinhibition are some of the first symptoms a person experiences when suffering from lack of sleep. If a person does not sleep after exhibiting the initial symptoms, they may experience apathy and decreased motivation, slowed speech, impaired memory, the inability to multitask, poor decision-making abilities, and even depression and a decrease in the function of their immune system. As a person reaches the point of falling asleep, they will fall into micro sleeps lasting about 5–10 seconds, causing attention lapses, nodding off during activities like driving (or perhaps flying!). This poses obvious risks to aviation activities.

Stages of Sleep

Sleep consists of two basic states: **rapid eye movement (REM) sleep** and **non-rapid eye movement (NREM) sleep**. According to information found on WebMD, while a person is sleeping the body shifts between NREM and REM sleep. Normally, the sleep cycle begins with a period of NREM sleep followed by a very short period of REM sleep. NREM sleep consists of stages 1 through 4 (discussed below). Each stage will occur in a series of cycles around every 90–100 minutes during the sleep period. Each of the four stages will typically last between 5 and 15 minutes. A complete sleep cycle consists of a progression from stages 1 through 4 before REM sleep occurs; then the cycle will begin again.

NREM Stages **Stage 1:** A reduction in activity between wakefulness and stage 1 sleep, when drowsiness sets in. The eyes are closed, the body is relaxing, pulse and respiration decrease, brain waves slow down. Stage 1 may last for 1 to 10 minutes. Many people may experience the feeling or even briefly dream of falling during stage 1, which may cause a sudden muscle contraction.

Stage 2: Brain activity will increase during stage 2. This is a stage of light sleep during which heart rate slows and body temperature decreases as the body prepares to enter deep sleep. Muscles continue to relax, and respiration becomes steady and even. Stage 2 may last around 10 minutes.

Stage 3: This is a deeper stage of sleep during which the body is in a state of continued relaxation featuring slow, even respiration and slower pulse. The sleeper is basically completely at rest, and it becomes difficult to awaken someone during this stage. If a person is awakened they may not feel rested and will likely experience some grogginess. Stage 3 may last around 5 minutes.

Stage 4: This is considered to be the deepest of the sleep stages and is more intense than stage 3. The body is relaxed and brain wave activity is slow, to the point that a person in this stage may actually be considered unconscious. Stage 4 lasts around 30–45 minutes, the longest sleep cycle, and is considered to be the most restorative to the sleeper. If woken from sleep during stage 3 or 4, a person will experience disorientation and significant grogginess, a condition known as **sleep inertia.** Sleep inertia can be very hazardous to safety as a person is not fully alert, responsive, or functional.

It is during the deep stages of NREM sleep (stages 3 and 4) that the body regenerates tissue, builds bone and muscle, and may build up its immune system. Thus, these stages are especially important to human health and overall well-being. As a person ages, they tend to sleep lightly and experience less restorative deep sleep.

REM Sleep Typically, a person will enter REM about 60–90 minutes after going to sleep, concurrent with NREM stages 1 through 4.The first period of REM typically lasts 10–30 minutes, with each recurring REM stage increasing in duration; the final one may last up to 60 minutes or so. Vivid or intense dreaming will typically occur during REM sleep as increased brain activity occurs. During some REM stages, the sleeper's pulse rate and blood pressure will fluctuate, the larger muscle groups remain relaxed, but smaller muscles like toes and fingers may twitch, and the eyes move rapidly in different directions, hence the name rapid eye movement. As each cycle is repeated around every 90 minutes or so, less time may be spent in stages 3 and 4 and more time in REM. The influence of anything that disrupts REM (stress, illness, alcohol, drugs, etc.) will result in sleep that is non-beneficial to the person. Generally, about 25 percent of the normal sleep cycle is spent in REM.

Amount of Sleep Required

Contrary to what many have come to believe through common knowledge, there is no hard and fast rule governing exactly how much sleep every person requires. In fact, required sleep amounts are, to a degree, variable. That is, some people can function on much less sleep on average, while others require more sleep to function normally. The amount of sleep a person needs depends on the individual. The need for sleep depends on various factors, one of which is age. The 8 hours of sleep per night of the common-knowledge norm is not necessarily universal; some people can sleep less and function well, while others require more than 8 hours. However, on average most adults require between 6 and 9 hours of sleep per day in order to ensure proper rest. However, what is important here is the issue of self-assessment. Each person should determine their own individual sleep requirements for optimal benefit. The quality of the sleep experienced is actually much more important that the quantity of sleep.

Fatigue

Fatigue and its symptoms are very common to many people both inside and outside aviation. As such, it is very much a normal part of life. For many people, fatigue may not pose a great danger, depending on the situation and person. However, when a fatigued person is flying, working in maintenance, functioning as an air traffic controller, or involved in other activities in aviation, fatigue can and does pose a very real threat to aviation safety.

Fatigue is another term that is not easy to define due to the large number and variety of things that cause fatigue. In simple terms, **fatigue** may be defined as being tired, sleepy, or weary. However, from an operational perspective applicable to aviation, fatigue has been defined by the FAA as a condition characterized by increased physical and/or mental discomfort with reduced capacity for work, reduced efficiency in task accomplishment, and loss of motivation or capacity to respond to stimulation, typically accompanied by weariness and tiredness. Fatigue is the common element in virtually all symptoms related to sleep deprivation and many other physiological and psychological stresses in aviation. Factors such as these have led some aviation organizations to adopt an active **fatigue risk management** program in order to monitor and track fatigue-related issues and to attempt to avert fatigue in their employees. An effective fatigue risk management program will provide specific training and education on fatigue and sleep, and will encourage employees to report any events in which fatigue may have played a role.

Common Causes of Fatigue in Aviation

In the aviation industry, there are a wide range of issues that cause fatigue. The following is a list of some common causes in a number of positions in aviation. Note that this is only a partial list; there are many more potential fatigue inducers than are considered here.

- Sleep deprivation or lack of quality sleep
- Times zone changes among pilots, maintenance technicians (called into the field for aircraft repairs, etc.), and other personnel required to travel as a part of their jobs
- Disruption of normal circadian patterns and sleep cycle (time zone variations, jet lag, etc.)
- Boredom (low mental stimulation)
- Over-familiarity with job and related tasks
- Choices and activities in personal life
- Periods of excessive mental stimulation
- Flight schedule (long flights; short, frequent flights; erratic flight schedule; etc.)
- Single-pilot flights
- Shift work
- Extended periods of time performing similar tasks
- Stress and pressure (work and personal life)
- Overexertion due to physically demanding activities
- Dehydration
- Noise
- Improper nutrition
- Personal illness
- Medications
- Vibration (often associated with propeller-driven aircraft)

Symptoms of Fatigue in Humans

Similar to the causes of fatigue, there are a number of different symptoms indicating that a person is experiencing some level of fatigue. The following is a partial list of some common fatigue symptoms in humans.

- Change in personality (irritability or grouchiness, giddiness, etc.)
- Feelings of indifference in task completion; loss of initiative
- Complacency
- Sleepiness, drowsiness
- Change in short-term memory
- Erosion of task performance
- Impaired decision making and judgment
- Difficulty concentrating
- Fixation
- Loss of situational awareness

- Increase in commission of errors
- Erosion of reaction time
- Depression
- Hallucinations (in some excessively fatigued persons)

According to the FAA, a number of studies have shown that persons experiencing fatigue consistently underreported how they were actually feeling. To make matters worse, the fatigued individual does not always fully understand the degree of personal impairment they are experiencing due to fatigue. No amount of experience, job knowledge, years in position, medication, or caffeine is enough to overcome the impacts of fatigue on the body and mind. Fatigue can only be managed through quality sleep, proper nutrition and hydration, and lifestyle choices.

Managing Fatigue

The truth is that fatigue can never be eliminated from any person; it is a normal part of life. In fact, to a degree fatigue may actually be a good thing, serving as a warning that we are in need of rest or some change in activity or lifestyle. Some key components to the management of fatigue are self-assessment (understanding and recognition of one's own condition, patterns, and fatigue symptoms) and education (teaching people about fatigue and the importance of its management). Here is a partial list of some strategies for managing fatigue in both professional and personal life.

- Understand your job requirements and situation as much as possible before employment. In other words, know what you are getting into. New or transitioning jobs often require changes in life style and/or patterns. You must be prepared to adapt ahead of time insofar as possible.
- Get plenty of quality sleep and rest; avoid products and situations that are not conducive to proper sleep.
- Maintain a healthy, balanced diet.
- Keep yourself properly hydrated. Remember, not all beverages are created equal from the standpoint of hydration.
- Avoid excessive amounts of caffeine in your diet.
- Avoid medications that impede sleep before and during the times you need to sleep, and avoid medications that make you drowsy before and during job-related activities. Read the warning labels!
- Use alcohol in moderation and only at times that will not impede you on the job. Remember, this means you need to stop drinking long before you are supposed to report to work. This should be at least 8 hours ahead of time.
- Make physical exercise a regular part of your life. Try to set up an exercise regimen of at least a few times a week.
- Identify the things in your life that are causing stress. Do your best to determine ways to alleviate and manage excessive stress in your life.
- Do not be afraid to seek medical and/or psychological assistance as needed. Unresolved physical and mental issues will only get worse if left untreated.
- Finally, be honest with yourself about yourself and your condition. Conduct a self-assessment on a daily basis. A useful tool for self-assessment that is well known in aviation is the **IMSAFE Checklist** (see **Figure 6–7**).

Figure 6–7. IMSAFE Checklist.

1. *Illness:* Am I sick? Do I have any symptoms? If so, what are the symptoms?
2. *Medication:* Have I been taking prescriptions or over-the-counter drugs? What are the possible side effects?
3. *Stress:* Am I under psychological pressure from my job or personal life? Do I have money, health, or family problems?
4. *Alcohol:* Have I been drinking within 8 hours? Within 24 hours?
5. *Fatigue:* Am I tired and not adequately rested? What are my sleep requirements and patterns?
6. *Eating:* Have I eaten enough of the proper foods to keep me adequately nourished during the flight and/or work activities?

Source: Federal Aviation Administration 1991

Chapter Summary

Human factors constitutes one of the most important components of aviation safety for a number of reasons that were discussed in this chapter. HF is a multifaceted discipline consisting of elements such as engineering, ergonomics, physiology, and psychology. HF is by far the leading causal factor in the majority of aviation accidents and incidents. Many organizations have already implemented or are in the process of implementing HF programs and initiatives in order to protect their employees and assets and to reduce the costs associated with accidents, incidents, fatalities, and injuries. Some HF initiatives in the aviation industry have produced measurable positive results. Many key aviation bodies such as the FAA, NTSB, and ICAO have long recognized the importance of HF in aviation safety and are strong proponents of HF training and programs for organizations and their employees.

HF models are useful, taking theoretical concepts and putting them into a format that is applicable to the real world. Both Reason's Model and the SHEL Model are widely recognized and used within the aviation industry to operationalize some HF elements. Some of the key components of HF and its applicability to aviation safety are found in human judgment, decision making, stress and stress management, situational awareness, sleep, and fatigue. Each of these components has the capacity to affect and be affected by each of the others. It is important that each person involved in aviation operations be cognizant of their individual physical and mental condition before becoming involved in aviation activities.

Past and current trends clearly demonstrate that HF will continue to occupy a prominent place in the aviation industry. Humans operate and function within the industry, and therefore HF must continue to be given strong consideration as we work to improve safety in aviation.

Chapter Concept Questions

1. What are some of the most common causal factors for stress in many people's lives? Identity the primary things in your life that cause you stress. What are the main stress symptoms that you experience? What can you do to reduce stress in your life?

2. Describe the basics of NREM and REM sleep, including the four NREM stages of sleep. Describe your own sleep cycle: How much sleep do you average per night? How would you describe the quality of sleep that you usually experience?

3. Describe in detail the following human factors models: Reason's Model (the Swiss Cheese Model) and the SHEL Model. Describe in detail a specific case or example in aviation where each of the models can be accurately applied.

4. Respond in detail to the following statement: *Technology is not sufficient to overcome human factors problems, accidents, and issues in aviation.* Be as specific as you can in your response.

5. Define each of the following independently of the other: (a) judgment; (b) decision making. Briefly discuss the importance of judgment and decision making in aviation-related activities and in life. Identify at least five of the many different factors that may influence how a person makes judgments and decisions.

6. What is situational awareness (SA)? Briefly describe the importance of remaining situationally aware in aviation activities. Overall and honestly, how would you describe your personal level of SA? What specific things can you do to improve SA in your life?

7. What is the systems approach? Discuss the potential usefulness of the systems approach in an organization's safety programs and policies.

8. Be honest in answering this question: Which of the five hazardous attitudes do you display most frequently? In what types of situations do you display these attitudes? What types of things can you do to minimize the presence and impact of these attitudes in your life?

9. Describe the basics of the FAA's aeronautical decision making (ADM) process.

10. You have been tasked by divisional leadership in your aviation company with developing a human factors program for your specific area (i.e., flight operations, aircraft maintenance, ramp operations, FBO, etc.). What are some of the most important elements that need to be present in your human factors program? What specific strategies would you use to address each applicable element? Example: What would you do specifically to get the employees in your division/department onboard with the HF program? In answering this question, speak generally about the basic aspects of virtually any human factors program (i.e., support of leadership and employees, an HF "champion," investigation process, etc.), and then be as specific as you are able in applying to a department or division within an aviation company.

Chapter References

Adams, C. 2009. Human Factors: Beyond the "Dirty Dozen." *Aviation Maintenance.* http://www.aviationtoday.com/am/issue/cover/Human-FactorsBeyond-the-"Dirty-Dozen%22_33730.html.

Alteon, A Boeing Company. n.d. Maintenance Human Factors Program Training for Managers. PowerPoint Presentation.

American Psychological Association. 2010. Why Sleep Is Important and What Happens When You Don't Get Enough. http://www.apa.org/topics/sleep/why.aspx#.

AviationKnowledge. 2010. Human Factors Models (in Aviation). http://aviationknowledge.wikidot.com/aviation:human-factors-models.

Edwards, E. 1988. *Human Factors in Aviation.* San Diego, CA: Academic Press.

Endsley, M. 2000. Theoretical Underpinnings of Situational Awareness: A Critical Review. In M. R. Endsley and D. R. Garland (Eds.), *Situation*

Awareness, Analysis, and Measurement. http://zonecours.hec.ca/documents/A2007-1 1399574.TheoricalUnderpinningsofSituationAwareness_ACriticalReview.pdf.

Federal Aviation Administration. n.d. Fatigue in Aviation. http://www.faa.gov/pilots/safety/pilotsafetybrochures/media/Fatigue_Aviation.pdf.

Federal Aviation Administration. n.d. Maintenance Human Factors Presentation System: Human Error. https://hfskyway.faa.gov/hfskyway/Presentations/Default.aspx.

Federal Aviation Administration. 1991. *Aeronautical Decision Making.* Advisory Circular 60-22. http://rgl.faa.gov/Regulatory_and_Guidance_Library/rgAdvisoryCircular.nsf/0/CCDD54376BFDF5FD862569D100733983?OpenDocument.

Federal Aviation Administration. 1995. National Plan for Civil Aviation Human Factors: An Initiative for Research and Application. www.hf.faa.gov/docs/natplan.doc.

Federal Aviation Administration. 2000. *FAA System Safety Handbook*, Chapter 15: Operational Risk Management. http://www.faa.gov/library/manuals/aviation/risk_management/ss_handbook/media/chap15_1200.pdf.

Federal Aviation Administration. 2000. *Maintenance Resource Management Training.* Advisory Circular 120-72.

Greenhut, N., Andres, D., & Luxhoj, J. T. 2006. *Graphical Enhancements to the Executive Information System (EIS) for the Aviation System Risk Model (ASRM).* American Institute of Aeronautics and Astronautics. http://www.rci.rutgers.edu/~carda/Docs/AIAA%20Paper_ver%203.pdf.

Hawkins, F. 1993. *Human Factors in Flight* (2nd ed.). Brookfield, VT: Ashgate Publishing Company.

Helmreich, R., and Foushee, H. C. 2010. Why CRM? Empirical and Theoretical Bases of Human Factors Training. In B. Kanki, R. Helmriech, and J. Anca (Eds.), *Crew Resource Management* (2nd ed). San Diego, CA: Academic Press.

Kilroy, C. 2008. Special Report: United Airlines Flight 232. *AirDisaster.com.* http://www.airdisaster.com/special/special-ua232.shtml.

Latorella, K. A., Prabhu, P. V. 2000. A Review of Human Error in Aviation Maintenance and Inspection. *International Journal of Industrial Ergonomics* 26, 131–161.

Luxhoj, J. T., and Maurino, M. 2001. An Aviation System Model (ASRM) Case Study: Air Ontario 1363. *The Rutgers Scholar* 3. http://rutgersscholar.rutgers.edu/volume03/maurluxh/maurluxh.htm.

National Transportation Safety Board. 1990. Aircraft Accident Report, United Airlines Flight 232. http://www.airdisaster.com/reports/ntsb/AAR9006.pdf.

Orasanu, J. 2010. Flight Crew Decision-Making. In B. Kanki, R. Helmriech, and J. Anca (Eds.), *Crew Resource Management* (2nd ed.). San Diego, CA: Academic Press.

PilotFriend. 2010. United 232. http://www.pilotfriend.com/disasters/crash/united232.htm.

Reinhart, R. 1996. *Basic Flight Physiology* (2nd ed.). New York: McGraw-Hill.

Shaver, E. 2009. A Short History of Human Factors and Ergonomics. *The Human Factor Advocate.* http://www.thehumanfactorblog.com/2009/01/06/a-short-history-of-human-factors-and ergonomics/.

Shaver, E., and Braun, C. 2008. *What Is Human Factors and Ergonomics?* Benchmark Research & Safety, Inc. http://www.benchmarkrs.com/_uploads/What-is-Human-Factors-and-Ergonomics.pdf.

Sheridan, T. B. 1988. *Human Factors in Aviation.* San Diego, CA: Academic Press.

Sogg, S. n.d. An Integrated Systems Approach to Human Factors in Commercial Aviation

Maintenance Systems. https://hfskyway.faa.gov/HFTest/Bibliography%20of%20Publications%5CHuman%20Factor%20Maintenance%5CAn%20integrated%20systems%20approach%20to%20human%20factors%20in%20commercial%20aviation%20maintenance%20systems.pdf.

Summary of the Various Definitions of Situational Awareness. n.d. http://www.raes-hfg.com/crm/reports/sa-defns.pdf.

Trollip, S. R., & Jensen, R. S. 1991. *Human Factors for General Aviation.* Englewood, CO: Jeppesen Sanderson, Inc.

Waloncik, D. S. 1993. General Systems Theory. http://www.survey-software-solutions.com/walonick/systems-theory.htm.

WebMD. 2010. Coping with Excessive Sleepiness: Sleep 101. http://www.webmd.com/sleep-disorders/excessive-sleepiness-10/sleep-101.

Wenner, C. A., & Drury, C. G. 2000. Analyzing Human Error in Aircraft Ground Damage Incidents. *International Journal of Industrial Ergonomics* 26, 177–199.

Wiegmann, D., Faaborg, T., Boquet, A., Detwiler, C., Holcomb, K., & Shapell, S. 2005. *Human Error and General Aviation Accidents: A Comprehensive, Fine-Grained Analysis Using HFACS.* Federal Aviation Administration, Office of Aerospace Medicine. citeseerx.ist.psu.edu/viewdoc/download?doi=10.1.1.74.7529&rep.

Wikipedia. 2010. Link Trainer. http://en.wikipedia.org/wiki/Link_Trainer.

Wilson, B. T. 2007. Situation Awareness: A Cognitive Definition. *FlightCog.com.* http://flightcog.com/essays/sadef.html.

Wood, R. H. 2003. *Aviation Safety Programs: A Management Handbook* (3rd ed.). Englewood, CO: Jeppesen Sanderson, Inc.

7 Ground Safety

Chapter Learning Objectives

Upon completion of this chapter, the reader should be able to:

- Understand the basic elements of ground safety and how they relate to various areas of an aviation organization.
- Describe the common causes and the categories and classifications of ground damage.
- Explain some of the main safety programs applicable to ground safety.
- Describe and discuss regulatory standards as they relate to the various programs described in this chapter.
- Describe some of the main hazards associated with various job functions within an aviation organization.

Key Concepts and Terms

Aircraft Ground Damage
Aircraft Ground Handling
Air Transport Association (ATA)
Americans with Disabilities Act (ADA)
Auxiliary Power Unit (APU)
Cargo Handling
Customer Service Agent
Deadman Switch
End-Range Motion
Fall Protection
FOD Walk
Foreign Object Damage
Foreign Object Debris (FOD)
Fuel Cell Entry
Hazard Communication Program
Hearing Conservation Program
Labeling System
Lockout/Tagout Standard
Marshaller
Material Safety Data Sheet (MSDS)
Permit Required Confined Spaces
Portable Hydrant Pumps
Shelf Life
Spoilage
Squib
Standard Threshold Shift
Stationary Hydrant Fueling System
Wing Walker

Introduction

The scope of ground safety in the aviation industry is broader than the average person outside the industry could ever imagine. Until the mid- to late 1990s, it was widely accepted within the industry and government that the Federal Aviation Administration (FAA) had primary jurisdiction over compliance. Perhaps this is why the air transportation industry was quickly becoming the industry with the highest injury rates. After all, the FAA does not focus on employee safety. Its primary focus is flight safety and protection of the flying public through regulatory actions.

In recent years, the Occupational Safety and Health Administration (OSHA) has become more important in the various worksites of many airlines, providing oversight and working to ensure workplace safety and regulatory compliance. The increased presence of OSHA combined with the costs associated with injuries to employees and damage to aircraft, equipment, and other assets have prompted most airlines to devote much more attention to ground safety than in years past. Airline operators have done this by providing funding, hiring safety professionals, and developing standards and procedures for ground safety, including providing specific training in ground operations and safety. Not only are these factors applicable to ramp and terminal employees, but arguably even important to aircraft maintenance because of the safety-critical functions of these operations. Aircraft maintenance is one of the most important operational components of aviation organizations due to the fact that so many stakeholders are relying on safe, well-maintained aircraft to transport people and goods. Further, though they are often much smaller in scale by comparison, the importance of ground safety at fixed-base operations (FBOs), corporate aircraft operations, and contract operations should not be overlooked.

This chapter will provide insight into some of the various elements of ground safety (ergonomics, aircraft ground damage, hazardous materials, etc.) and how they relate to various operations throughout an aviation organization. Many of these elements will be discussed in the context of the specific job areas in which they operate (e.g., lockout/tagout in maintenance, noise in ramp operations, etc.). Because of the complexity of the standards and how they relate between different aviation organizations, the various programs, regulations, and other risk elements will be discussed only at a high level rather than in comprehensive detail.

Aircraft Ground Damage

One of the most significant factors affecting ground safety in the aviation industry is the occurrence of **aircraft ground damage**. As implied by the phrase, aircraft ground damage occurs when an aircraft receives damage while on the ground. The **Air Transport Association (ATA),** now known as **Airlines for America,** defines aircraft ground damage as "damage to the exterior of an aircraft caused by equipment or personnel requiring any corrective action beyond inspection and sign-off that is considered under the control of ground operations personnel at the time damage occurs."

Source: Airlines for America

The ATA definition also includes aircraft damages caused by jet blast from another aircraft. An example of jet blast damage would be an object such as a ladder being blown into an aircraft by jet blast. Under the ATA definition, the inadvertent deployment of an aircraft emergency evacuation slide would also be considered aircraft ground damage.

There are many situations in which aircraft receive damage on the ground; some of the more common of such events occur during aircraft servicing (loading/unloading of baggage/cargo, etc.), when ground support equipment is being operated around aircraft, during maintenance, and while an aircraft is being towed or pushed back from its parking position or from the gate, in the case of air carrier operations. According to the ATA, some of the most common causes of aircraft ground damage events are as follows:

- Damage inflicted during aircraft towing and pushback from gate
- Collisions between aircraft and ground support equipment (GSE) such as belt loaders, tugs, baggage carts, and ground power units
- Inadvertent deployment of evacuation slides and tail cones
- Damage related to the use of passenger loading bridges due to operator error or collision during aircraft movement

Source: Airlines for America

Aircraft ground damage may be further defined as being controllable, as causing a flight delay or cancellation, and as inflicting damage to an aircraft exceeding a certain dollar amount in parts and repair.

Although most are manufactured to exacting standards and are inherently structurally sound, aircraft are actually rather easy to damage, so that inflicting significant damage to an aircraft does not always require a major collision or impact. Some seemingly minor impacts can cause an aircraft to be damaged beyond acceptable specifications and tolerances, requiring maintenance before being returned to service. Such situations often result in delayed or canceled flight operations, as in many cases the aircraft must be, at minimum, inspected by a qualified maintenance technician, and in some cases it will be taken out of service until repairs can be made and it is authorized to return to service. Thus, aircraft ground damage events are often very expensive to the aircraft operator.

Some may be surprised to learn that aircraft ground damage events are extremely common occurrences, particularly in the airline industry. While the majority of damage events are minor, some occurrences are much more serious, requiring additional maintenance beyond basic inspection to determine the degree of damage and to make repairs. The following section covers the basic categorization of ground damage events.

Aircraft Ground Damage Categorization

In the airline industry, ground damage events are commonly classified into various categories based on the type of damage to the aircraft or what (or who) caused the event to occur. There are a number of damage categories that may be used by some air carriers to classify the various types of aircraft damage. The following categorization system is an example of how an air carrier might choose to classify its aircraft ground damages. Please note that this is an example only and does not represent the systems used by air carriers as a whole.

Category	Description
1	Major damage. In general, this is the most serious type of damage, as it requires maintenance to the aircraft beyond basic inspection, often requiring flights to be delayed and/or canceled. Many Category 1 events are very expensive to the aircraft operator due to lost revenue and the associated costs of repair.
2	Damage to the interior of the aircraft (flightdeck, cabin, cargo bins, etc.)
3	Minor damage
4	Foreign object damage (FOD)

5	Damage related to weather (hail, lightning strike, etc.)
6	Damage to aircraft caused by another airline
7	Damaged caused by mechanical failure of equipment (ground support equipment, etc.)
8	Old damage (previously discovered and recorded)
9	Damage caused by maintenance personnel
10	Other damages (those not readily fitting into another category)
11	Holding area until further classification can be completed

The classification of aircraft ground damage events is generally conducted following the initial inspection by a qualified person and may be modified as the event is investigated. Thus, the final damage category may by different from the initial classification assigned to the event. For example, at first glance a small dent in the engine nacelle of a regional jet caused by contact with a bag or ground support equipment may seem to be minor and thus initially be assigned to Category 3 (minor damage). However, upon further inspection the actual damage may prove to be more serious than originally thought, and the aircraft may need to be removed from service for more extensive inspection and repair. The final classification of the event might then be changed to Category 1 (major damage).

Airline ground damage events may be further categorized by using classifications such as:

- *Rolling Stock:* A stationary aircraft is impacted by wheeled ground support equipment (GSE), either motorized or non-motorized, self-propelled or towed.
- *Loading Bridge:* Damage inflicted when the aircraft is impacted by a passenger loading bridge or passenger assistance device being positioned for enplaning or deplaning of passengers.
- *Marshalling Error:* Damage occurring when the aircraft is moving and under the guidance of ground-based support staff.
- *GSE Disconnection Error:* Damage caused by failure to disconnect GSE from aircraft and GSE pulling away from the aircraft.
- *Pushback or Towing Damage:* Damage that occurs during pushback and/or towing operations, involving procedures, personnel, or equipment.
- *Other:* Damage events not fitting into one of the above categories.

Common Causal Factors in Ground Damage Events

Although there are a large number of factors that contribute to the occurrence of aircraft ground damage events, here are several of the more common causal factors present in some form in many events. This list has been created based on the experience of the authors and other safety personnel in the aviation industry with firsthand knowledge. This list is rather general, as there are many specific types of situations which could fit into one or more of these areas, and there is overlap between some of them.

- Failure to follow established procedures and policies properly
- Loss of situational awareness

- Hurrying, rushing while performing tasks in and around aircraft; time pressure
- Driving/positioning ground support equipment too close to the aircraft
- Failure to ensure or maintain adequate aircraft wingtip clearance
- Complacency
- Fatigue
- Negligence
- Environmental factors
- Presence of a chain of events leading to the occurrence
- Adoption of "tribal knowledge" ("this is how we do things here")
- Lack of consistent daily operational checks of ground support equipment
- Failures in communication between employees
- Horseplay
- Stress

The Costs of Aircraft Ground Damage

Even cursory consideration given to the issue of aircraft ground damage reveals that events of this nature are often extremely expensive to firms operating aircraft, such as airlines and corporations utilizing aircraft for business purposes. Some estimates have placed the cost to the global airline industry due to aircraft ground damage at over $3 billion annually. Many airlines use a figure of approximately $500 per minute in determining the cost of their delays associated with aircraft ground damage events. Also, Many U.S. airlines estimate their average direct costs attributable to preventable aircraft ground damage flights in the range of $50,000 to $75,000 per event, and indirect costs often drive these figures higher in terms of overall economic impact. Obviously, many of the costs associated with ground damage events are very high as a result of replacement or remanufacture of parts and the cost of labor paid to the maintenance technicians performing the repairs.

Lost Revenue From the perspective of the airlines, one of the most significant costs of aircraft ground damage is the revenue lost due to flight cancellations. An aircraft sitting on the ground after being damaged is not able to generate revenue for the air carrier. Most airlines maximize the number of flight segments that can be operated by a given aircraft on a daily basis to achieve maximum possible revenue generation. The seats on an aircraft operating in revenue service not occupied by paying passengers have entered into a state of **spoilage**, meaning that the potential revenue that would have been generated by the seats has expired or "spoiled" and cannot be realized. In some events, the damage to the aircraft is so significant that the aircraft must be taken out of service for several consecutive days or even weeks while it is being repaired. Assuming the air carrier does not have an aircraft available to fly as a spare in place of the damaged plane (which is sometimes the case), the loss of revenue due to the damage event may be significant. Thus, lost revenue constitutes a very real cost of aircraft ground damage events.

Delays Another cost often associated with ground damage events is found in delayed flights. Even in cases in which flights are not canceled and the aircraft is permitted to remain in revenue service, quite often damage events will cause at least some flight delay to occur due to the need for the aircraft to be inspected and authorized to remain in service by a qualified individual. If the actual delay is significant in terms of the actual departure time as compared to the scheduled departure time, it may be difficult for the aircraft to get back on schedule. This may in turn cause downline flights to be delayed, causing a cumulative delay affecting a number of flights scheduled to be flown with the damaged aircraft.

Foreign Object Debris/Damage (FOD)

Another serious issue common throughout the aviation industry that regularly causes damage to aircraft operating on the ground is **foreign object debris (FOD)**. As the name implies, FOD is any type of foreign object on the ground in ramps, aircraft parking areas, taxiways, runways, and other areas frequented by aircraft. FOD may consist of rocks, parts of aircraft that have fallen off, pieces of luggage (zippers, locks, wheels, etc.), or any other type of objects that should not be in these areas and may cause damage to aircraft. If foreign object debris causes damage to an aircraft, it is then called **foreign object damage**.

Each year, FOD-related events cause a significant number of ground damages to occur. While the undercarriage, tires, and landing gear of aircraft may be especially susceptible to FOD, other parts of aircraft are also at risk. FOD is sometimes ingested into jet engines and can cause very serious damage to the fan disk and other parts of an engine.

FOD Prevention and Management Although FOD is extremely common on the airfield, there are a number of ways to prevent and manage FOD effectively to minimize the possibility of damage to aircraft. Many airports conduct daily FOD inspections of the runways, taxiways, and other areas of an airfield, and FOD is typically removed shortly after it is found. Such FOD inspections can be made a part of the daily airfield inspections required by FAR 139 for commercial service airports. Airports not having commercial service also benefit from performing daily airfield and FOD inspections.

Another common practice engaged in by a number of airports and other aviation organizations is the **FOD walk**. An FOD walk involves a group of people spreading out in a line across ramps, taxiways, and runways to seek and remove FOD from the airfield. FOD walks are very effective ways of locating and removing FOD that may have been missed during other types of airfield inspections, because the people involved are very close to the surfaces of the airfield and have a greater chance of spotting FOD.

Training employees in FOD removal and reminding them of the importance of this practice is another means of preventing and managing FOD. Airline ramp areas often produce significant amounts of FOD, and employees need to be vigilant in removing FOD and disposing of it as soon as possible to minimize the likelihood of damage to an aircraft. However, employees who work at airports and operations of all sizes need to be alert and participate in FOD identification and removal.

Prevention of Aircraft Ground Damage

Although there are a number of strategies that may be employed in the effort to prevent aircraft ground damage, the following are general recommendations for aircraft ground damage prevention.

- Provide adequate training for all employees in job procedures and company policies.
- Adhere to all established policies and procedures; avoid shortcuts.
- Establish and maintain continual proper communication among all flight and ground personnel.
- Establish an active internal campaign within the aviation organization to raise awareness on the problems associated with ground damages.

- Ensure that local and corporate (as needed) organizational leadership is present and active in monitoring operational activities.
- Create and establish a consistent observation process for operations taking place around aircraft.
- Keep all ground support equipment in good operating condition through regular maintenance.
- Conduct daily inspections of ground support equipment. This is actually an OSHA requirement as well as good safety practice.
- Local leadership should remind employees to work safely and follow procedures and policies.
- Conduct daily FOD inspections of all ramp and operations areas; follow up with local airport leadership to ensure daily airfield inspections are being conducted by qualified personnel.
- Encourage employees to work with efficiency, but not to "hurry" or "rush" in performing their tasks.
- Do not approach ground support equipment closer than five feet to any aircraft, with the exception of belt and container loaders and lifting devices used to enplane/deplane passengers with disabilities, as necessary.
- Drive slowly when operating equipment in the proximity of aircraft; maintain clearances between equipment and aircraft.
- Avoid horseplay and negligent behaviors.
- Remain alert at all times.
- Finally, report any known or possible damage to aircraft, no matter how minor it may seem, *immediately* to a supervisor, maintenance technician, or flight crew member. This is extremely important!

Ramp Operations

The ramp operations occurring at many airports present a variety of challenges and hazards for employees. In the airline industry and other areas of aviation, ramp operations are where the majority of serious injuries occur. This section will discuss the types of operations that occur on the ramp, some of the typical hazards that are present, and control measures that can help to minimize the hazards to employees and passengers. But before addressing the hazards associated with each individual job function, there is one hazard exposure that may impact all ramp employees: noise. The following section will discuss ramp noise in more detail.

Noise

Noise is an inevitable issue for aircraft operators, maintenance technicians, and ramp workers. Many aircraft are equipped with an **auxiliary power unit (APU)**, a small turbine engine that provides power for some of the systems of an aircraft. The APU is often running during ground handling, and will produce a 102–109 decibel output. Exposure to this amount of noise for more than about 30 minutes daily could cause permanent hearing loss to those employees working in proximity to an operating APU. **Table 7–1** shows OSHA's permissible noise exposure levels. The ramp operations areas at many airports feature a working environment that produces a continuous amount of noise, necessitating the creation and implementation of a **hearing conservation program**. Employers must administer a continuing, effective hearing conservation program whenever employees are exposed to noise levels at or exceeding the eight-hour time-weighted average (TWA) of 85 decibels (dB) measured on the "A" scale (see Table 7–1).

Table 7–1. The permissible noise exposures, or permissible exposure limits (PELs), for which exposure should not be exceeded.

Duration Per Day (Hours)	Decibel Sound Level
8	90
6	92
4	95
3	97
2	100
1½	102
1	105
½	110
¼ or less	115

Notes: 1) When the daily noise exposure is composed of two or more periods of noise exposure of different levels, their combined effect should be considered, rather than the individual effect of each. 2) Exposure to impulsive or impact noise should not exceed 140 decibels peak sound pressure level.

Source: 29 CFR 1910.95, Table G-16. OSHA.

Periodic monitoring and testing of affected employees is required. In addition, hearing protection must be provided to affected employees (at no cost to the employee). The hearing protection provided must have a noise reduction rating (NRR) high enough to attenuate the employee's exposure level below a 90-dB TWA. The Occupational Noise Exposure Standard (29 CFR 1910.95) provides further details for monitoring and testing.

Employers are required to record work-related hearing-loss cases when an employee's hearing test shows a marked decrease in overall hearing. If an employee's hearing test (audiogram) reveals that the employee has experienced a work-related standard threshold shift (STS) in hearing in one or both ears, and the employee's total hearing level is 25 decibels (dB) or more below audiometric zero (averaged at 2,000, 3,000, and 4,000 Hz) in the same ear(s) as the STS, the case must be recorded on the OSHA 300 Log. Employers can make adjustments for hearing loss caused by aging, seek the advice of a physician or licensed health care professional to determine if the loss is work-related, and perform additional hearing tests for verification.

A **standard threshold shift** is defined in the occupational noise exposure standard [29 CFR 1910.95(g)(10)(i)] as a change in hearing threshold, relative to the baseline audiogram for that employee, of an average of 10 decibels (dB) or more at 2,000, 3,000, and 4,000 hertz (Hz) in one or both ears. In this case the STS need only be reported to the employee. Refer to **Table 7–2.**

If the employee has shown an STS, the employee's overall hearing ability in comparison to audiometric zero must then be examined. Using the employee's current audiogram, average the hearing levels at 2,000, 3,000, and 4,000 Hz to determine whether he employee's total hearing loss exceeds 25 dB from audiometric zero. In this case the STS must be reported to the employee and recorded on the OSHA 300 Log. Refer to **Table 7–3.**

Table 7–2. Example of a Standard Threshold Shift (STS) That is Not OSHA-Recordable

Frequency (Hz)	Baseline (dB)	Current Audiogram (dB)	Difference (dB)
2000	10	20	10
3000	5	10	5
4000	15	30	15
Average	10	20	10

Baggage Handling

Baggage handling contributes to the majority of injuries throughout the airline industry and also accounts for many of the most debilitating and costly. Typically, baggage is first handled by ramp employees in a bag room or at planeside. Complicating the handling process is the fact that bags come in so many shapes, sizes, and weights. For regional aircraft and even some heavy aircraft operators, the baggage handling process is further complicated by physical limitations of aircraft cargo compartments. In many aircraft, cargo bins are only a few feet high and thus do not permit employees to keep their backs straight and lift/move bags and cargo from an optimal posture. Many back and shoulder injuries arise from baggage operations.

Furthermore, many ramp workers must engage in significant amounts of repetitive motion each day due to the nature of the work they perform and the amount of bags and cargo they must handle during a shift. Airlines operate under significant time constraints due to the rigidity of their schedules, and thus ramp workers are often operating under the constant pressure of getting the flights turned around as quickly as possible. This time factor is yet another potential contributing factor to injuries because of the urgency to turn the aircraft around swiftly and the reality that people often take shortcuts in order to do their jobs more efficiently. Deviation from established operating procedures is a common factor in employee injures and aircraft ground damage events.

Around the time of this writing, many airlines have begun to charge fees for checked baggage in addition to the fees that were previously been implemented for excess weight (i.e., over 50 lbs.). These measures have discouraged some travelers from bringing excess baggage. The result certainly means fewer bags, sometimes fewer jobs, but also fewer injuries to employees.

Table 7–3. Example of an OSHA Recordable Standard Threshold Shift (STS)

Frequency (Hz)	Baseline (dB)	Current Audiogram (dB)	Difference (dB)
2000	20	30	10
3000	30	35	5
4000	10	25	15
Average	20	30	10

Figure 7–1. Baggage handler loads an aircraft.

Ramp employees such as the baggage handler in **Figure 7–1** can reduce the chance of injury from baggage handling by testing the weight of baggage and cargo before the lift and by not jerking the bag. For heavy items, team lifting or the use of assist devices is recommended. An assist device may be a dolly or lift truck, or using the corners of bag belts or carts as leverage to lift the baggage. When loading to a cart, loading heavier items first to the lower shelves is preferred.

Ground Handling of Aircraft

Aircraft ground handling is usually described as the reception and dispatch of aircraft upon arrival and preparation for departure. However, there are also occasions where aircraft must be transferred from gate to gate (hangar to gate is typically handled by aircraft maintenance personnel) and/or between terminals. With some advance notice, ramp personnel are notified via radio of the aircraft's approach. For efficiency, final preparations and equipment placement should be done prior to arrival, but doing so also reduces rushing and the shortcuts that often lead to injury or ground damage.

The movement of the aircraft can be one of the most dangerous tasks in the airline industry. But with proper training, communication, and oversight, the risks can be minimized. Ramp

Figure 7–2. Ground crew escorts aircraft into position.

employees communicate with flight crews via hand signals and sometimes radio. A proper understanding of which signals to use and when to use them is vital for the protection of lives and assets. When receiving an aircraft, a designated parking spot large enough to handle the given aircraft is used. Positions are usually identified by ground markings, with a "T" indicating the appropriate stop position. One marshalling employee is typically used with two **wing walkers**. A **marshaller** or aircraft guide (see **Figure 7–2**) is a person trained to visually and (sometimes) audibly direct the movement of an aircraft on the ground. The marshaller should always being in direct visual contact with the pilot in command of the aircraft and with both wing walkers. Wing walkers are employees trained to assist the marshaller with aircraft movement by appropriately signaling to the marshaller if the way is clear for the travel path of the aircraft wings.

Cargo Handling

Cargo handling is similar to baggage handling in that the packages and materials handled vary in size, weight, and shape. While some cargo may be handled using equipment within the facility, manual handling of materials will always be necessary when an aircraft is involved. Sometimes this manual handling may occur several times before the cargo makes its final destination. Handlers must be trained and skilled in the technique of not only maximizing the use of the space for storing cargo for transport, but also stabilizing it during the storage process to minimize struck-by/against hazards before and after transportation due to falling materials.

Most cargo transported by an aircraft will either be ad hoc (i.e., not part of the planned cargo) and stowed in the cargo area alone or along with passenger baggage, or in storage containers called ULDs or unit load devices (see **Figure 7–3**) or air cargo pallets.

Figure 7-3. Employee moves containerized cargo into transport position.

Cabin Services

Cabin Services handles the food and beverage services for passenger aircraft. Although many passenger airlines contract this service to a third-party vendor, some handle their own service internally. Cabin Services employees restock beverage, snack, and food items on aircraft and will often provide cleaning service between departures and at the end of the day. Services include a complete dressing of the cabins, which could include replacement of pillows, blankets, and magazines and sanitization of seat-back tray tables.

Some of the hazards to this group of employees include:

- *Lifting/lowering of food and beverage products.* This is one of the most frequent and severe sources of injury for this job type. Injuries typically occur when, during the lift, materials are not kept close to the body or the employee bends at the waist rather than the knees.
- *Struck-by potential from protruding metal of ovens, drawer edges, or metal latches.* Employees must be careful to replace drawers, doors, and latches to prevent cut injuries.
- *End-range motions due to awkward location of equipment.* Due to limited space, some equipment and supplies may be stored in locations that require poor or undesirable posture. An **end-range motion** is when an extremity (elbow, wrist, shoulder, etc.) is straightened to or near is maximum potential. In addition to the extra force (or weight) created by the lifting or lowering of materials away from the body, potential trauma to the joints can occur if the end limits are exceeded. This

can occur if, while in an end-range posture, a foreign element creates a sudden jerk or stoppage of the anticipated movement of the body.

- *Infection potential from used food and drink utensils.* Employees should take caution when handling food and drink materials by using protective gloves or, at a minimum, following proper hygiene practices. Proper hygiene includes limiting the handling of items that were touched by the mouths of passengers, avoiding touching the face while servicing, and frequently washing the hands with soap to reduce exposure potential. The airline can also reduce risk by providing hand-held foods or disposable food and beverage utensils. Although airlines have eliminated many services because of costs, doing so also reduces exposure.

Aircraft Fueling

Transport-category aircraft consume significant amounts of fuel and will often require refueling at each stop. The high degree of flight activity associated with airline operations creates the need for regular refueling, particularly at airline hub facilities. Aircraft refuelers face two primary hazards: fuel spillage and fire. Only trained and qualified individuals should be allowed to perform fueling operations. Only refueling equipment that has been recently inspected and found to be in good operating condition should be used.

Most refueling operations involve one of a few different methods. Depending on the location, refueling services for many airlines are performed with a **stationary hydrant fueling system** that siphons

Figure 7–4. Fueling employee connects nozzle to fuel from truck.

fuel directly from the storage tanks to the aircraft. Regional aircraft are typically fueled from **portable hydrant pumps** or refueling trucks (see **Figure 7–4**), while general aviation aircraft are typically fueled from trucks.

To avoid fuel spills, the fueling operator should inspect daily all refueling equipment for good seals in the line and ensure that a good connection is maintained. Knowledge of current fuel levels and the amount to be added is essential as well. Most spills tend to occur with fueling trucks away from the aircraft and when the trucks are being loaded. The scenario usually includes misuse of the **deadman switch**, a device consisting of a cable and switch, sometimes at the truck and sometimes in the user's control. When the cable or switch is released, the fuel flow stops. Spillage often occurs when the operator is inattentive to fuel level in the vehicle or has bypassed the switch.

Fires from fueling service can come from a number of different sources, but all can be avoided with proper training and use of appropriate safety equipment. Again, the best prevention tool is to perform tasks with trained and qualified individuals using appropriate equipment in good operating condition. The equipment should be inspected daily for damage and wear.

For fueling vehicles, the parking position is the first decision in the safety process. The vehicle should be away from the aircraft APU (auxiliary power unit) and positioned where it can be quickly moved in the event of an emergency. The vehicle should be chocked and, if passengers are on board during the fueling process, the cabin door must remain open. To minimize the buildup of static electricity, the proper sequence for grounding and bonding should be followed. More details on the requirements of aircraft fueling can be found by referring to the National Fire Protection Association (NFPA) 407, "Standard for Aircraft Fuel Servicing."

Disability Services

Since the enactment of the **Americans with Disabilities Act (ADA)**, most (if not all) airport facilities have been equipped with assist devices and personnel to ensure the comfort and mobility of disabled passengers. Among these aids are special nearby parking, elevators, handicap-accessible restrooms, wheelchairs, and so on. Airports and airlines also provide personnel (see **Figure 7–5**) to assist with passenger movement from gate to gate, terminal to terminal, and in and out of aircraft. This is sometimes provided by internal employees, but often by contract service.

The main hazards to this employee group include lift/lower and push/pull hazards. Potential for injury exists when lifting and lowering while passengers are moved to/from wheelchairs and aircraft boarding chairs.

What complicates the matter of assisting the movement of people is the fact that they are not wrapped up in a tight, neat package easily transported like a box. Instead, individuals come in many sizes and weights. Sometimes the disability

Figure 7–5. A disabled passenger is assisted into the terminal.

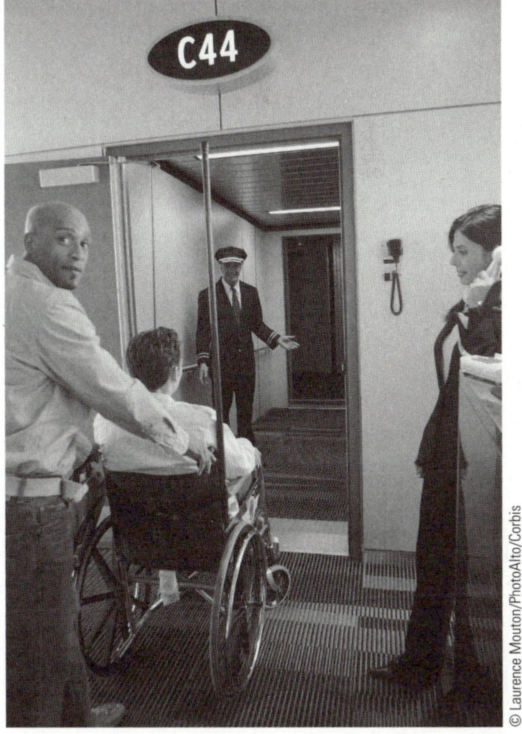

itself can produce an added challenge for assist personnel. Assist personnel should understand their limits and employ proper lifting/lowering methods to reduce the potential for injury. This includes understanding the assist equipment and its capabilities, knowing the best positions to maximize power during a lift/lower, and completing the task without causing injury or serious discomfort to the passenger.

The other part of the equation is understanding when to push versus when to pull (e.g., a wheelchair). The human body has much more power and control when pushing than when pulling. In addition, pulling often will result in an end-range motion of the arms or elbows. However, there are times when pulling is appropriate. To move a wheelchair or boarding chair over a hump or entrance to the aircraft, the assist employee should pull the chair to the obstacle then pull over it without jerking.

Ground Service Equipment (GSE) Maintenance

The GSE maintenance group provides preventive and emergency repair of ground support equipment such as vehicles, power units, belt loaders, and anything else used on the ramp or in a hangar. Larger organizations may have personnel who are specialized in one or more pieces of technical equipment; smaller organizations may employ select vendors or maintenance contractors to provide these services. Regardless of the organization type or size, GSE maintenance personnel should keep in mind a few considerations for the end users (ramp operations employees) of their product:

- Elimination of protruding edges or cut surfaces
- Condition of tires and wheels
- Operation of steering, braking system, lights
- Fluid levels (engine oil, coolant, transmission, etc.)
- Ease of movement or operation of the equipment
- Condition of latches and hinges
- Reduction of end-range motion requirement for use of equipment
- Condition of hoses, cables, belts, bumpers, etc.

One of the most important aspects of a successful GSE shop is organization. To keep a GSE shop running smoothly, it must have an effective preventive maintenance program that is well organized, scheduled, and adhered to by the entire organization. Most individuals change their oil every 3,000–5,000 miles. This is a form of preventive maintenance. And, while most people drive their vehicles for a limited time each day, the equipment on the ramp or in a hangar is usually used for several hours per day and often placed in stressful situations (heavily loaded, operated in extreme temperatures, etc.). Because of the varying equipment and dynamic features, preventive maintenance is much more complicated than simply changing the oil every few thousand miles. GSE will often handle the changing of transmissions and engines in tugs, components in ground power units, and wheel assemblies for baggage or maintenance carts, as well as many other requirements. For this, the shop is required to store and use many different types of chemical products. Aside from aircraft maintenance, the GSE shop will maintain more chemical products than any other aviation operation. And, because each of these items must be a part of the **hazard communication program** for the organization, it must be maintained in accordance with the standard (29 CFR Part 1910.1200). In many organizations, the GSE shop will often be cited for hazard communication violations because a shop attendant will purchase a product from a local automotive parts supply that is not on the organization's list of approved products and a **material safety data sheet** (**MSDS**) cannot be provided.

In addition to preventive maintenance, emergency situations occasionally arise. Oftentimes the organization has few or no spares for downed equipment and will need it serviced and returned quickly. This makes it extremely important for the handlers of equipment to perform daily checks before and after their shifts to recognize, document, and notify GSE of any discrepancies before they become an emergency case. In general, some of the things that should be checked include fluid levels, indications of leaks, foreign materials, shiny metal indicating wear, and worn belts or chains. Operators should also be attentive to problems while operating the equipment, listening for peculiar noises and looking for evidence of leaks or smoke.

Customer Service Functions

Customer service agent is typically the term given to employees who work the ticket counter (see **Figure 7–6**), gate, or service counter of an aviation organization. Although there are fewer hazards associated with customer service positions, they should not be excluded from the organization's health and safety program. Though usually only for limited distances, many of these employees handle baggage and cargo.

Figure 7–6. Customer Service Agent prepares to transfer baggage to conveyor.

Most ticket counters are organized so that there is a small step between the passenger and agent area, usually containing a scale to weigh the bag. A transfer belt will typically be located behind the agent. It is important for the agent to request that the passenger lift the bag to the step or scale area and avoid overreaching. It is equally important for the agent to pivot the feet when transferring baggage to the transfer belt to avoid twisting the trunk. As with ramp baggage handlers, agents should test the weight of baggage before lifting and avoid jerking. For heavy baggage or cargo, team lifting or the use of assist devices is recommended.

Aircraft Maintenance

There are many different job functions within aircraft maintenance. As a result, there are also many different exposures to injury. The majority of chemical exposures will be found here, along with varied material handling exposures, fall potential, hazardous energy sources, and so on. There are other job functions that experience these types of exposures, and they will be covered in more detail here because of the frequency of exposure by the aircraft maintenance job function.

General

There are thousands of parts that go into the manufacture of an aircraft. While many aircraft components are built by different manufacturers around the world, many will be replaced at some point during the life of each aircraft. Most of these replacements will take place in a hangar environment and, usually and preferably, during a scheduled maintenance visit rather than in a part failure situation. When scheduled maintenance is conducted in a hangar, many shops become involved in the effort. This could include the Paint Shop, Composite Shop, Upholstery Shop, Engine Shop, NDT (Non-Destructive Testing) Shop, Sheet Metal Shop, and Battery Shop, among others.

The Paint Shop is usually used for aircraft parts and components such as landing gear or cowlings. Many of these parts come to the Paint Shop after they've been stripped for non-destructive testing or when composite repairs are required. In addition to the Paint Shop, the Composite Shop uses many solvents for numerous processes. The Composite Shop also uses many two- and three-part products for composite development. All of these products are time controlled, and some must be temperature controlled as well. Including this as part of the hazard communication program and training or as a separate module is a must to ensure these controls are addressed. Although there are strain hazards from lifting and lowering in these shops, the main hazards are from chemical exposures. The use of ventilation and personal protective equipment such as gloves, aprons, and respirators is often required. See **Figure 7–7** for an example of one of the hazardous tasks that might take place in an aircraft maintenance shop.

Managing Chemical Products

Solvents, cleaners, resins, and epoxies, among other items, are frequently used chemicals in aircraft maintenance. There are hundreds of chemical products required for the maintenance of each aircraft, most of which are specified by the aircraft manufacturer's maintenance manuals. Communication of the hazards of these chemicals to employees is governed by the Occupational Safety and Health Administration (OSHA) under Title 29 CFR 1910.1200, otherwise known as the Hazard Communication (HazCom) Standard.

The HazCom Standard requires employers to develop a written program that describes how chemical products at the facility are received, labeled, stored, and handled, for example, to reduce the risk to employees exposed to the chemical hazards involved. As with any other hazards to

Figure 7–7. Welding conducted in one of many maintenance shops that may be found in any airline hangar facility.

employees, the hierarchy of risk minimization should be followed. For chemical products, this means the elimination of chemicals more dangerous to health and the environment by substitution of a less dangerous product or elimination of the process altogether. In some cases, allowing third-party vendors who specialize in certain maintenance functions using the more dangerous chemicals can be cheaper both monetarily and through risk reduction. One of the best examples of this is that most regional carriers and many major carriers will contract out work for repainting of aircraft. The facilities and chemicals involved in stripping are usually too expensive for these operations to be viable for internal maintenance.

For each chemical product to be used at the facility, a list should be organized and maintained (updated) with the most current information for each product. While not required, many companies choose to add detailed information to this list as an added benefit to management and affected employees. This information might include details of how the product should be labeled, stored, and shipped, or which shops the product is used or stored by. In addition to this inventory list, a material safety data sheet (MSDS) must be obtained for each product on the list. MSDSs will accompany many shipments of products, but if a new product is received without one, it should be requested before the product is allowed to be distributed for use.

For the program to work, each affected employee (a person who uses or could be expected to use the specific products in his/her operation or job function) must be trained on information regarding how to find and use the inventory list and MSDSs, read and understand the labeling system employed by the organization, and specific hazards and injury/illness prevention measures for the chemical constituents to which they are exposed. In other words, the employee must be able to read the label, understand the hazards involved with the product, and determine what personal protective equipment (PPE) is required to minimize exposure. Employees should also know how to safely handle the material, follow the organization's requirements for spills, and dispose of the product as necessary. In some cases, detection equipment may be necessary to measure exposure levels in the area, including oxygen and carbon monoxide.

Matériel (parts) employees are usually tasked to label and store chemical products upon initial receipt. These employees must be able to understand the **labeling system** well and know which chemical products should be segregated from others. Oxidizers, for example, such as those commonly found in aircraft sealants, should be kept away from oxygen cylinders or oxygen generators. Exposure to excess oxygen increases the rate of oxidation and can cause extreme heat from the rapid decomposition (chemical reaction) of the product. Employees who need additional information about a chemical product have the right to request and receive the MSDS for the product within a reasonable period of time.

Explosive devices are also present in aircraft maintenance and must be stored and handled carefully. In fact, there are specially made storage cabinets for such devices. For aircraft, in addition to flares, there are two other types of devices that are considered explosive. The first is called a **squib**; it is used to activate the engine fire-extinguishing system. The other is called an initiator, which serves as a "blasting cap" for fire bottles located near the fuel tanks used to release the extinguishing system in the fuel tank's vent or wing surge tanks. Both types of devices should be controlled by the **shelf life** date stamped on the device by the manufacturer. Once the device has reached the shelf life date, it should be removed and disposed of in accordance with federal, state, and local regulations. The local police authorities will often accept and safely dispose of these units.

Fall Protection

Fall protection is often a difficult and expensive requirement for airlines, particularly for maintenance operations outside the hangar environment. Within hangars, a system of staging platforms and flexible lifelines may be used to reach virtually every area of an aircraft.

But out on the parking ramp of a hub facility or the tarmac of a small airport facility, such solutions are usually not available. To safely accomplish tasks requiring fall protection in these situations, the use of a bucket truck or and anchoring system that uses a vacuum or suction is preferred. The anchoring system, which is often a solution for in-hangar maintenance as well, can be applied by single users or used in combination to attach a flexible lifeline that can span the length of the hull or wings.

Fuel Cell Entry

Fuel cell entry presents another unique hazard to the airline industry. Entry into aircraft fuel cells is considered a confined space entry as defined by the 29 CFR 1910.146, Standard for **Permit Required Confined Spaces.** Because most of these spaces are very small, entry presents a significant challenge. An additional consideration is the hazard associated with the fuel from the wings and center tanks of aircraft. Non-sparking tools and monitoring equipment are necessary to keep fire and health hazards under control.

Prior to work in a fuel cell, the fuel must be completely removed and the tanks adequately vented to remove remaining vapors. Venting is accomplished through an integrated venting system or the use of an air-flow system. Under the permit system, only authorized and trained employees may enter the permit space. An attendant is also required outside the space. The attendant's duties are to monitor the entrants and perform all attendant duties assigned in the permit space program. This overview of a permit program or fuel cell entry program provides only a few of the many details that cannot be covered by a few paragraphs in this text. Specific procedures for a program should be tailored to the needs of the organization.

Control of Hazardous Energy (Lockout/Tagout)

The OSHA Standard for the Control of Hazardous Energy, otherwise known as the **Lockout/Tagout Standard**, is found in 29 CFR 1910.147. This standard is most applicable in aircraft maintenance, but is also important in ground service equipment maintenance. The purpose of this standard is to specify requirements for isolating sources of energy for purposes of maintenance or work in areas where energy sources present a danger to employees. These energy sources include electrical, hydraulic, pneumatic, mechanical, steam, and so on. The energy is turned off at the source and controlled by a set of locks and tags to ensure no other employee can return energy to the system.

In aircraft maintenance, lockout/tagout is mainly used in the shops for equipment and machines used to work on aircraft parts, but a tagout system is usually used to control energy inside the aircraft having to do with aircraft systems where employees will work. A tag may be placed on operating levers, inspection door, or switch, or a system fuse may be removed and replaced by a tag identifying the location where maintenance is occurring. The company must demonstrate that all authorized personnel have been appropriately trained in the use of the system and can properly isolate energy sources, and that affected employees have been appropriately trained to identify the system and refrain from removing tags or attempting to re-energize affected systems. Locks and tags may only be removed by the persons who apply them.

Aircraft Jacking

The jacking of an aircraft can be a dangerous task if not handled with the appropriate precautions. The procedures vary by aircraft type, and the manufacturer's prescribed procedures must be followed. Therefore, only general safety precautions and procedures related to jacking will be discussed.

Aircraft jacks, usually of a tripod style, are used to lift aircraft for maintenance purposes. Before each use, the jacks should be thoroughly inspected to ensure they can handle the capacity (weight) to be lifted, the proper functioning of the safety locks, condition of the pins, and an overall serviceability check. Fuel may need to be transferred or removed for appropriate weight and balance requirements. No equipment should be around the aircraft and no one should be on board, unless the maintenance manuals specifically require such practice to observe leveling instruments in the aircraft. The aircraft should be positioned on a sturdy, level surface protected from the wind. The jacking positions on the aircraft are specified by the aircraft manufacturer's maintenance instructions.

An overall assessment of the area and configuration of the system to check for hazards should be completed prior to lifting the aircraft. Jack pads ensure that the aircraft load is properly distributed

Case Study: Implementation of a Behavior-Based Safety System

Introduction

This case study demonstrates the effectiveness of the implementation of a behavior-based safety system within an organization. The setting is a large organization that includes a major global commercial cargo and freight air carrier service.

Before Behavior-Based Safety

The organization has been around for many decades and has rapidly become a leader in safety over the last 20 years. With programs and processes being implemented to reduce injuries, facilities, and equipment, the organization had made great gains through engineering of equipment and facilities that reduced material handling in some cases and, in others, eliminated it altogether through systems automation. However, there was still a large human element in the equation, which accounted for at least 92 percent of the injuries experienced on an annual basis. And, while the training programs established were good and continuously evolved for improvement, something had to be done to address the human element. Injury rates were in the double digits, and compensation costs were astronomical.

The First Round

The company depended on their insurance carrier for much of their loss prevention and turned to them for help in developing and implementing a behavior-based safety system. The corporate training function collaborated with insurance carrier consultants as well as front-line safety committee members. Their goal was to design each element of the process, review root causes to determine contributing behaviors, and develop effective training for implementation. The initial installation included forms for each loss source type (slips/falls, lift/lower, etc.). While in theory this was a good approach, there were many issues that led to much-needed improvements. Furthermore, for each loss source form, there were several behaviors included (8–12). The wording for each behavior was too lengthy, and having so many behaviors on the form made it cumbersome for the observers. The result was that many didn't follow the process, and of those that did, quality was usually lacking.

The training that was provided had included some good elements, such as an explanation of why the process was needed, a look into human behavior, and the methods for conducting observations and providing effective feedback. There were even workshops included for developing feedback skills. However, because of the size of the organization, the training was only provided to selected personnel. This typically included managers and safety committee chairpersons. Unfortunately, first-hand training was not consistently provided to the full- and part-time supervisors who would be conducting the bulk of the observations in the operations.

All wasn't bad, though, as this was at least a good start to a daunting task for such a large organization. As the saying goes, "It takes a lot longer to turn around a big ship than a small boat." The organization did experience some positive results, with slight downward trends in injury frequencies. But it wasn't quite the result that was anticipated and it wasn't an acceptable result for the time and dedication devoted to this process.

The Revised Process

After several seasons of slow but continuous improvement, the organization's leadership wanted to see a larger return on their investment. Again collaborating with its insurance carrier, the organization revised its process to focus on more specific behaviors which would help streamline it for the observer, improving quality and efficiency. A review of each facility and each operation's worksite analysis would be conducted to determine their most frequent and severe loss sources. Looking even deeper at specific root causes for each job type and other aspects such as the day of the week, equipment involved, and so on would further improve the behavior selection. Shortening the behavior statement and limiting the number of behaviors to 4–6 made the process more efficient, allowing observers to focus on and understand a few key behaviors that were contributing to their losses.

The organization also started providing training to front-line supervisors, who conduct the majority of the observations. While most would receive a two-hour classroom training session, including an observation practicum and feedback workshop, many would also receive one-on-one training in the operation. Quality checks on observation forms were integrated, and those identified as "unacceptable" were provided additional training or mentoring.

The Result

A few big positives came from implementation of this process. First, it got everyone focused on key behaviors and safety expectations. Second, while initial results were small, over the first ten years of implementation, the organization experienced an overall reduction in injury frequency of almost 61 percent. Morale improved and safety committee members were empowered through the process of reviewing their worksite analysis, developing the observation forms (with input from the supervisors in their area), and conducting observations. While the huge cost savings could be cited, the real savings were in the number of employees who were able to return to their homes in the same condition as when they left.

at the jack points. If other work is being performed on the aircraft, a check should be conducted to ensure stress panels or plates are in place to avoid structural damage. The jacks should be extended until they contact the jack pads and checked for proper alignment. Most accidents occur because of misaligned jacks.

When all is ready for lifting, a maintenance technician should be positioned at each jack with a crew leader positioned visible by each technician. The crew leader's primary responsibility is to ensure the jacks are operated simultaneously to keep the aircraft level and to avoid overloading any of the jack points. Before the aircraft is lowered, the crew leader ensures that all personnel, work stands, and equipment are clear of the area around the aircraft and that the landing gear is down and locked, with all locking devices properly installed.

Chapter Summary

This chapter discussed some of the basic elements of ground safety such as ergonomics, aircraft ground damage, and hazardous materials, and the role they play in the primary job functions typically found in the aviation industry, most of which are in the commercial airline setting. Some of these elements overlap several job functions, but were discussed in some detail in the subsection dealing with the job function for which they are typically more important. For example, hazardous materials, although used in a few different job functions, are typically most relevant to aircraft maintenance. For each of the job functions, the most common risks and hazards were discussed in some detail, with key terms noted and defined. Key equipment used to minimize risk was also described.

Flight Safety Programs

Daniel Krueger, Michael Ferguson, and Brandon Wild

Chapter Learning Objectives

After completing this chapter, the reader should be able to:

- Explain the regulatory requirement for Part 121 air carriers to have a Director of Safety.
- Describe the role of the Director of Safety at Part 121 air carriers.
- Discuss the primary components of a Flight Operational Quality Assurance Program (FOQA).
- Describe the basics of an Aviation Safety Action Program (ASAP).

Key Concepts and Terms

Aviation Safety Action Program (ASAP)

ASAP Manager

Director of Safety

Event Review Committee (ERC)

Flight Operations Quality Assurance (FOQA)

FOQA Manager

Memorandum of Understanding (MOU)

Non-Sole Source Report

Sole Source Report

Introduction

In the end, flight is the goal and ultimate end product of the majority of aviation-related support activities. Consider this: In aviation, even most of the simplest and seemingly unconnected activities are ultimately conducted in support of flight operations. The action of an administrative assistant writing a memo for a superior may not have a direct connection to flight, but in aviation the lion's share of activities will usually be conducted in support of actions that in some way will be connected to producing flight. Since flight is the goal, it naturally follows that safety in flight should be absolutely paramount. Thus, *flight safety* is arguably the ultimate goal of aviation activities. This statement does not negate the importance of safety in aviation ground activities, but rather is an acknowledgment of everything working toward ensuring the safety of flight and in all supporting activities.

What does flight safety mean? Most obviously, this term can be defined to mean that aircraft accidents (crashes) do not occur and that people are not killed or injured in such events. Beyond question, avoiding and/or preventing aircraft accidents is the most significant aspect of flight safety. However, while the prevention of accidents is certainly a significant portion of the definition, flight safety is more than just avoiding accidents involving aircraft. Flight safety includes the policies, procedures, safety programs, training, organizational culture, and other important elements designed to prevent undesired occurrences such as aircraft accidents. **Figure 8–1** depicts an important flight deck procedure to ensure safety. Flight safety is a means of incorporating safety in such a way that is permeates the entire organization. This represents a direct connection between flight safety and Safety Management Systems (SMS).

Figure 8–1. A flight crew goes through their pre-departure checklist.

© Bernard van Berg/Getty Images

In general terms, within many aviation organizations such as airlines, flight safety is a separate department within the overall flight operations organization. This is not always the case, but it is rather common. A flight safety department often possesses and develops specific safety standards designed to prevent accidents from occurring. As stated earlier in this book, events and hazards that are unreported may lead to undesired occurrences that may have been prevented if they had been reported and investigated. A flight safety department cannot prevent what it does not know about. When hazards and events are reported, there must be appropriate mechanisms in place for responding to them. This point drives toward the heart of the main purpose of flight safety: the prevention of accidents and other negative events in order to make aviation organizations safer and more profitable.

The purpose of this chapter is to discuss some of the most common flight safety features and programs present in many aviation organizations. Although most of these programs and features are consistent with those of FAR 121 air carriers, some of these programs can be applicable to a wide variety of aviation companies in support of flight safety activities. While some of these programs and positions are required by federal aviation regulations, others are strictly voluntary but are nonetheless present in many aviation organizations. The fact that this is so results from a recognition that programs of this nature are beneficial to an aviation organization, its people, its assets, and even members of the general public. Flight safety must be given a high degree of priority in any aviation company as it conducts its day-to-day activities and operations.

Director of Safety

Many aviation companies such as major and regional airlines, corporate aircraft operators, and other organizations have a specifically designated person serving in the position of **Director of Safety**. For airlines operating under FAR 121, the Director of Safety is a position that must be present in order for the company to operate. FAR 119.65 states the following:

> *Sec. 119.65 — Management personnel required for operations conducted under part 121 of this chapter.*
>
> *(a) Each certificate holder must have sufficient qualified management and technical personnel to ensure the highest degree of safety in its operations. The certificate holder must have qualified personnel serving full-time in the following or equivalent positions:*
>
> *(1) Director of Safety.*
> *(2) Director of Operations.*
> *(3) Chief Pilot.*
> *(4) Director of Maintenance.*
> *(5) Chief Inspector.*
>
> *(b) The Administrator may approve positions or numbers of positions other than those listed in paragraph (a) of this section for a particular operation if the certificate holder shows that it can perform the operation with the highest degree of safety under the direction of fewer or different categories of management personnel due to—*
>
> *(1) The kind of operation involved;*
> *(2) The number and type of airplanes used; and*
> *(3) The area of operations.*
>
> *(c) The title of the positions required under paragraph (a) of this section or the title and number of equivalent positions approved under paragraph (b) of this section shall be set forth in the certificate holder's operations specifications.*
>
> *(d) The individuals who serve in the positions required or approved under paragraph (a) or (b) of this section and anyone in a position to exercise control over operations conducted under the operating certificate must—*
>
> *(1) Be qualified through training, experience, and expertise;*
> *(2) To the extent of their responsibilities, have a full understanding of the following materials with respect to the certificate holder's operation—*
> *(i) Aviation safety standards and safe operating practices;*

Source: U.S. Department of Transportation/FAA

(ii) 14 CFR Chapter I (Federal Aviation Regulations);
(iii) The certificate holder's operations specifications;
(iv) All appropriate maintenance and airworthiness requirements of this chapter (e.g., parts 1, 21, 23, 25, 43, 45, 47, 65, 91, and 121 of this chapter); and
(v) The manual required by §121.133 of this chapter; and
(3) Discharge their duties to meet applicable legal requirements and to maintain safe operations.
(e) Each certificate holder must:
(1) State in the general policy provisions of the manual required by §121.133 of this chapter, the duties, responsibilities, and authority of personnel required under paragraph (a) of this section;
(2) List in the manual the names and business addresses of the individuals assigned to those positions; and
(3) Notify the certificate-holding district office within 10 days of any change in personnel or any vacancy in any position listed.

Source: U.S. Department of Transportation/FAA

To paraphrase, FAR 119.65 outlines in general terms the requirements for an air carrier to have a Director of Safety and other required positions and briefly discusses the basic requirements of the position. It is expected (and required) under the regulations that the Director of Safety possess a clear understanding of the applicable FARs, company safety policies and procedures, and the overall operation of the organization.

The typical Director of Safety at an airline or other company is a highly experienced, knowledgeable person chosen for the position based on a combination of their working background, operational knowledge, and subject-matter expertise. In some cases, this person may be a pilot with an extensive background of flight qualifications and aviation safety experience. In other cases, the Director of Safety may have worked in a non-flying capacity in operations, safety, security, or another area. Beyond the basic requirements outlined in the FARs, each company may have its own unique minimal qualification standards for selecting a Director of Safety in accordance with their organizational and operating requirements.

In aviation organizations with such a position, the Director of Safety has the primary authority over and responsibility for many (if not all) safety programs in the company. This may include but is not limited to Safety Management Systems (SMS), Flight Operations Quality Assurance (FOQA), Aviation Safety Action Program (ASAP), occupational health/injury prevention initiatives, and other safety programs within the organization. Depending on the size of the organization, the Director of Safety will usually have a number of managers or other employees charged with specific areas of oversight and management for various safety-related programs.

Aviation Safety Action (ASAP) Programs

A common and important program present in many aviation companies is an **Aviation Safety Action Program (ASAP)**. Fundamentally, ASAP is a voluntary safety program employed by participating organizations as a means of reporting and recording information related to possible and actual safety issues and hazards. An ASAP is designed to be used by airlines operating under FAR 121 and 135, and by major domestic aviation repair organizations certificated under FAR 145. An ASAP is a partnership between the FAA, the company, and the participating employee group, and the entire program is bound in a signatory document called the **Memorandum of Understanding (MOU)**. ASAP programs are segregated by employee groups; they are most commonly available for Flight Operations, Dispatch, Maintenance, In-Flight, and Ground Operations. Each program is managed separately and requires an MOU for each employee group.

As described by the FAA in AC 120-66B, the primary purpose of ASAP is to encourage the employees of air carriers and aviation maintenance repair stations to voluntarily report safety information and potential hazards that may be critical to identifying potential issues that could lead to accidents (see **Figure 8–2**). The FAA and participating organizations have found that identifying accident "precursors" is essential to preventing accidents from occurring. Precursors are issues, problems, and hazards present before an accident occurs, which, if identified and mitigated, could prevent an accident. Rather than taking disciplinary action against an individual employee or company that voluntarily reports a problems or event, ASAP uses corrective action to address issues and provides the reporting employee certificate protection. This approach to safety improvement represents a positive shift toward changing the internal safety culture of a participating organization, as discussed in Chapter 1 of this book.

Another positive attribute of ASAP is that it provides a mechanism for the reporting, collection, analysis, and retention of the safety data obtained voluntarily. This information is used to determine the causal factors of events and issues and to assist in the creation of specific, effective corrective actions. All ASAP activities are designed to educate the appropriate persons to prevent similar events from occurring in the future.

Participation in an ASAP can be likened to a party agreement entered into voluntarily by the FAA, a certificate holder, and, if appropriate, other parties. ASAP is basically a form of safety

Figure 8–2. As pertaining to flight safety: ASAP programs rely upon reports submitted by flight crews.

partnership between the airline (and other participating certificate holders), the FAA, and some specific third parties, such as the flight operations department and/or an employee group's labor organization. Should any of the participating groups no longer be agreeable to the conditions of the MOU, they may withdraw from the ASAP at any point.

Since an ASAP is a voluntary program, it is important to assure employees they can participate in reporting safety events and concerns without fear of reprisal. Thus, ASAP is designed to correct identified problems rather than to take action against a reporting employee, who thus does not need to be concerned that the FAA or the company will use reports accepted under the program to take enforcement action against them.

ASAP Reporting and Data Collection

Most organizations having an ASAP have a specific form of electronic reporting for use by their employees. These electronic reporting systems are made readily accessible for applicable employees to use at their disposal to report safety events concerns and hazards. The data collected is usually maintained in a master repository for analysis and to determine whether events are of immediate concern due to the presence of possible significant risk and/or systemic ramifications.

The FAA requires that several criteria be satisfied in order for a report to be included in ASAP. These are outlined in AC 120-66B:

- The report must be submitted in a timely manner (generally within 24 hours of the completion of the day in which the event occurred) or whatever criteria are agreed upon between the carrier and the FAA for determining timeliness, or within 24 hours of an event that may have been outside of regulatory compliance.
- If the ASAP report involves a possible regulatory violation, the event must have been unintentional and must not appear to involve intentional disregard for safety or regulations.
- The report must not contain any false information or concern substance abuse or criminal activity.

Source: U.S. Department of Transportation/FAA

In addition, AC 120-66B indicates that the following types of reports are not included in an ASAP:

- Reports containing information on events that may be the result of intentional actions, or that appear to involve an intentional disregard for safety
- Reports containing information on substance abuse/controlled substances, falsified reports, or possible criminal activity
- Reports not submitted in a timely manner
- Reports of events that did not occur when the submitter was acting as member of the company (certificate holder)

Source: U.S. Department of Transportation/FAA

The ASAP Manager

Like any organizational initiative, in order to be effective and consistent an ASAP must have a responsible manager in charge of the entire program. In most participating organizations, the **ASAP Manager** is the primary authority for managing the program and ensuring compliance of the program with the standards established in the MOU. The ASAP Manager is not necessarily a certificated airman; however, the manager should have a good working knowledge of flight operations and how different departments within the organization work in conjunction with one another to ensure safe and compliant operations. It is also not uncommon for the organization to appoint the ASAP Manager to manage multiple ASAP programs within the organization.

An ASAP Manager is also responsible for disseminating ASAP reports and findings to company leadership and for providing timely feedback to employees who have submitted ASAP reports. The manager or a designee appointed by the manager maintains the ASAP database and also creates reports on the program to distribute to those in the company who need to know about the specifics of the program. The manager also provides oversight of any investigation under the ASAP, and also serves as one of the primary contact persons representing the organization to the FAA, NTSB, and other organizations when addressing ASAP issues. The ASAP manager will often have other duties and responsibilities in addition to those already discussed. The manager is also responsible for maintaining each program MOU, as they are generally only valid for a period of 18–24 months, and an ASAP program will become void should the MOU expire. The Certificate Holding District Office (CHDO) for the organization also has the authority to grant an extension for any given MOU should it become necessary to allow the organization additional time to prepare and seek approval of a new MOU.

The ASAP Manager is also responsible for preparing quarterly reports for the FAA, for which a template is available in AC 120-66B. The quarterly report summarizes the activity of the program, specifying the number of reports accepted into the program, and also highlights key safety issues identified in the program.

ASAP Event Review Committee

Another key component of ASAP is the **Event Review Committee (ERC)**. AC 120-66B identifies an ERC as a group that is usually comprised of a management representative from the company (certificate holder), a representative from the employee group (such as a pilot or maintenance technician), an FAA representative from the Certificate Holding District Office (CHDO), and in some cases, a representative from an employee union. The ASAP Manager typically will chair meetings of the ERC but is not considered a voting member of the committee.

The ERC will determine if an ASAP report submitted should be accepted into or rejected by the ASAP program. The ERC will review each report *de-identified* (reporter name redacted) to ascertain if the reported event was unintentional and did not involve any intentional disregard for safety. Once a report is accepted into the program, and all recommended corrective actions have been accepted and completed or rejected with explanation, the ERC will then close the report with an appropriate risk assessment.

The primary purpose of the ERC is to review and analyze reports submitted under an ASAP, identifying safety issues and hazards, and working to generate possible corrective actions to problems identified through the reports. The ERC may choose to share information pertinent to creating effective solutions to identified problems and issues. There may also be occasions where the ERC will review reports that involve different departments, and it may become necessary to invite subject-matter experts (SME) to assist the ERC in fully understanding a reported event . Anyone acting as an SME should be required to sign a confidentiality agreement to ensure the integrity of the ASAP program. SME participants can offer the ERC greater knowledge of operations and procedures concerning maintenance, dispatch, ground operations, air crew scheduling, and other aspects of the organization.

Once an ASAP report has been accepted into the program, the ERC is also responsible for placing each report into either a **sole source** or **non-sole source** category. A sole source report is defined as a report for which the reporter is the single source of the information contained within the report, and had the reporter not filed a report the event would have been unknown to the FAA and likely the company as well. It is not uncommon for 90–95 percent of an organization's ASAP program reports to be sole source, and it is typically for this reason that both organizations and the FAA participate in the program.

Non-sole source reports are those where there is existing evidence of the event outside of ASAP, and the FAA would have been aware of the event without the ASAP program. An example of this would be if air traffic control (ATC) filed a report on an event that was also reported under ASAP. The ATC report would have been filed regardless of whether an organization had an ASAP program, and it is in these non-sole source type of events that the certificate protection benefits established by the ASAP program protect the reporter, so long as the report is accepted by the ERC. Without the ASAP program, an FAA inspector would likely open an investigation and ultimately the certificated employees involved in a particular event would be at the disposition of the FAA investigation, which could potentially lead to certificate action (e.g., suspension or revocation) against the employee.

There is one other difference between sole source versus non-sole source reports as outlined in the AC: the FAA response for each category of report. The AC states that for ASAP reports that are sole source, the FAA may not respond at all because they were previously unaware that the event occurred, and so it simply wouldn't be fair for the FAA to respond to events that would have gone unreported. For ASAP reports that are non-sole source, the FAA is allowed to respond and may exercise "administrative action" against the reporter. Administrative action is very different from formal certificate action in that administrative action may only involve one of three actions against the reporter:

1. Letter of No Action
2. Letter of Warning
3. Letter of Correction

These letters are typically held on file locally at the CHDO and are expunged after a period of two years. Employees can often become confused about the difference between sole source and non-sole source, especially employees involved in non-sole source reports, so it is imperative that organizations that have ASAP programs educate their employees on the details of their respective MOU and AC 120-66B.

Flight Operational Quality Assurance

Flight Operational Quality Assurance (FOQA) is a program that uses recorded data from aircraft to improve safety and efficiency. Aircraft that are able to participate in FOQA programs are normally equipped with a digital recording device that is separate from the flight data recorder, and records similar parameters of the aircraft's flight (see **Figure 8–3**). These parameters, usually measured from engine start to engine shutdown, include, but are not limited to, airspeed, altitude, and engine RPM. Data are recorded on either a PCMCIA card or optical disk, which are removed from the aircraft during regular maintenance. With new technological advances, some newer aircraft, such as the Boeing 787, transmit data via wireless data link. The recorded flight parameter data are downloaded into a secure server and analyzed by an organization's safety department for safety of flight issues and trends. Today, most major U.S. airlines and many regional airlines have active FOQA programs.

Figure 8-3. FOQA programs utilize recorded aircraft flight performance and parameter data to improve flight safety.

FOQA and AC 120-82

In 2004, the FAA issued Advisory Circular 120-82, which offered airlines certain protections in exchange for having an FAA-approved FOQA program as well as sharing de-identified safety trends. The following sections describing the basic function of FOQA programs for U.S. airlines are taken directly from AC 120-82:

a. FOQA is a voluntary safety program that is designed to make commercial aviation safer by allowing commercial airlines and pilots to share de-identified aggregate information with the FAA so that the FAA can monitor national trends in aircraft operations and target its resources to address operational risk issues (e.g., flight operations, air traffic control (ATC), airports). The fundamental objective of this new FAA/pilot/carrier partnership is to allow all three parties to identify and reduce or eliminate safety risks, as well as minimize deviations from the regulations. To achieve this objective and obtain valuable safety information, the airlines, pilots, and the FAA are voluntarily agreeing to participate in this program so that all three organizations can achieve a mutual goal of making air travel safer.

b. A cornerstone of this new program is the understanding that aggregate data that is provided to the FAA will be kept confidential and the identity of reporting pilots or airlines will remain anonymous as allowed by law. Information submitted to the FAA pursuant to this program will be protected as "voluntarily submitted safety related data" under Title 14 of the Code of Federal Regulations (14 CFR) part 193.

(1) In general, aggregate FOQA data provided to the FAA under 14 CFR part 13, section 13.401 should be stripped of information that could identify the submitting airline prior to leaving the airline premises and, regardless of submission venue, should include the following statement:

WARNING: This FOQA information is protected from disclosure under 49 U.S.C. 40123 and part 193. It may be released only with the written permission of the Federal Aviation Administration Associate Administrator for Regulation and Certification.

(2) However, if an airline voluntarily elects to provide the FAA with aggregate FOQA data that includes airline identifying information, then it should include an additional statement that it is the proprietary and confidential property of [Airline Name].

Source: U.S. Department of Transportation/FAA

This information from AC 120-82 demonstrates that FOQA is a voluntary program, a form of partnership between the FAA and participating airlines in order to improve safety. Confidentiality is a key component of FOQA.

In-Focus: Flight Operational Quality Assurance

Brandon Wild is the Flight Operational Quality Assurance (FOQA) Manager for a large airline based in the southern U.S. He is an expert in FOQA and has guided his company's program successfully through periods of change and significant growth. Here in his own words, Brandon discusses the importance of FOQA at his company.

The great aspect of a FOQA program is that it allows our safety department to see everything that happens during a flight, conveniently recorded and downloaded for our analysis. This data allows our company to be proactive when it comes to solving small safety of flight issues before they become larger problems. After analyzing the recorded data, we are able to "get the word out" to our pilots through training programs, fleet bulletins, or flight plan remarks. FOQA is a big part of the overall Flight Safety Department, which helps our company to maintain a high level of safety.

Besides our first and foremost reason for having a FOQA program, which of course is safety, we are able to utilize data compiled from our aircraft for other purposes as well. We use FOQA data to troubleshoot engine issues for Technical Operations. We can then use these solutions to analyze archived data to detect if the same issue might be about to occur on another aircraft engine. Our airline also uses the data for fuel burn issues, such as measuring the amount of flights using single engine taxi, detecting the number of flights that are not flying at their optimum altitude for fuel burn, or finding aircraft that are burning more fuel due to engine or airframe issues.

Another great aspect of FOQA is that we use it to help improve safety and efficiency, not just at our airline, but in the industry as well. We use FOQA in conjunction with other data sources (FAA, other airlines, etc.) to do comparisons of approaches to airports using fuel efficient arrival techniques. We also partner with other airlines and Air Traffic Control to improve approaches and departures at specific airports that show a higher rate of less than optimal safety issues.

FOQA enables my airline to conduct studies and data analysis that might otherwise not be able to be accomplished. Our FOQA program is a very valuable tool that is used for safety as well as making us a much more efficient airline.

Courtesy of Brandon Wild

Figure 8–4. FOQA data provides valuable information on aircraft flight parameters which is used to predict/prevent future occurrences.

FOQA programs first began in the 1970s and quickly caught on as a proactive safety tool with international airlines. Due to security of data and legal concerns, U.S. airlines did not develop FOQA programs until the mid-1990s, and even then these carriers were rather slow to recognize the value of FOQA programs. FOQA programs are usually operated by the airline in conjunction with the union or association that represents the pilots (for companies having a labor organization representing its pilots).

Airline FOQA Personnel

Airlines generally have several key employees responsible for the management and oversight of the FOQA program, as well as for the analysis of data collected. FOQA program personnel typically include a designated **FOQA Manager**, a data analyst (or multiple data analysts, depending on an airline's size), and several pilot association representatives, who serve as "gatekeepers." Airlines that have a very complex fleet structure (operators of multiple aircraft types) may choose to have fleet-specific representatives (B777, B737, A320, etc.). These individuals will usually include a senior instructor from the particular fleet type to serve as the fleet expert and work on issues and events specific to their aircraft type.

Some airlines also choose to have a technical operations maintenance representative who can handle complex issues dealing with aircraft digital recording devices. These individuals are also able to act as liaison with the maintenance and engineering departments of the airline. The scope of an airline's FOQA program structure depends on the size of the airline, the number of aircraft equipped with FOQA recording devices, and how much information and analysis an airline chooses to gain from their program.

Use of FOQA Data

As the FOQA data are collected and analyzed, the specific flights from which the data were derived are de-identified. FOQA data is used to solve systemic issues, and does not concentrate specifically on individual flights. At some airlines, if more information is needed from a particular flight due to an issue or event that is recorded on the flight, the pilot gatekeeper, who has access to the identified data, may contact the flight crew to find out more details about the event or issue. These crew contacts are done in a de-identified manner, with the gatekeeper being the only person who knows the identities of the crew. This is to ensure the confidentiality of the flight crew in order to protect the integrity of the FOQA program.

FOQA has become an integral part of many airlines' safety programs. These programs provide greater insight into aspects of the airline's operations. FOQA is a major portion of the overall flight safety picture, enabling airlines to have a high level of safety (see **Figure 8–4**).

Chapter Summary

In this chapter, we have briefly discussed some of the more common flight safety programs that are widely used in the airline industry. We have also discussed the regulatory requirement for airlines in the United States to have a Director of Safety, and have outlined the fundamental areas of responsibility played by this person.

As mentioned at the start of this chapter, safety in flight is paramount, and both ASAP and FOQA are programs designed to enhance the safety of organizations through the collection of safety reports, data, and other pertinent information. Organizations that have active FOQA and ASAP programs have in place programs that when properly used and managed create significant benefits to the company, its employees, and the general public.

Chapter Concept Questions

1. In your own words, describe the position of Director of Safety and a Part 121 air carrier.
2. What are some of the main areas of responsibility of the Director of Safety?
3. Define and describe in detail an Aviation Safety Action Program.
4. Describe the role of an ASAP Manager.
5. What is a Memorandum of Understanding under ASAP?
6. What is an Event Review Committee? How does the ERC process work?
7. Define and describe in detail an Aviation Safety Action Program.
8. Describe the role of a FOQA Manager.
9. In your own words, how might ASAP programs make an air carrier safer?
10. In your own words, how might a FOQA program make an air carrier safer?

Chapter References

Federal Aviation Administration. 2002. *Aviation Safety Action Program (ASAP)*. Advisory Circular 120-66B.

Federal Aviation Administration. 2004. *Flight Operational Quality Assurance*. Advisory Circular 120-82.

RisingUp Aviation. 2010. Federal Aviation Regulations. http://www.risingup.com/fars/info/part119-65-FAR.shtml.

Airport Safety

William Towle and Michael Ferguson

Chapter Learning Objectives

After completing this chapter, the reader should be able to:

- Discuss some of the most important considerations around safety at airports.
- Describe the basics of FAR Part 139 and its applicability to commercial service airports.
- Discuss the significance of wildlife management programs to airport safety and describe some of the methods used to manage wildlife in and around airports,
- Explain the importance of daily airport self-inspections and some of their primary elements.
- Discuss the basics of airport driver training programs and their purpose.
- Explain the importance of airport familiarization.
- Describe the basics of airport rescue and firefighting.
- Discuss the importance of security and wildlife fencing in airport safety.

Key Concepts and Terms

Airport Emergency Plan (AEP)
Airport Familiarization
Airport Operating Certificate
Airport Operations
Airport Operations Area (AOA)
Attractants
Capital Improvement Program (CIP)
Electronic Flight Bag (EFB)
FAR Part 77
Field Condition Report
14 Code of Federal Regulations (CFR) Part 139
Frangible
Movement Area
National Incident Management System (NIMS)
Notices to Airmen (NOTAM)
Pavement Condition Index (PCI)
Porous Friction Course (PFC)
Runway Incursions
Runway Safety Area
Spalling
Snow and Ice Control Plan (SICP)
Wildlife Hazard Management Plan
Wildlife Hazard Assessment

Introduction

With respect to airports, safety is a wide-ranging and very important topic encompassing virtually all of what is called **airport operations**. Most airport operations tasks are completed in order to maintain and operate the airport as safely as possible. Safety is an issue for all airport managers across the United States, and the FAA is constantly working on ways to enhance safety at airports. To ensure a safe airport, there are many things that need to be taken into consideration. Although only commercial service airports are required to comply with FAR Part 139, all airports should abide by the safety standards outlined in FAR 139. This chapter will provide a brief overview of some of most prominent elements that should be present to ensure safety at airports.

FAR 139

In the United States, safety standards for commercial service airports are governed under **14 Code of Federal Regulations (CFR) Part 139**, also referred to as FAR Part 139. Under Part 139, the FAA issues operating certificates to U.S. airports that feature commercial air carrier service. Specifically, the FAA issues operating certificates to the following: airports having scheduled and unscheduled air carrier service employing aircraft containing more than 30 seats; airports that have scheduled air carrier operations in aircraft with more than nine seats but fewer than 31 seats; and any other airports the FAA Administrator requires to possess an operating certificate. Thus, the key principle in determining the applicability of FAR Part 139 to a given airport is the presence of some level of commercial air carrier service.

The main purpose in the issuance of an **Airport Operating Certificate** is to ensure safety in aviation operations. Since every successful flight begins and ends at an airport, it is very important that airport operators and managers maintain at least minimally acceptable levels of safety at their respective facilities and airfields. To be awarded an operating certificate by the FAA, an airport must adhere to and maintain specific operational and safety standards, including daily airfield inspections; maintaining aircraft movement areas such as runways, taxiways, and ramps; airport rescue and firefighting units, and many other areas. The minimal requirements for airport certification under FAR 139 vary depending on the size of the airport and the type of operations conducted.

Because of the status of FAR 139 as a regulatory set of standards, it is important that all commercial service airport managers and operators take measures to ensure compliance. Also, FAR 139 standards are intentionally designed to make the airport operating environment as safe as possible. Following are some of the specific practices that need to be in place for FAR 139 certificated airports.

Airport Emergency Plans

Although every airport should have an **Airport Emergency Plan (AEP)** in place, commercial service airports operating under FAR Part 139 are required to have an active AEP in place. An Airport Emergency Plan is a document that addresses essential emergency-related actions planned to ensure the safety of and emergency services for the airport populace and the community in which the airport is located. An emergency or an incident at an airport can occur anytime and anywhere. Depending on the type of event, the emergency may be short-lived, lasting only a few minutes, or it may go on for several days; similarly, incidents may vary greatly in severity. The important consideration is that while emergencies can seldom be exactly predicted, they can and must be prepared for. The following sections cover some of the main elements of an effective AEP.

A good AEP begins with planning to better understand the needs of an airport emergency and what the role and responsibilities are of each responding agency. Depending on the type and extent of the event, there are many agencies that may be involved in responding to an emergency, especially at a commercial service airport. These agencies will often include the Federal Aviation Administration (FAA); the National Transportation Safety Board (NTSB); members of local law enforcement, firefighting, and rescue; the American Red Cross; and local hospitals. If the event is suspected of being security-related or the result of possible criminal activity, the Transportation Security Administration (TSA) and the Federal Bureau of Investigation (FBI) may be involved in the response and subsequent investigation. The AEP planning process brings these groups and agencies together to discuss what resources each will bring to an emergency.

Effective in 2011, the FAA requires that the **National Incident Management System (NIMS)** be integrated into the planning of an Airport Emergency Plan. NIMS provides a systematic, proactive approach to guide departments and agencies at all levels of government, nongovernmental organizations, and the private sector to work seamlessly to prevent, protect against, respond to, recover from, and mitigate the effects of incidents, regardless of cause, size, location, or complexity, in order to reduce the loss of life and property and harm to the environment.

During the planning process, agencies become familiar with a specific airport and how it operates. It is extremely important that the agencies involved, especially law enforcement and fire, understand the layout of the airport and where various entrance and exit points are located so that swift ingress and egress during an emergency can be expedited. If the airport is fenced, gaining access through locked gates should be discussed and made part of the plan. An airport emergency action plan should include the roles and responsibilities of each participating agency for various emergencies such as aircraft accidents, bomb threats, or hazardous chemical spills. The plan should also include how to handle the media and friends and families of accident victims.

Depending on their size, commercial service airports are required to conduct a full-scale onsite emergency exercise either every year, or a full-scale exercise every third year and a "tabletop" emergency exercise every year. It is advisable that smaller airports follow a similar schedule so the participants in an airport emergency will be familiar with the airport layout and are prepared ahead of time. More information on AEPs can be found in the FAA's Advisory Circular 150/5200-31C, *Airport Emergency Plans*.

Airfield Inspections

While some hazardous conditions may develop virtually instantaneously, others can occur more gradually. It is important that the airport operator have an airport safety self-inspection program that monitors specific airfield conditions in order to identify unsatisfactory conditions and develop prompt, specific corrective actions. Airfield self-inspections are required under FAR 139 and must be conducted daily. At a minimum, an airfield self-inspection should include a visual observation of the ramps, runways, taxiways and other aircraft movement areas to locate any foreign object debris (FOD) that may pose a damage threat to aircraft (see Chapter 6 for more information on FOD). However, during a more detailed inspection of an airport, consideration is given to pavement condition, airfield lighting, signage, runway, taxiway and ramp markings, and many other critical areas. Not only are daily airfield inspections required for compliance under FAR 139, they are an inherent part of ensuring that airports are as safe as possible for all aircraft operations.

Daily self-inspections are extremely important. A self-inspection program should be established to ensure the airport is operating safely. In addition to the items discussed above, there are several other things to consider when conducting daily inspections. The **Runway Safety Area**, the areas off the ends and sides of runways, should be inspected daily to ensure no ruts or erosion is present due to a vehicle or aircraft exiting the runway. Also, the only objects that should be present in the runway safety area are required lights and signs; no other object should be present in these areas. The daily inspection should also include examining all markings, lights, and signs to ensure they are highly visible and are not obstructed by snow or vegetation.

All on-field pilot aids such as wind cones, precision approach path indicators (PAPIs), and the airport rotating beacon should be inspected to ensure they are working properly and are obstruction free. The inspection of these types of aids may have to be done at night in order to ensure the visibility of lights.

In cold weather climates, snow and ice can become an issue at airports. Thus, a daily airport self-inspection should include looking for any snow and ice accumulations that might affect the safety of aircraft operations. If it is found that snow or ice is present on runways, a braking action test should be conducted using an FAA-approved friction meter, if such a device is available. Additionally, a field condition report should be completed and conveyed to airport tenants indicating the current conditions of the airfield, such as the depth and type of any snow and ice. Re-inspections should be performed and updates issued anytime field conditions change rather than doing this only once a day.

Using a well-developed checklist specific to the airfield in question is a good practice to ensure a proper self-inspection. Documentation of each inspection is required under Part 139 as evidence that the inspection took place and for liability reasons. An airfield inspection should include noting the number, type, and location of wildlife that might be present in and around the airport. This type of recordkeeping demonstrates safety proactivity on the part of the inspector. Actions of this nature will help in the protection of aircraft, pilots, and passengers at an airport.

Airfield Lighting, Signs, and Markings Another critical component of airfield self-inspection is assessing the condition of all airfield lighting. This includes but is not limited to runway and taxiway lights, approach lighting systems such as those associated with instrument landing systems (ILS), and ramp lighting. The safety of aircraft and vehicle operations during twilight and night operations is enhanced by ensuring that all lighting systems on the airfield are fully functioning and adequate in nature. Airfield lights and lighted signs are critical to wayfinding at airports since they are the primary night visual aids to help pilots and others identify where they are and where they are going on the airfield. Lights and signs have specific colored lenses that differentiate between taxiways and runways, so it is imperative that pilots be able to see these lights and that each contains the correct color lens. If a light is burned out or inoperable, the airport operator must repair the light as soon as possible. Lights should be inspected to make sure they are operable and that no fixtures are broken. Inspection of lights may have to be conducted at night, when they are most visible.

There are two basic types of lights present on airfields: inset pavement lights and those mounted on **frangible**, or "breakaway," mounts. All lights and signs mounted above the ground must be secured with a frangible mount, a coupling that secures the light in place at the ground, but is made to break away easily in the event an aircraft deviates off a runway or taxiway. FAA Advisory Circular 150/5220-23 describes frangible mountings in more detail.

Figure 9–1. Example of an airfield sign.

Airport signs are color-coded to indicate the location of taxiways, runways, and other important areas of the airfield, as well as to serve as positional and directional indicators (see **Figure 9–1**). There are a variety of different signs used on most airports to assist pilots and others in navigating the airfield safely and efficiently. Although colored lights and signs may be redundant, redundancy, in concert with pavement markings, provides an additional level of safety in the event a light or a sign is broken, temporarily out of service, or inadvertently covered by snow or grass.

Painted airfield surface markings are another component of wayfinding on an airport. Runway markings are painted white and taxiway markings are painted yellow, thus indicating to a pilot what type of surface they are on. A good self-inspection program will include night inspection to ensure that painted surface markings are visible. Pavement markings should be installed with glass beads so they are reflective at night. Markings that cannot be seen by pilots and others operating on the marked surfaces are useless. This is true of lights and signs as well. In the northern climates, snow and ice should always be removed from and around the lights and signs. Weeds and grass should be trimmed so the lights and signs do not become obstructed from view. Over time, lights will dim and sign panels will fade. Lights that have lost their effectiveness but have not yet burned out should be proactively changed, as should sign panels that have faded.

Airfield Pavement Self-inspections and Maintenance Another important area of consideration is the condition of airport pavement areas. The surfaces of runways, taxiways, ramps, and aircraft parking areas should be inspected daily for cracks, holes, and any type of **spalling** or other visible deterioration. Spalling refers to the physical deterioration of paved areas such that the surface of the material is in a state of "crumbling" or is fractured appreciably. As a part of any daily airport safety inspection, cracks, areas of spalling, or other problems on paved surfaces should be logged and reported right away so that proper repairs can be made in a timely manner. Paved surfaces such as runways are especially important to keep in a state of proper repair due to aircraft takeoff and landing operations.

Inspections of pavements should also be conducted during and after a rain event to see if any ponding has occurred. Water or "birdbaths" on the runway after a rain event can be extremely unsafe and can cause an aircraft to hydroplane and possibly lose directional control. **Porous friction course (PFC)** and grooving can be done to the pavement to help eliminate the possibility of hydroplaning during a rain event. PFC allows water to drain through the pavement; grooving channels the water through grooves below the grade of the runway. Both of these systems allow water to drain from the runway, thus eliminating ponding. If a runway is grooved, the airport operator should inspect the grooves to ensure they are not blocked with sand, rubber deposits, or paint that can hinder the flow of water. When inspecting pavements while it is raining, the inspector needs to ensure that water is able to run off properly and is not ponding at the edges. Vegetation on pavement edges should be low enough to allow proper drainage of pavement surfaces, but the pavement edge should not be more than three inches higher than the adjacent ground.

Airfield surface maintenance activities such as crack repair and sealing and seal coating of asphalt are cost-effective ways to prolong the life of airport pavements and are part of a good pavement maintenance program. Spending a nominal amount of money annually on preventive maintenance, including crack repair and sealing, may reduce the need to pay larger amounts of money to replace or resurface asphalt runways, taxiways, and ramp surfaces sooner than may have been necessary. These types of actions will allow the pavement to perform better over a longer period of time for less cost. Such a program should include an inspection every year to determine the **Pavement Condition Index (PCI)**, which evaluates the condition of an airport's pavements so the airport operator can determine what pavement needs attention. This practice also helps the airport operator in estimating future maintenance costs and in planning which pavement areas need improvement the most. For example, an airport can plan to crack-seal every other year and monitor the PCI, which will allow the operator to better identify when the pavement needs to be replaced. Pavement life will most likely be extended by an effective pavement maintenance program.

Airport Support Equipment Maintenance

All airport support equipment used for operations, snow removal, mowing, and other important functions should be inspected prior to being placed into service for the season (see **Figure 9–2**). In addition, any equipment that is used regularly needs to be inspected daily as a means of ensuring the equipment is maintained in a safe and consistent manner and to keep it in good working order. On occasion, it will be necessary to replace aging equipment or to obtain new equipment. For planning purposes, an airport manager should develop a **capital improvement program (CIP)** to plan for equipment replacement. A CIP is an economic tool used to plan for the acquisition of

Figure 9–2. A deicing unit in operation.

expensive equipment. Due to the high cost of new equipment, an airport operator should create a plan for procurement of and payment for new or replacement equipment, and the plan should also identify when existing units need to be taken out of service based on age, cost, and usage. This can all be accomplished with a properly developed and managed CIP.

Recordkeeping

It is extremely important for an airport operator to keep accurate, up-to-date records of everything. All self-inspections, field condition reports, employee training, and other important activities should be carefully documented to ensure accuracy and compliance. Keeping accurate records is especially critical in the event of an accident at the airport. Recordkeeping helps reduce the probability of the airport operator's liability in the event of being challenged in court. Aside from liability protection, recordkeeping also enables the airport operator to keep track of when tasks were completed in order to track when a certain procedure needs to be performed again, whether daily, weekly, monthly, or annually. Self-inspections should be documented and the documentation maintained at the airport for a period of at least 12 calendar months from the time the inspection is conducted.

Wildlife Management

Management of wildlife and wildlife attractants on or near airports is extremely important to ensuring the safe operation of aircraft operating on or in the vicinity of an airport. There are many wildlife hazards on airports, but the most common are those posed by the presence of birds. According to the 2005 edition of the FAA's *Wildlife Hazard Management at Airports*, birds were involved in over 97 percent of all wildlife strikes by aircraft, and less than 20 percent of all wildlife strikes are reported. Bird strikes are among the greatest safety concerns to both airport operators and pilots, and it is the responsibility of the operator to keep birds and other wildlife away from landing and departing aircraft. Although one bird ingested into an aircraft engine may not be enough to cause it to fail, a whole flock certainly has this capability (see **Figure 9–3**). The size of the bird is another consideration when determining how great a hazard it is to an aircraft. A four-pound bird traveling at 200 mph (relative to aircraft speed) has approximately the impact force of a 2½-ton truck. Mammal strikes are less common, but can typically cause more damage. Thus, airports located near deer populations should construct fencing since deer/aircraft collisions can be very serious or even fatal.

There are many different ways to keep wildlife away from airports; one of the most important is to keep wildlife **attractants** away from airfields. As implied by the name, attractants are anything that attracts wildlife to an airfield, typically food, water, and shelter. Managing attractants is an important component of wildlife management; it can range from reducing agricultural activities around airports that draw birds and mammals, to reducing the amount of water located on or near an airport. According to the FAA's Advisory Circular 150/5200-33B, *Hazardous Wildlife Attractants on or Near Airports*, hazardous wildlife attractants should not be permitted within a five-mile radius of the airport to protect the approach, departure, and circling airspace.

Figure 9–3. Geese, such as these seen at Ronald Reagan National Airport, can be very hazardous to aircraft taking off and landing.

However, it may be impossible to keep hazardous wildlife attractants from the airport, and the goal is not to make the areas at and around the airport a wildlife desert, devoid of any animals. Rather, the goal is to manage and mitigate the hazards posed by wildlife as much as possible. Typically, a **wildlife hazard assessment** will be completed to help understand the different types of wildlife that are present at and around a given airport.

Under FAR 139.337, a wildlife hazard assessment must be conducted by a wildlife damage management biologist who has professional training and/or experience in wildlife hazard management at airports, or by a person working under direct supervision of such an individual. The wildlife hazard assessment must contain at minimum (paraphrased from 139.337):

(1) An analysis of any events at the airport that prompted the assessment
(2) Identification of all wildlife species observed and their numbers, locations, local movements, and daily and seasonal occurrences
(3) Identification and location of attractants at or near the airport
(4) A description of wildlife hazards to air carrier operations
(5) Recommendations for reducing identified wildlife hazards to operations

Source: U.S. Department of Transportation/FAA

Once the wildlife assessment is completed, the airport will complete a **wildlife hazard management plan,** which is a document that will incorporate best practices on how to control the wildlife on the airfield. When wildlife is found on the airfield, there are many different ways to deal with or control the hazards. Using dogs to chase off birds, treating grass with chemicals, using "cracker shells" to frighten wildlife, covering ponds with grid-wire systems, removing habitat, and trapping are just a few of the methods that airports around the country have to control wildlife. Euthanizing is another way to control wildlife, but this is typically a last resort when other forms of control have failed. Permits are often required before euthanizing birds or mammals. The U.S. Department of Agriculture's Wildlife Services Agency is a good resource for information pertaining to managing wildlife at airports.

In-Focus: Airport Wildlife Management

John Ostrum is the Manager of Airside Operations at a large hub airport in the Midwest. He is a renowned expert in the field of airport wildlife management. In this brief excerpt from an interview conducted in the summer of 2010, John describes some important elements of wildlife management at airports.

One reason wildlife management is important at a Part 139 airport is simply because it's a requirement. The FAA defines very clearly what they call *triggering* events. Examples of triggering events are multiple aircraft strikes with wildlife, damaging strikes, numbers of wildlife capable of causing these, and as far as I'm concerned there is no airport in the country that doesn't meet those standards. So that means the FAA is saying to certificated airports that they have to have wildlife hazard assessment and from that comes the wildlife management plan. The FAA clearly defines the requirements of a wildlife hazard management plan. Once your airport has a wildlife assessment (which is a 12-month look at all the things happening at your airport and around your airport up to five miles out), you develop a wildlife management plan. The management plan is what the airport is telling the FAA how they are going to mitigate any hazards identified in the wildlife assessment and in the management plan.

As far as some of the components of a wildlife management plan, you're looking at wildlife types, populations, and attractants, not only on your airport but around your airport. So obviously if you look at the issue from a Wildlife 101 perspective, the first thing you have to ask is: What problems do I have? And the second question is: What problems may I have? And so when you look at those factors they are part of the processes or surveys used in developing your management plan. Basically the FAA wants you to look at your airport year-round. Whether it be a once a week survey or every two weeks, the FAA wants it done on a regular basis. The FAA requires regular assessment activities by qualified airport wildlife management biologists. The FAA has identified education requirements and experience standards. The basic requirements are that you have to meet the OPM Office of Personnel Management 486 Biologist requirements. The individual has to have gone through a training program (a minimum of three days). They also have to have completed at least one wildlife management assessment under a qualified biologist and they have to know the different species and patterns.

Wildlife look for three things: food, water, and shelter. If you have wildlife on your airport they are there for one or more of the three. Wildlife can be a hazard and do pose a risk, but they are looking to survive. Understanding why they are at your airport is important. Once you understand these things, then the biologist's job is to make recommendations. What can you do to mitigate these hazards? The recommendations made should take into consideration population management, habitat management, and methods of managing problem species. The wildlife survey also considers the airport's wildlife strike history based on information contained in the National Strike Database. The wildlife assessment is a one-year picture in time of the status of your airport from a wildlife perspective.

The intention of a wildlife management program is not to make the airport a biological desert. That is impossible. You are never going to sterilize the airport to where you have no animals. Develop a plan, identify the problems and how to deal with them, and determine what resources you have for dealing with the problems. That is a wildlife management plan in a nutshell. Wildlife management is an integrated approach, because no one tool will do everything.

Courtesy of John Ostrum

Airport Familiarization

It is very important for an operator and any users of an airfield to be familiar with the surroundings at an airport. **Airport familiarization** is the practice of learning the layout of the airfield and where things are located within what is called the **Airport Operations Area (AOA)**. The AOA consists of the runways, taxiways, ramps, and roadways located in the operations portion of an airport. The driver of a vehicle or the flight crew of an aircraft operating in the AOA must be familiar with the layout of the runways and taxiways so as not to become disoriented and end up in an area where they are not supposed to be (see **Figure 9–4**).

Additionally, a Part 139 certificated airport must provide specific training to anyone who will operate a vehicle within the **movement area** of the airport, which consists of runways and taxiways at towered airports that a vehicle operator must receive permission to enter. The required training consists of airport familiarization, movement and safety areas, airport communications, and other subject areas required under Part 139. In an effort to further minimize runway incursions, the FAA continues to work to find ways to make airports less complicated and easier to navigate so a vehicle

Figure 9–4. Some airports, such as LAX shown here, can be confusing to the unfamiliar pilot.

does not end up on a runway with the potential for a collision with an aircraft. Anyone driving on the AOA needs to become familiar with the runway and taxiway layouts. Vehicle operators also need to understand airport markings and lighting, as well as where the navigational aids are on the field. Drivers should be able to find landmarks in the event they become disoriented so they can re-orient themselves while looking at a map. The next section briefly covers required training for drivers operating in movement areas.

Airport Driver Safety Training

One of the most important safety programs at an airport is the driver training program. This program provides annual training to anyone with driving privileges in the AOA of an airfield. Since airport operations employees often need to drive vehicles on runways and taxiways to conduct airport safety self-inspections, repair lights and signs, or manage wildlife, it is imperative they know how to drive safely on these surfaces. In an effort to minimize runway incursions, surface incidents, and ultimately accidents, the FAA requires every person who is allowed to drive on the movement area of an airport to complete an annual driver training program.

The training program should consist of airfield familiarization to ensure that each driver knows the layout of the airfield. This will help them avoid becoming lost on the airport and stay oriented at all times, thus enhancing their situational awareness. The program should also include training on airfield markings, lighting, and signs so the driver can identify the wayfinding markings on the airfield. There are special procedures for entering a movement or safety area, and these need to be included in the training so that drivers do not inadvertently enter the movement area without prior permission from controllers.

Another important piece of the driver training program is airport communications. Proper use of the radio and aviation jargon is extremely important when communicating with the air traffic control tower at a commercial service or general aviation airport. Thus, the basics of proper airport radio communications will be a required portion of any complete driver training program.

Airport Safety Areas (FAR Part 77)

Many things can affect the airspace above and around airports. Objects affecting navigable airspace may not necessarily be located on the airport. **FAR Part** 77 identifies various imaginary surfaces that are required to be kept clear by the airport operator to ensure the safety of an airport and its associated airspace. If an airport has a precision approach instrument landing system (ILS), the imaginary surfaces defined in FAR 77 might be as far as 50,000 feet (9.46 miles) away from the airfield. It is up to the airport operator to ensure that no obstructions that could pose a hazard affecting the safety of aircraft penetrate these imaginary surfaces. Since most applications for tower and antenna construction are submitted to municipalities and/or counties, the operator should meet with the applicable city and county jurisdictions to ensure they are aware of the airspace requirements. This will help reduce the number of undesired tower construction projects that could affect an airport. The daily self-inspection process can help identify potential hazards relating to the imaginary surfaces before they become real problems. Obstructions such as trees growing off the end of a runway may take years to penetrate the imaginary surface, but can be a real hazard if their growth is left unchecked.

Runway Incursions

One of the most serious safety threats at airports involves **runway incursions**. According to the International Civil Aviation Organization (ICAO), a runway incursion is "any unauthorized intrusion onto a runway, regardless of whether or not an aircraft presents a potential conflict." The FAA adopted this same definition in 2008.

Under the ICAO definition, runway incursions can take many forms. Basic examples would be an aircraft, vehicle, or person moving onto a runway or taxiway without receiving clearance from air traffic or ground control prior to entry. Runway incursions can be highly dangerous, and numerous fatal accidents have occurred as a result of such events. While most runway incursions do not result in accidents, they represent a very real threat to aviation safety, and the FAA has made incursion prevention a priority.

Incursion Categories

According to the FAA website on runway safety, there are four categories of runway incursions:

- *Category A:* A serious incident where a collision is narrowly avoided
- *Category B:* An incident in which separation decreases and there is a significant potential for collision, which may result in a time-critical corrective/evasive response to avoid a collision
- *Category C:* An incident characterized by ample time and/or distance to avoid a collision

Source: U.S. Department of Transportation/FAA

- *Category D:* An incident that meets the definition of runway incursion such as improper presence of a single vehicle/person/aircraft on the protected area of a surface designated for the landing and takeoff of aircraft but with no immediate safety consequences

Source: U.S. Department of Transportation/ FAA

Prevention of Incursions

Because aviation operations at airports involve aircraft, vehicles, markings, signage, and people working in a number of different capacities, it is important to note that incursion prevention involves a concerted effort on the part of all parties operating at airports. There is no "one-size-fits-all" solution to runway incursions due to the variety of operations, situations, issues, and people involved. Thus, preventing runway incursions is a shared responsibility among pilots, aircraft operators, air and ground traffic controllers, and vehicle drivers. Automated warning systems, education, and situational awareness are some of the primary keys to preventing incursions.

The FAA's Runway Safety Management Strategy In their efforts to prevent runway and taxiway incursions, the FAA focuses on a number of key elements in the runway safety management strategy. This is a multifaceted approach to runway safety featuring outreach, awareness, and infrastructure and technology improvements. *Note*: The majority of the following was derived and paraphrased from the FAA website on runway safety.

Outreach to Pilots The majority of runway incursions are caused by pilots deviating from regulations and air traffic control instructions. FAA research of taxi clearances found that clearer and more explicit instructions are needed from controllers to pilots. As a result, the FAA has issued new requirements for controllers to give explicit directions to pilots on precise routes to taxi from gate to runway. The new requirements also include a directive that all aircraft must have crossed all intervening runways before receiving their takeoff clearance.

The FAA also encourages airport operators to construct perimeter roads around the airfield to minimize the number of times vehicles are driven across taxiways and runways, thus reducing the probability of incursions by vehicles. The FAA has also updated standards for runway markings and signs.

The FAA has published a booklet for pilots discussing safe communication procedures at towered and non-towered airports, and has also partnered with the Aircraft Owners and Pilots Association in creating two online courses to help better educate pilots on runway safety. The FAA has also expanded the role of the Flight Service Station by providing runway safety information to pilots at towered and non-towered airports. FAA safety inspectors also verify that pilots have current surface movement charts (airport diagrams) available and that these charts are actually used by flight crews.

Situational Awareness To promote greater situational awareness about runway safety, the FAA has taken several important courses of action. They have produced DVDs highlighting safe surface operations and communications procedures for general aviation and commercial pilots. The FAA is also developing re-creations of actual incursions to enhance discussion and understanding of these events among air traffic controllers. Some airport managers and fixed-base operators (FBO) actively participate in Runway Safety Action Teams to address runway safety issues pertinent to their specific airport. The FAA also requires driver training programs for all airport operators and employees who access airfield movement areas at commercial airports.

Technology The FAA has made several technological improvements to enhance runway safety at U.S. airports, including the following:

- Airport Movement Area Safety System (AMASS), a radar-based system that tracks ground movements and generates automatic visual and audio alerts to controllers when the potential for a collision is detected on runways and taxiways. As of 2009, the FAA had installed AMASS at 34 of the busiest airports in the United States.
- Airport Surface Detection Equipment, Model X (ASDE-X) is a precise surface detection system. The positions of aircraft and vehicles equipped with ASDE-X can be detected and observed on displays by controllers, providing real-time location information. ASDE-X integrates data from a variety of sources, including radars, transponder multilateration systems, and Automatic Dependent Surveillance–Broadcast (ADS-B) to provide accurate target position and identification information. As of 2009, ASDE-X is being implemented at 35 of the busiest airports in the United States.
- Runway Status Lights (RWSL). The FAA developed RWSL to increase situational awareness for aircrews and airport vehicle drivers and to serve as an additional component of runway safety. RWSL derives traffic information from surface and aircraft approach surveillance systems and automatically illuminates red in-pavement lights to alert flight crews to potentially unsafe situations. At the time of this writing, the FAA is set to deploy RWSL at the following U.S. airports: Atlanta (ATL), Boston (BOS), Charlotte (CLT), Chicago O'Hare (ORD), Dallas-Ft. Worth (DFW), Denver (DEN), Detroit (DTW), Ft. Lauderdale (FLL), Houston George Bush (HOU), Las Vegas (LAS), Los Angeles (LAX), Minneapolis (MSP), New York Kennedy (JFK), New York LaGuardia (LGA), Newark (EWR), Orlando (MCO), Philadelphia (PHL), Phoenix (PHX), San Diego (SAN), Seattle (SEA), Washington Dulles (IAD), and Baltimore-Washington (BWI).
- Final Approach Runway Occupancy Signal (FAROS). This system is designed to provide visual alerts on runway status to pilots planning to use a runway. FAROS provides aircraft approaching for landing with alerts by flashing the Precision Approach Path Indicator (PAPI) lights. At the time of this writing, FAROS is being tested at Dallas-Ft. Worth and Long Beach/Daugherty Field.
- Another technological enhancement is the **Electronic Flight Bag (EFB)**. As described by FAA Advisory Circular 120-76A, an EFB is an electronic display system for flightdeck or cabin use that displays a variety of aviation data and performs basic calculations on performance data and fuel calculations. EFB also contains important electronic data such as aeronautical charts, flight route information, weather information, and checklists. EFB provides efficient access to important data for flight crews, reduces their workload, and minimizes paperwork in the flightdeck. These factors help to enhance situational awareness for flight crews.

The Runway Safety Council Another initiative to combat runway incursions is the formation of the Runway Safety Council in the fall of 2008. The council is a joint effort between the FAA and the aviation industry to research and identify the root causes of runway incursions. The council is comprised of representatives from various parts of the aviation industry and the FAA. A working group within the council examines investigation data from severe runway incursions, conducts a root cause analysis, and presents their findings and recommendations to the council on ways to improve safety. The council reviews the recommendations, and if accepted, the recommendations are assigned to the office within the FAA and the agencies within the aviation industry best able to control the root cause and prevent further runway incursions. The council tracks recommendations to ensure appropriate action is taken and to monitor the effectiveness of corrective actions.

Airport Facilities

Ultimately, it is the responsibility of the airport to protect the users of the airport as well as the general public in the vicinity of the airfield from hazards associated with the operations taking place at the airport. This responsibility includes ensuring that all publicly accessible buildings such as passenger terminals are free of hazards and in good operational condition. At Part 139 commercial service airports, facilities maintenance personnel are to make sure terminals are clean, functional, and well maintained. At smaller commercial service airports, maintenance personnel will often be "jacks of all trades," completing tasks that range from checking on the heating and air conditioning system to fixing a faulty door handle. These personnel will also be available when electrical and plumbing issues arise, but may only have a limited ability to address them. After assessing the issue, maintenance personnel may arrange to bring in a contractor for issues requiring specific expertise. Larger commercial service airports will often have an entire staff of electricians, plumbers, and other skilled persons to keep the facility functional and safe.

Security Fencing

At many airports, especially commercial service airfields, security fencing is installed around much of the perimeter of the airport property to deter people from entering sensitive areas on the airport. Fencing of this type is typically eight feet high, may have barbed wire along the top, and will surround the entire airport. At smaller commercial service airports, fencing might only be in proximity to the airline terminal building or the general aviation area, where there are typically higher concentrations of people and aircraft. Such airports may not fence more remote areas of the airport where it may be less likely, though not impossible, for a person to enter the grounds.

Wildlife Fencing

Wildlife fencing at airports is somewhat different from security fencing. Wildlife fencing is typically ten feet high and does not use barbed wire. The idea is to keep wildlife out, not people, so fence height is required rather than sharp deterrents such as barbed wire. Nevertheless, wildlife fencing still acts as a security fence by virtue of creating a barrier to sensitive areas. To be effective, fencing needs to surround the entire airport. Any gaps in the fence that can allow wildlife to enter the airfield will reduce the effectiveness of the entire fence. In short, fencing is only as strong as its weakest link.

Another important consideration is that animals will sometimes dig under fencing. Thus, not only should the fence be ten feet high in order to keep deer and other large animals out; it should also extend approximately two feet into the ground to keep out animals that may dig under, such as coyotes and dogs. When considering wildlife fencing, an airport operator must keep in mind the type of wildlife that is intended to be kept off the airfield.

Airport Fueling Operations

Most airports that are considered important to the national airport system have some type of fueling operations located on-site. In this discussion, we are not referring to the many private airfields in the United States—those with grass strips and similar small airfields—but rather to other types of airports accessible to any aircraft operator or pilot. The majority of the airports that offer fueling operations provide at least 100LL (low lead) aviation gasoline; many others have Jet A fuel available for sale and dispensing into aircraft. Many of these airports feature one or more fixed-base operators (FBO) that provide fueling and other services. However, a number of airports have self-service systems where pilots fuel their own aircraft.

Because aviation fuels are flammable liquids, special care must be taken to ensure that all fueling operations are performed in a manner that is safe, efficient, and in compliance with all federal, state, and local regulations. Regardless of who is conducting fueling operations at the airport, the airport and fuel service provider must abide by the National Fire Protection Association (NFPA) 407, *Standards for Aircraft Fuel Servicing*, as well as FAA Advisory Circular 150/5230-4, *Aircraft Fuel Storage, Handling and Dispensing on Airports*. Compliance with these directives is accomplished through a contract agreement between the airport operator and the fueling operator. This agreement will spell out the responsibilities and requirements that the airport operator will place on the fueling operator to ensure the fueling operator is storing and dispensing fuel in a safe and compliant manner.

If fuel storage tanks are located on the airport property, it is the responsibility of the airport operator to have the fuel storage and operations areas inspected on a quarterly basis by the local fire marshal. This inspection typically consists of checking fuel tanks for leaks, and making sure that fuel lines are in sound operational condition and that fire extinguishers are serviceable and up to date. At commercial service airports, inspection of fueling systems is also conducted by the FAA Airport Certification Inspector as part of the annual inspection of Part 139 airports. Airports that have self-serve fueling facilities need to have their facilities inspected regularly as well.

Annual fire safety training is also required by the FAA for airport fuel operators. Every two years, fueling operator employees must complete an FAA-approved fuel training course. In the other years, fueling employees must complete a recurrent training course. This training must be documented in writing and the documentation kept on file by the fueling operator for 24 calendar months.

Airport Rescue and Firefighting (ARFF)

Airport rescue and firefighting (ARFF) is another requirement for Part 139 airports, based on the type and size of aircraft that operate at the airport. Part 139 requires that all commercial service airports have ARFF facilities and equipment located at a distance of not more than three minutes from the center point of the airfield for deployment in the event of an emergency. To satisfy this requirement, many airports have at least one ARFF station located on the airfield; many larger airports have multiple facilities. Some smaller airports that do not have an ARFF station directly on-site may satisfy this requirement by having an off-site station located in close proximity that still allows for three-minute center field access.

Part 139 specifies minimum types of ARFF equipment (trucks, etc.) required on-site to respond in the event of an accident, based on the size of aircraft and frequency of flights (see **Figure 9–5**). To determine ARFF requirements of an airport, FAR 139.315 contains a formula based on a combination of the length of the aircraft and the average daily departures of aircraft. The indexes range from the lowest (Index A) to highest (Index E). FAR 139.315 states:

> *. . . if there are five or more average daily departures of air carrier aircraft in a single Index group serving that airport, the longest aircraft with an average of five or more daily departures determines the Index required for the airport. When there are fewer than five average daily departures of the longest air carrier aircraft serving the airport, the Index required for the airport will be the next lower Index group than the Index group prescribed for the longest aircraft.*

Source: U.S. Department of Transportation/FAA

Figure 9–5. Typical ARFF fire engine.

Based on the index, the FAA has minimum requirements that are identified in Part 139.317, which specifies the number and types of firefighting vehicles needed, as well as the minimum amount of dry chemical and of water and aqueous film-forming foam (AFFF) agent that need to be carried on the vehicles.

Annual specialized training is required for ARFF firefighters, since aircraft fires are so different than structural. Part 139.319 identifies 11 subject areas in which ARFF firefighters need to train every 12 calendar months to remain compliant. Many small airports have their operations employees cross-trained as firefighters; other airports have firefighters from their local jurisdiction provide fire protection; still others contract this service out.

Notices to Airmen (NOTAMS)

The **Notices to Airmen** (**NOTAM**) system is an avenue for providing timely information to airport users regarding conditions around the airport that may affect aircraft operations. The use of NOTAMs is required by the FAA; sometimes these are the only communications that may exist between an airport and aircraft operators. Anything that could be helpful for a pilot to know should be disseminated in the NOTAM system. Issues such as lights being out, an air show taking place, or airport markings being unclear can be communicated through NOTAMs. Other NOTAMs might include notices regarding anything that could be hazardous to an aircraft, such as birds in the vicinity of the airport or personnel and equipment on or near runways. NOTAMs are critical to a pilot's flight planning so that they are not surprised by something when they are on or near the airfield, so it is extremely important that the airport operator issue timely and accurate NOTAMs.

An airport operator issues NOTAMs through the FAA Flight Service Stations, the offices that pilots communicate with when they are conducting flight planning. NOTAMs need to be recorded by the airport operator and are required to be kept on file for 12 calendar months.

Winter Operations

Winter operations pose a special set of challenges to airports operators and users. With any winter weather event comes the possibility of snow, sleet, or ice (see **Figure 9–6**). If a certificated Part 139 airport is located in an area that can potentially receive a winter storm, the airport is required to have a **Snow and Ice Control Plan (SICP)**. Depending on the size of the airport, the SICC plan can be rather basic or fairly in-depth. The SICC identifies the snow removal equipment available to the airport operator and lays out who will be responsible for snow removal operations and when

Figure 9–6. Digging out of the snow at Chicago O'Hare.

they will occur. A typical SICC will have a map delineating areas of priority and outlining the areas where snow will be removed first, second, third, and so on. The plan will describe how the snow will be removed and will also ensure that no snow piles will be left on the airfield that could pose a potential hazard to aircraft operations.

During winter weather events, the FAA requires that commercial service airports disseminate a **field condition report** to airport tenants, including the air carriers, fixed-base operators, and the air traffic control tower. This report will identify any existing airport conditions other than clean and dry. Field condition reports are critical and need to be completed anytime conditions change, which can be quite often during a winter weather event. The field condition report will specify what type of contaminant, such as snow or ice, is on runways and taxiways, and whether the airport surfaces have been plowed or sanded. Some airports provide web-based field condition reports; the onus is then on the airport user to go to a website to obtain the most current field conditions.

Chapter Summary

This chapter briefly discussed some of the most common and important safety requirements for airports certificated under FAR 139. FAR 139 contains the federal regulations under which commercial service airports are certificated by the FAA. In order to obtain Part 139 certification, airport operators must demonstrate that their airfield at least minimally satisfies the requirements of FAR 139. The operator must also ensure that these standards are maintained over time to ensure not only regulatory compliance, but the safety of the airfield.

In particular, Airport Emergency Plans, self-inspections, equipment maintenance, recordkeeping, wildlife management, airport familiarization, airport driver training, FAR 77, airport fencing, fueling operations, and airport rescue and firefighting were discussed as necessary features of the overall safety of airports.

Airport safety includes many different elements under the responsibility of the operator to ensure that an airport is as safe, compliant, and efficient as is realistically feasible. Because aircraft operations can affect pilots, passengers, and members of the general public, the safety of airports is paramount, and the airport operator must be proactive in ascertaining the condition of the airport at all times and correcting issues before they develop into problems that could compromise safety.

A comprehensive airport safety program will include each of the above elements. Since all flight operations have their origin and completion at an airport, it follows that airport operators have the responsibility of regularly inspecting and conducting ongoing and consistent maintenance of the entire airfield, including runways, taxiways, aprons, and all associated buildings and facilities under their purview.

Chapter Concept Questions

1. Describe the basics of FAR 139 and its purpose in ensuring airport safety. What are some of the most important components of 139?

2. Why is it important for airport operators and their employees to conduct daily airport self-inspections? Name five specific areas of airports that should be inspected daily.

3. Why is it important for an airport to develop and maintain an effective wildlife management program? What is a wildlife hazard assessment? Why is it important for an airport operator to manage wildlife attractants?

4. What are the basic requirements under 139 for airport rescue and firefighting operations?

5. Describe the elements that should be present in a comprehensive and effective Airport Emergency Plan.

Chapter References

Aircraft Owners and Pilots Association. 2011. Airport Signs and Markings. http://www.aopa.org/asf/publications/flashcards/flashcards.pdf.

Federal Aviation Administration. 2003. *Guidelines for the Certification, Airworthiness, and Operational Approval of Electronic Flight Bag Computing Devices.* Advisory Circular 120-76A.

Federal Aviation Administration 2009. *Airport Emergency Plan.* Advisory Circular 150/5200-31C.

Federal Aviation Administration. 2009. Fact Sheet: Runway Safety. http://www.faa.gov/news/fact_sheets/news_story.cfm?newsId=10166.

Federal Aviation Administration. 2010. *Aircraft Fuel Storage, Handling and Dispensing on Airports.* Advisory Circular 150/5230-4A.

Federal Aviation Administration. 2010. FAR Part 139: Airport Certification. http://www.faa.gov/airports/airport_safety/part139_cert/.

Federal Aviation Administration. 2010. *Hazardous Wildlife Attractants on or Near Airports.* Advisory Circular 150/5200-33B.

National Fire Protection Association. 2007. *Standards for Aircraft Fuel Servicing.* NFPA 407.

10 Emergency Response

Chapter Learning Objectives

After reading this chapter, the reader should be able to:

- Understand the various facets of emergency response planning.
- Describe the roles of communication for an effective emergency response.
- Rank emergency response events into different levels for classification and communication purposes.
- Outline a basic emergency response plan for multiple risk (response) types.
- Describe some of the key positions in an emergency response and their basic roles during a response.

Key Concepts and Terms

Aviation Family Assistance Act (AFAA)
Care Teams
Chart of Events
Contained Spill
Contingency Plans
Emergency Command Center (ECC)
Emergency Response Exercises
Emergency Response Manager or Coordinator
Emergency Response Planning
Family Assistance Teams
Fire Risks
Incident Command Center (ICC)
Incident Commander
Independent Sites
Large Quantity Generators (LQG)
Levels of Notification
Loss Type Potential
Mutual Aid Agreement
Non-Contained Spill
Nonspecific Threat
Small Quantity Generators (SQG)
Specific Threat
Spill
Threat
Triaged
Weather-Related Risk

Introduction

With regard to emergency response in the aviation industry, the first thing that may come to mind is an aircraft accident. However, while proper response to an aircraft accident is certainly important, there are many other areas of focus in aviation for which preparation is necessary to minimize injury or loss. While all commercial air carriers are required to develop an emergency response plan in the event of an aircraft accident, it is also important for every aviation organization to plan for local emergencies as well. These **contingency plans** help to minimize the effects of many types of local emergencies such as tornadoes, hurricanes, bomb threats, sabotage, and hazardous chemical spills by developing policy and assigning responsibilities before, during, and after various emergency events. Apart from reducing the negative publicity that comes with a poor response to emergencies, every aviation organization has a moral and ethical obligation to provide maximum protection for passengers, employees, and the general public.

Emergency Response Planning

Emergency response planning includes the organization of emergency response efforts on the federal, state, and local levels. It is a process that defines public health priorities for effective consequence management and explains the concepts of communication and the Incident Command structure. The emergency response plan must take into consideration many factors important to successful planning. These include, but are not limited to:

- The size of the organization and each of its independent sites
- Resources of the organization
- Location of each site
- Loss type potential (life, aircraft, fire, weather-related, terrorism) for worst-case scenario (largest aircraft and number of seats, largest potential fuel spill, etc.)
- Local emergency resources (fire, chemical response, etc.)
- Emergency drills (see **Figure 10–1**)
- Plan responsibility, that is, which departments and positions are responsible for plan maintenance and revision
- Plan distribution and control, that is, who receives a copy of the plan and maintains updates, who is responsible for ensuring the latest revisions are applied and communicated, and how each of these is accomplished

Once the organization has assessed the risks associated with its operations, a plan can be developed to account for the risk factors (as those described above) in detail. Details should include a description of the risk type, the ways in which the risk can affect the operation (including passengers and the public that may be affected), and measures in place to respond to an event. Every measure should be taken—through engineering and administrative controls—to prevent such events from occurring in the first place. That is, prevention of undesired events is the ultimate goal. Nevertheless, an emergency response plan is necessary when the measures taken fail or in the event of natural disaster, human error, or any other unexpected event.

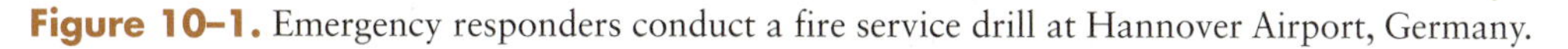

Figure 10–1. Emergency responders conduct a fire service drill at Hannover Airport, Germany.

© Holger Hollemann/dpa/Corbis

Organization Size

Regardless of the size of any organization, planning should be flexible enough to provide protection for passengers, employees, and the general public when exposed to any risks associated with the operation. The size of the organization will be a factor in determining how expansive an emergency response plan should be considered. By way of example, an independent FBO (fixed-base operator) would only need to plan for emergencies on a local basis, perhaps only in coordination with local and/or state agencies. On the other hand, a major or regional air carrier would need to deal with much broader demographics and multiple departments as well as federal, local, and state agencies, and perhaps international agencies in some cases. **Independent sites** (destination or outlying cities or towns as opposed to larger "hub" sites) or field cities/stations often have far fewer resources at their disposal. For these sites, planning with nearby emergency response agencies and/or quick mobility of essential resources may be necessary. The organization, with the assistance of the local operations, would develop a local emergency response action plan specific to that site incorporating information on assistance available from local agencies.

Organization Resources

Resources of the organization are important to an emergency response for obvious reasons. If the organization cannot provide adequate resources, the response effort could be compromised or insufficient. A poor response effort is usually the product of poor communication and preparation, which can lead to negative media attention. The media can sometimes be quick to point out the failure of an emergency response effort as it is likely to be "visible" or "newsworthy," which can add to public speculation as to the organization's preparedness and commitment to safety.

Resources are also important in performing a proper investigation, responding to the needs of the families of accident victims, and for efficient restoration to operational status, since an airline needs to keep operating, even following an accident. Larger and most mid-size organizations typically have an **Emergency Response Manager or Coordinator** and a budget for response equipment, supplies, and drills. This position is responsible for the maintenance and revision of the written program as well as the organization and/or oversight of local emergency response plans and drills, and other related administrative duties in support of the emergency response program.

Most of the duties associated with the Emergency Response Manager (ERM) and staff are concerned with the planning effort. In the event of a disaster, the role becomes an effort in the management of resources. Managing emergency response resources includes ensuring communication protocols are met, equipment and supplies are appropriately organized and transported, and the needs of the Incident Commander and support staff are met. To support this effort, the ERM will usually be located at the organization's headquarters in an assigned incident command center (defined in the Communications during an Emergency section later in this chapter). To ensure proper communication protocols are met, the ERM will check to see that the proper notification tree has been activated, that any call-out equipment employed has worked effectively, and that each required response position has an active representative who is participating as planned. Equipment and supplies are usually stored near the headquarters or the lead aircraft maintenance facility, or in multiple facilities. These products may need to be delivered to the event site to support the response effort, which may require coordination of the ERM with flight control and other interdepartmental functions to acquire aircraft assets to use for this effort. Also, any issues that arise either in the Command Center or in the field may involve special needs for which the Incident Commander may need resources, which would usually be coordinated by the ERM.

Response equipment and supplies should be stored and maintained in a state of readiness to ensure efficient response. The amount of equipment and supplies stored should be adequate to respond to a worst-case scenario. For example, an air carrier should be prepared to provide appropriate supplies in response to the maximum number of passengers on their largest aircraft or enough spill control materials to handle an event involving the largest hazardous material container in use. These supplies should be packaged in such a manner that they can be quickly and easily transported to the event site. Such could include, but are certainly not limited to flashlights, cameras, batteries, blankets, towels, rain gear, notepads and pens, satellite phones, pop-up cover tents, portable chairs and tables, and tarps.

In-Focus: Airline Emergency Management

Toshia Marshall has been an airline industry professional for over ten years. During the previous seven years of her career, she has been the emergency management professional at two different airlines, including her present position with an airline based in the western United States. In this brief excerpt from an interview conducted in the summer of 2011, Toshia shares her thoughts regarding the importance of emergency management and how the profession relates to the airline industry.

In my opinion, one of the biggest risks in running an airline is the potential of an aircraft accident. The risk isn't so much in the probability of this happening—as agencies, authorities, and air carriers alike have put an extreme amount of energy into practicing and creating safe processes and procedures; but more so the risk lies in the impact that an accident will have on an air carrier and all involved as a result. At the core of an accident or any disaster, history has repeatedly revealed the impact to be greatest on the people that are involved in such a catastrophe. In 1996 and as part of the Federal Family Assistance Plan, this factor has been recognized in the passing of legislation known as the Aviation Disaster Family Assistance Act (ADFAA). Prior to this legislation, surviving passengers, family, and friends left to mourn their loved ones were impacted further by poor handling of the aftermath of an accident, not limited to the airline, local, state, and federal governments. Citizens demanded change and the airline industry, as a whole, was not moving fast enough on its own.

Emergency management has become vital to airlines. Both domestic and international carriers work consistently and continuously to improve upon meeting Victim Support Tasks (VST) stated in the ADFAA, minimizing secondary assaults and effectively managing the components of an activation, supporting efforts to provide family assistance and perform an investigation. For almost as long as there has been air transportation, there has been emergency management. However, in the last couple of decades the field has formalized, and higher education curriculum has been developed to meet this growing area.

The foundation of emergency management includes a plan for mitigating hazards, preparing for activation, responding to the expected and unexpected, and maintaining flexibility towards recovery. Command, control, communication, and coordination are critical functions to an emergency management program as well. Identifying a management process, making resources available and streamlining the approach for information flow is imperative as commonly structured groups make decisions towards life safety, stabilizing the situation, preserving the environment, property preservation, and evidence, and ultimately to do what's right for the greater good of people, while meeting the need of better order in the midst of chaos.

Courtesy of Toshia Marshall

Communication

Communication is one of the most important key resources of an event response. It is critical to the response effort itself, but also important to loved ones of passengers and employees who may be involved. While we live in an age of exceptional technology with the advent of smart phones, there are still some areas where this type of equipment may be rendered useless after an event. Also, with the possibility of the media releasing inappropriate information as well as the possibility of sensitive information being transported through compromised advanced devices, the use of smart phones and similar devices should be avoided whenever possible. Satellite phones or landline telephone services may offer a more effective and secure means of communication. Other forms of technology such as long-range video and microphones can be used to get information that can be appropriately handled by the organization.

Communications during an Emergency

Caution should always be taken by the members of any aviation organization when communicating any information related to an emergency. In fact, some aviation organizations provide specific training to some of their employees on effective and appropriate means of communicating with the

Site Location

The location of each site where a local plan is developed is important for many reasons. Several considerations must be taken when developing each element of the plan. They include asking the following questions:

- Where is the location geographically?
  - What are the typical weather challenges for the area (tornado, hurricane, ice/snow, flooding, etc.)?
  - Are there large bodies of water nearby?
  - Is the area prone to or does it have a history of earthquakes?
- What is the distance from the organization's headquarters or Incident Command?
  - Is the time to get to this location (and points in between) acceptable?
  - If not, is there another organization that can be contracted to assist in the event of an emergency?
  - If applicable, what are the procedures for taking an aircraft out of service to fly key response personnel to the accident site? (applicable primarily to airlines)
- What agencies have jurisdiction over the location (local, state, federal, and/or international)?
- Is the site and surrounding area of operation predominantly industrial, residential, or rural?

Loss type potential simply means the type of loss exposures that may exist. For aviation, what is considered a "worst-case scenario" is usually an aircraft crash. However, there are other types of losses including ground damage, serious injury, chemical spill, fuel spill (usually listed separately from chemical spill because of size and frequency potential), fire, weather-related (hurricane, tornado, volcano, flood, winter storm, etc.), bomb threat, sabotage or terrorism, and biological exposures (blood-borne pathogens, epidemics, etc.). Each of these requires individual response planning and preparation (contingency plans), as mentioned previously. While there may be some similarities in response technique, most will require individual or specific attention. This chapter will not cover details for all types of contingencies, but will give a basic example for some of the more prominent risks (tornado, fire, fuel spill, etc.) in later sections.

Local Emergency Response and Resources

Local emergency resources must be determined to make alternate resourcing plans where necessary. First, any organization will fare better if it develops a positive relationship with the local fire department and perhaps other responding agencies. The fire department is a valuable resource that can help with training and understanding of applicable federal, state, and local regulations. Furthermore, inviting fire responders to tour the facilities of a company not only helps in identifying potential risks, it gives them an opportunity to become familiar with the facility, the operations and scope of the organization, and the chemical products that are used and stored there. This knowledge could make a response much safer and faster, perhaps even saving lives and property in the event of an emergency. Finally, for those facilities or operations with limited resources (either equipment or personnel), mutual aid agreements may bring increased benefits to those organizations involved. A **mutual aid agreement** is simply an agreement between two or more organizations to render aid in the event of specified emergencies. For example, an airline in a small field city may require personnel and equipment resources from the local FBO to move a disabled aircraft.

Fire Event and Response Readiness

A study of the **fire risks** associated with a site will help determine the amount and size of extinguishing equipment necessary for a local response as well as any training needed for personnel. This can be as simple as gathering information on the types of products at the facility, understanding their characteristics (how flammable are the products?), and determining the maximum quantity that will be stored. Making these determinations helps to identify the proper size and types of fire extinguishers and other equipment necessary to handle the hazards in each area. Also consider: If fire response personnel are across the tarmac at an airport or hangar facility, perhaps minimal equipment and training is all that would be needed. However, if response takes 5–10 minutes or more, it would make more sense to have a more sophisticated system including better-than-standard fire suppression equipment and personal protection for on-site personnel trained to attend to a fire. In either case, it is extremely important for the site to provide specific awareness training on the exposures in each operation as well as individual responsibilities for response (even if the response is to evacuate). The following is an example of the provisions that might be developed for responding to a fire and/or explosion in a building or hangar setting:

Employee Responsibility:
In the event that an employee discovers a fire or explosion, he/she should:

- Attempt to extinguish a small fire, but only if there is backup support and only if he/she has been trained in the proper use of the extinguisher.
- If the attempt fails to extinguish the fire immediately, the employee should immediately notify the shift supervisor [this term is used for the supervisor on duty at the facility or area involved] via one of the following methods:
  - Dial the supervisor's extension
  - Two-way radio
  - Hangar public address (PA) system
  - Fire alarm pull box (place follow-up call to the shift supervisor as soon as possible)
  - Provide the following information:
  - Exact location of the fire
  - Type of fire (e.g., electrical, flammable liquid, or combustible material)
  - Whether or not the fire is near a critical system (e.g., major equipment or aircraft, fuel or gas lines, etc.)

If a fire is announced, all other employees should follow the instructions of the shift supervisor. If the pull alarm is used for facility notification, employees are expected to evacuate the facility according to the emergency evacuation plan.

Shift Supervisor Responsibility:
When advised of a fire/explosion, the shift supervisor will ascertain the nature, extent, and location of the fire/explosion and formulate response tactics. **Figure 10–3** illustrates how the media can quickly respond to the scene of aircraft incidents. Poor response and handling of

Figure 10–3. News crews capture the evacuation of a Boeing 747.

© STR/epa/Corbis

an event could prove devastating to public perception of an airline. Fires don't always happen to aircraft or the airlines. Sometimes the airport itself, or portions thereof, can be affected (see **Figure 10–4**). Position titles may vary from this example.

- The shift supervisor will direct emergency responders (first aid attendants, EMT, fire department, etc.) to the impacted area to begin response activities.
- The shift supervisor will request any additional assistance or equipment as necessary when advised by the emergency responders.

Upon notification of a fire/explosion, the shift supervisor will notify the local fire department by dialing 9-1-1. As necessary, the shift supervisor will initiate evacuation procedures. During off hours, the shift supervisor will contact appropriate management personnel per company protocol.

The shift supervisor is to respond by directing a person or persons to proceed to any gates or locked doors/entry points to act as a liaison between facility personnel and the local fire and police departments (allowing quick direction and information).

Designated management and technical support personnel should report to the designated briefing/conference room or alternative location as designated by the incident commander and provide support as required.

Emergency Response Personnel [trained to use fire suppression equipment or trained as first aid responders]:

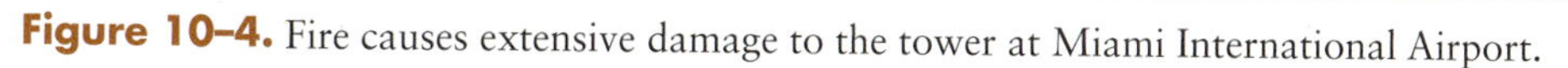

Figure 10–4. Fire causes extensive damage to the tower at Miami International Airport.

Specific fire extinguishing tactics are to be used on a case-by-case basis; however, the following general principles apply:

- No actions will be taken that will subject any personnel to unreasonable risk.
- Rescue and first aid take priority to fire suppression.
- The possibility of explosion should always be considered.
- Sprinkler systems are not to be shut off except by order of the fire department commanding officer after all evidence of fire is gone.

Weather-Related Events (Tornado, Hurricane, Etc.)

A study of the **weather-related risk**(s) common in an area will help determine the types of preparation necessary to respond to each. While coastal areas may be susceptible to hurricanes or other tropical storms, other parts of the country may be plagued by tornadoes, severe thunderstorms, or flooding, for example.

Tornadoes are one of nature's most deadly storms (see **Figure 10–5**). They can wipe out neighborhoods in a matter of seconds with winds that may reach in excess of 300 miles per hour in the case of extremely violent and destructive tornadoes. The path of a tornado can reach in excess of

Figure 10–5. Workers repair storm damage on the roof of the main terminal at St. Louis Lambert International Airport , which was struck by a tornado in April 2011.

one mile wide and 50 miles long. Although some states in the Midwest tend to receive the majority of storms that spawn tornadoes, every state is at some level of risk for the development of severe weather under the proper atmospheric conditions. According to the Federal Emergency Management Agency (FEMA), tornadoes are most likely to occur between 3 P.M. and 9 P.M., but they can occur at any time of day. The following is an example of the types of provisions that may be incorporated for response to the threat of a tornado:

> *If a "tornado watch" is issued by the National Weather Service (NWS), specific personnel should be assigned to monitor weather conditions, listen for broadcast warnings, and report on threatening conditions. If a "warning" is issued by the National Weather Service [meaning that a tornado or funnel cloud has actually been sighted in the area or NWS Doppler radar has detected a severe thunderstorm capable of producing a tornado], the following steps are to be taken:*
>
> - *Personnel will be notified by the public address system. If this option is not available or not functioning properly, each area supervisor will be notified and required to inform all affected personnel verbally as soon as possible.*
> - *Emergency responders should be placed on alert.*

Source: Federal Emergency Management Agency (FEMA)

- *Hangar and terminal personnel are to seek shelter in the base floor "Storm Shelter" location identified on the Emergency Evacuation Map, if time permits. If not, seek shelter near an inside wall, preferably under a table or desk to help protect against falling or flying debris.*
- *If possible, all nonessential utilities should be shut off.*
- *Secure all loose equipment. Aircraft maintenance hangar crews will be instructed to follow aircraft maintenance instructions on securing aircraft and closing hangar doors.*

Move outdoor activities inside and close appropriate doors. After the passing of the tornado and the all-clear is given, personnel should inspect their areas for damage. If the hangar or terminal was struck, emergency responders will begin rescue and first aid and damage control activities. Damage assessment, clean-up and restoration, and other recovery activities should follow. Where a storm has resulted in devastation either at the company facility or employee homes and neighborhoods, assistance should be provided to address an emotional recovery as well.

Source: Federal Emergency Management Agency (FEMA)

Threat Event

Title 49 (Transportation) Code of Federal Regulations (CFR) Part 1544.301 requires aircraft operators to adopt a contingency plan and implement it when directed by the Transportation Security Administration (TSA). The aircraft operator must update the plan on an annual basis and participate in airport-sponsored exercises. Part 1544.303 describes how the aircraft operator should respond to a bomb or air piracy threat. Bomb threats against the security of a flight or facility, or information regarding the act or suspected act of air piracy must be reported to the TSA.

A **threat** (see **Figure 10–6**) is an expression of intent to harm another human being or property. A bomb threat may be the most common occurrence, but awareness of and training for other forms of threat are also important, even if the response may be virtually the same. Other forms of threat include air piracy, threats of physical violence (terrorist to customer, employee to employee, customer to employee, or employee to customer), and threats of property destruction (which could occur by numerous means, as when an individual flew his small aircraft into the IRS building in Austin, Texas, in 2009, or car bombing attempts). Threats are classified as specific or nonspecific. A **specific threat** allows for positive target identification or is made by a person claiming to be a member of an organization or group that is judged credible (known terrorist groups or organizations). A **nonspecific threat** may be related to one or more targets, but there is doubt as to their credibility because of inaccurate locale descriptions, no known affiliations, and so on.

The U.S. Bureau of Alcohol, Tobacco, Firearms and Explosives (ATF) identifies two reasons why a person would report a bomb threat. They are:

- The caller has definite knowledge or believes a bomb has been or will be placed and wants to minimize personal injury or property damage.
- The caller wishes to create an atmosphere of anxiety and panic in an effort to disrupt normal activities at the facility where the device is reportedly placed.

Source: U.S. Department of Justice/ATF

The majority of bomb threat calls are proven to be false. To avoid creating alarm and panic, threats should be handled with discretion. However, a threat received should always be treated as "real" and must be forwarded immediately to designated personnel (security management, the operation's control center manager, etc., trained in threat response and assessment) within the organization to assess the validity of the threat. The employee and/or responding agent receiving notification may also seek guidance from law enforcement agencies and/or airport authorities.

Figure 10–6. Armed police officers guard the main route into Heathrow International Airport in London, UK, July 1, 2007. Police arrested five people after an SUV drove into Glasgow International Airport's terminal entrance and caught fire in a terrorist attack.

© Bloomberg/Getty Images

Threats may be conveyed through many avenues, including through personal contact (person-to-person), handwritten or typed notes, or by telephone. Telephone threats are the most common form of threat communication. A threat response form is usually designed for gathering information conveyed via phone. However, the same form may be used to record information received through other means. Threat report forms should be located within reach of every telephone capable of receiving calls from an outside source. The person receiving the call should remain calm and courteous to the caller to avoid triggering a negative reaction. Although it may not be all-inclusive or achievable in all cases, the following is an example of information that can be collected by the receiver (if the threat is other than a bomb, some information may be noted as not applicable, or "NA"):

- What is the identity of the caller and where is s/he calling from?
- Where is the bomb?
- When is the bomb scheduled to go off?
- What is the purpose of the threat?
- If an aircraft is involved, what is the flight number, aircraft number, departure city and time, and destination city and time?
- If an airport or building is involved, where is the facility and location of the bomb within the facility?

After the call is concluded, answer the following:

- Was the caller male/female?
- What was the caller's approximate age?
- Did the caller have a European, Asian, Middle Eastern, or other accent? If so, identify.
- What were the voice characteristics of the caller (loud, soft, rasping, intoxicated, deep, pleasant, etc.)?
- What were the speech characteristics of the caller (fast, slow, stutter, slurred, crying, disguised, stressed, etc.)?
- What was the caller's manner (calm, obscene, irrational, angry, scared, etc.)?
- Describe any background noises (music, voices, animals, trains, aircraft, etc.).
- Include any other information pertinent to the call, including the exact wording of the caller.

If there is a specific threat (to a specific location or persons, etc.) and the threat cannot be determined definitively as a hoax, personnel should be briskly but calmly evacuated. If the threat involves some type of chemical, the ventilation system should be shut down to minimize the potential contamination area. Evacuation should also be ordered if a suspicious device is found. As part of the emergency response program, evacuation procedures should include evacuation routes from locations throughout each facility. These routes should include the quickest route to exit (and a backup) as well as meeting or assembly points outside the facility.

Spill Response

For the purposes of an emergency response plan, a **spill** is defined as the accidental release of liquid or solid material from its proper container, whether from container failure, upset, or unintentional drainage. Spill response equipment should be obtained for the various types of spills that could potentially occur in or at the facility. A spill will either be contained or not contained. A **contained spill** is one that is captured, either by secondary containment (an outside container or pallet capable of holding any spilled product) or by a diking system (a curbed or ramped area capable of containing the spill, usually seen around a fueling island). A **non-contained spill** is one without secondary containment, usually at a location other than where the product is stored or while in transit. For example, uncontrolled spills can occur during the fueling process, as a result of improper loading of a fuel truck, or by dropping chemical drums or containers during transit.

The specific requirements for spill and clean-up equipment will vary by chemical makeup. Some clean-up equipment is manufactured with materials that bond and jell with petroleum products; others react only with the chemicals they are made to absorb. Each individual site should take measures to prevent spills from occurring. Prevention may include barriers to protect liquid drums from traffic, spill prevention nozzles or transfer devices, and appropriate storage and secondary storage requirements (not allowing improper stacking or storage in improper containers). If these measures fail, the site must be prepared to respond to the maximum potential anticipated spill. For example, if it is determined that one location within the facility stores a 55-gallon drum containing a flammable liquid that could leak in the location, materials to dike, absorb, and stop the spread of the material should be kept in a nearby area. These may include booms, absorbent pads and mats, and spill berm containment dikes (to keep liquid product from reaching drains, water, ground, etc.). Also, non-sparking equipment and a salvage drum (or drums) large enough to collect the liquid and absorption materials will be necessary for clean-up. Some suppliers provide such equipment in ready-to-use overpack spill kits, which are drums that contain much or all of the material and equipment needed to contain and clean up a spill for which the pack is rated. Of course, anyone expected to respond to a chemical spill should be appropriately trained and should employ the appropriate personal protective equipment for the hazard exposure.

Aircraft maintenance facilities often have hazardous chemicals as part of certain production requirements. The use of many of these chemicals is often dictated by the aircraft manufacturer, and the use of substitute materials is often difficult to get approved. Many of these products have flammable or corrosive properties. If the on-site or nearby fire department does not have the capability of responding to a spill, a local contract with a capable and qualified company must be negotiated, or the facility must determine and train its own response personnel and provide the appropriate equipment for the maximum potential response.

The following is an example of the types of provisions that may be incorporated for response to spills of hazardous materials:

Small spills which pose no safety and health dangers and are not likely to adversely affect the environment are to be handled by trained area personnel with appropriate materials on hand. These personnel should:

- *Eliminate the source of the spill by closing valves, righting drums, turning leaking drums over, etc.*
- *Prevent the spill from spreading and going down drains by using absorbents or spill berm containment dikes.*
- *Add neutralizing agents, if applicable.*
- *Dispose of materials and absorbents properly.*

If personnel are unable to contain or clean up the spill:

- *Evacuate and secure the area; prevent others from contacting the material.*
- *Call the local spill response or Hazardous Materials (HazMat) Response Team; report the type of spill (oil, fuel, etc.), including approximate size of the spill, and have someone meet the Response Team at the door or location specified.*
- *Place spill containment berm dikes where possible to minimize drain, water, and ground contamination.*

Evacuation:

The shift supervisor should commence evacuation (refer to the company's evacuation procedures) of the area and/or facility if the following type of leak has occurred:

- A major leak; specified as:
  - A leak or spill of more than 20 gallons of acutely hazardous materials by inhalation [methyl ethyl ketone (MEK), toluene, etc.] in an area with little or no ventilation
  - A large fuel spill (50 gallons or more) within an aircraft hangar or within 50 feet of an occupied building or office facility
- A minor leak which personnel have trouble containing (continues to flow).

If the shift supervisor judges that a health and safety hazard exists, he/she may declare a full site evacuation of all nonessential personnel. The local fire and police departments should be notified by the operations manager or his designated representative whenever an evacuation has taken place. The safety manager (usually a position at the corporate level) is responsible for contacting all federal and state agencies requiring notification to inform them about the spill and its specifics. Upon arrival, the fire department has the authority to decide if a larger-scale evacuation is necessary.

Large Quantity Generators of hazardous waste must submit a copy of the facility's Hazardous Waste Contingency Plan to all local police departments, fire departments, hospitals, and state and local emergency response teams that may be called upon to provide emergency services. A Large Quantity Generator is a facility that generates 1,000 kilograms (about 2,200 pounds) per month or more of hazardous waste, or more than 1 kilogram (about 2.2 pounds) per month of acutely hazardous waste. **Small Quantity Generators** of hazardous waste are not required to submit such data. They must, however, attempt to make arrangements to familiarize police, fire departments, and emergency response teams with the layout of the facility, properties of the hazardous waste handled at the facility and associated hazards, places where facility personnel would normally be working, entrances to roads inside the facility, and possible evacuation routes. Small Quantity Generators generate more than 100 kilograms (about 220 pounds), but less than 1,000 kilograms of hazardous waste per month.

Aircraft Crash Event

Although air travel is the safest form of travel available, there is still the potential for a catastrophic loss at any location where the aviation company operates or along a route of flight (see **Figure 10–7**).

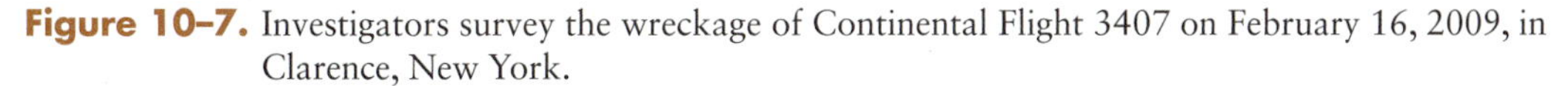

Figure 10–7. Investigators survey the wreckage of Continental Flight 3407 on February 16, 2009, in Clarence, New York.

For this reason, each location within the organization should be prepared with various contacts and specific responsibilities to ensure preparedness and an appropriate and timely response. These contacts and preparations include hotels, rental cars, churches and clergy for survivors and families of accident victims, and separate hotels for accident responders, meeting and media rooms, family assistance personnel, and so on. These basic preparations may be helpful in providing an efficient and caring response. An aircraft accident is a terrible occurrence, and preparatory steps need to be taken to ensure that the family members of accidents victims are cared for on a continual basis and that they are provided information as quickly as possible regarding the status of their loved ones. Also, an aviation organization needs to ensure that employees are provided with personal and emotional support to help them work through the event. The loss of an aircraft may be a very personal event for the employees of the company for any number of reasons, and providing emotional support for them is crucial.

The vast nature of responsibilities and the complexity of requirements for a crash response would warrant an entire text to cover in detail. In this chapter, only a basic overview has been provided, with general information given regarding some specific areas such as communications and resources. The subject is reiterated in this section to stress the fact that not only should there be a corporate plan in place, but it is equally important insofar as possible to have resources available and prepared in outlying field stations and even rural locations as needed.

Family Assistance: Caring for Families of Accident Victims

The **Aviation Family Assistance Act (AFAA)** was passed by Congress in 1996. It assigns the National Transportation Safety Board (NTSB) the responsibility to integrate resources of the federal government and other organizations to support the efforts of state and local governments and the airlines to attend to the needs of victims and their family members after an aviation disaster. Under the act, the Federal Family Assistance Plan for Aviation Disasters was developed to provide guidelines to companies within the United States and its territories.

Under the AFAA, the NTSB's primary task is to coordinate federal assistance and serve as a liaison between the airline and affected family members. The NTSB will coordinate the assistance efforts of state and local authorities. The NTSB also provides and coordinates briefings to family members, releases information to the media pertaining to family support, and maintains communication with affected families to meet their needs and provide continuous progress updates.

The airline has certain responsibilities under the act as well. In addition to immediate notification to the NTSB after learning of a crash, the airline has the responsibility to provide a reconciled passenger manifest to the NTSB, secure necessary facilities for families that may arrive at the departure, destination, or site city, assist family members with travel, food, and lodging, and provide notification to family members prior to releasing passenger names to the public. The airline is also responsible for coordinating American Red Cross assistance, providing contact person(s) from among trained airline personnel to support family member needs as they arrive, participate in daily activity and support meetings to ensure problems are resolved effectively, and ensure that the support function continues as smoothly and with as much care as possible.

There are certainly many other responsibilities for an airline in the event of an accident. The aforementioned examples are just some of the action items pertaining to the overall family assistance effort. More information on this topic can be obtained from the Federal Family Assistance Plan for Aviation Disasters on the NTSB Transportation Disaster Assistance website at www.ntsb .gov/family.

In addition to the responsibilities of the NTSB and the airline, the American Red Cross, the Department of State (DOS), the Department of Health and Human Services (DHHS), and the local medical examiner or coroner have certain responsibilities with regard to the response effort, which can be found on the NTSB website or on the various agencies' respective websites.

Evacuation Plans and Maps

In addition to the written emergency response plan, drills and exercises, and any associated awareness training, each facility is required to post evacuation plans and maps that describe the routes of exit, exits, shelter for severe weather, assembly areas for evacuations, and so on. The evacuation plan should include important phone numbers (fire department, security, first aid providers, etc.) and any other information (information such as zone or area leaders, employees trained in first aid, etc.) that is important to relay to personnel. The map can depict the entire facility or just the immediate area that includes the exit routes, exits, shelter for severe weather, and evacuation assembly area for that area. The map should include a symbol showing the "North" direction, a "you are here" identifier, and it should be positioned where the floor plan matches what the viewer would see.

The evacuation plan and map are typically on the same board or wall area. In many cases, the plan portion is included as part of the directions on the map or the notation of emergency contacts (perhaps just "911"). Better maps include useful details for various emergency situations, such as locations of fire extinguishers, automated external defibrillators (AEDs), first aid kits, and so on. Such maps can also be used as training aids for new employees and emergency responders. The key is that the map be clear and understandable to all who may view it.

Emergency Drills and Exercises

Finally, all of the information in this chapter is useless if drills are not conducted by an aviation organization to prepare for the "real thing." Many aviation organizations (including airlines) conduct **emergency response exercises** including mock aircraft accident drills on a regular basis in order to prepare their employees insofar as possible for their respective areas of responsibility in the event of an actual accident or other occurrence necessitating the activation of emergency response procedures. Conducting emergency response drills enables each responder to better understand their job tasks and how their actions affect other areas of the response (communication between departments and outside agencies, etc.). Drills involving aircraft incidents should use fictitious flight numbers, aircraft numbers, and victim names. Personnel involved in the drill should be informed of the "drill" status as it is carried out. Invite outside agencies (fire department, police, Red Cross, etc.) well in advance to allow for their scheduling needs.

Conducting drills will assist in making necessary alterations to the response plan in areas that may be deficient as evidenced by the exercise. They provide the organization the opportunity to see where and how they can make personnel and resource adjustments, reducing the potential for failure during an actual emergency.

What Happens When IT Happens?

The first priority for any type of accident or incident is to attend to those that have been injured. For a serious incident (such as an aircraft accident) involving multiple people, any victims should be **triaged**. That is, persons with the more serious injuries should be attended to first until adequate resources for all arrive. The second priority is for the protection of the accident site, aircraft, buildings, equipment and other assets, as well as the environment. It is very difficult to imagine every scenario possible at any given location. But careful planning and practice (drills) will minimize the negative effects of an incident.

Chapter Summary

Through sound processes of prevention, many potential tragedies, injuries, and damages can be avoided. However, there are still some that cannot be avoided and yet others that will happen due to oversight or negligence. In any case, an organization must plan for what could happen and prepare employees to respond in a timely manner and with care. Aside from the potential for excessive loss, a poor response can lead to a poor public image, which increases the indirect or hidden costs of an event. Any corporate communications representative would agree that negative press can have a huge impact on the success of a company. While this impact is usually short-term, its effects can be devastating to an organization in an industry that is highly visible to the public and greatly affected by a lagging economy.

This chapter does not cover in detail all loss source types. It does, however, provide selective detail on three loss source types (fire, bomb threat, and hazardous material spill response). While not every detail was included, these sections do provide some insight as to what should be included in the planning process.

Chapter Concept Questions

1. A document developed to help to minimize the effects of many types of local emergencies such as tornado, hurricane, bomb threat, sabotage, and hazardous chemical spills is often known as a ________________.

2. What would you consider to be the most important factor in an emergency response, and why?

3. What are some of the things to consider, geographically, when developing a local emergency response plan?

4. An agreement between two or more organizations to render aid in the event of specified emergencies is known as a(n) ________________.

    a. Contingency Aid Agreement

    b. First Response Contract

    c. ERP Agreement

    d. Mutual Aid Agreement

5. Describe the difference between a tornado "watch" and "warning."

6. A threat that may be related to one or more targets, but there is doubt as to their credibility because of inaccurate locale descriptions, no known affiliations, or other factor, is considered a ________________.

    a. Specific threat

    b. Nonspecific threat

7. Describe emergency situations for which a facility might order an evacuation.

8. Small spills which pose no safety or health dangers and are not likely to adversely affect the environment may be disposed of in a sewer drain. True or False

9. A facility that generates 1,000 pounds of hazardous materials each month is considered a Small Quantity Generator (SQG). True or False

10. In the event of a crash, the airline has the responsibility to provide a reconciled passenger manifest to the NTSB under the ______________ Act.

Chapter References

DiBerardinis, L. J. 1999. *Handbook of Occupational Safety and Health* (2nd ed.). New York: John Wiley & Sons, Inc.

Federal Emergency Management Agency (FEMA). 2010. Determine Your Risk and Plan for Emergencies. http://www.fema.gov/plan/prepare/.

Federal Emergency Management Agency (FEMA). 2010. Ready America: Prepare. Plan. Stay Informed. http://www.fema.gov/plan/prepare/.

National Fire Protection Association (NFPA). 2007. *NFPA 1600: Standard on Disaster/Emergency Management and Business Continuity Programs.* Quincy, MA: NFPA.

National Safety Council. 2001. *Accident Prevention Manual for Business & Industry: Administration & Programs* (12th ed.). Itasca, IL: National Safety Council.

National Safety Council. 2007. *Aviation Ground Operation Safety Handbook* (6th ed.). Itasca, IL: National Safety Council.

National Transportation Safety Board (NTSB). 2010. Information for Family and Friends: Accident Investigations. http://www.ntsb.gov/Family/family.htm.

National Transportation Safety Board (NTSB), Office of Transportation Disaster Assistance. 2008. Federal Family Assistance Plan for Aviation Disasters. http://www.ntsb.gov/Publictn/2008/Federal-Family-Plan-Aviation-Disasters-rev-12-2008.pdf.

U.S. Department of Transportation. 2009. Title 49 CFR, Part 1544, Sections 301 (Contingency Plan) and 303 (Bomb or Air Piracy Threats).

U.S. Environmental Protection Agency. 2010. Wastes: Hazardous Waste. http://www.epa.gov/waste/hazard/generation.

11 Health and Wellness

Chapter Learning Objectives

After reading this chapter, the reader should be able to:

- Understand the value of a wellness program to an organization.
- Describe the benefits of providing a wellness program to employees, both professional and personal.
- Understand the basic concepts of ergonomics and how it plays a role in the aviation industry.
- Describe effective injury prevention measures related to physical work stresses.
- Develop a basic plan to educate employees on the subject of wellness and encourage them and their families to participate.

Key Concepts and Terms

Annulus Fibrosus
Anthropometric Design
End-Range Motion
Ergonomics
Impingement
Instability
Intervertebral Disks
Musculoskeletal Disorders
Nucleus Pulposus
Optimum Posture
Spinal Cord and Nerves
Static Load
Study of Workplace Design
Wellness Programs
Work Conditioning

Introduction

Healthcare is about a $2.3 trillion industry today in the United States, essentially making it the biggest sector of the country's economy. Costs continue to rise each year due to factors such as advances in technology and research as well as inflation. The latest healthcare reform introduced by the Obama Administration will also likely add significantly to the rising cost for many U.S. citizens. Many employers across the country are faced with absorbing much of the additional cost of insurance coverage, and will likely pass some of that cost along to employees. Typically, an employee pays about 20 percent of the total cost for insurance coverage.

In a struggling economy, today's organizations are cutting back, doing more with less, and finding creative ways to curb spending and save money. Some of these organizations are beginning to realize the value of wellness programs as a way of reducing their healthcare costs. The rationale is that healthier employees do not require as much costly healthcare as a result of reducing their risk of injury on the job. Proponents suggest that wellness programs lead to healthier and happier workers, which leads to lower costs and improved productivity, including fewer missed workdays. Additional opportunities exist for benefits that are extended to family members of employees.

Key Reasons for a Wellness Program

It is no secret that the United States has become a "heavy" nation. With advances in technology and other innovations, it has become easier for many of us to become more sedentary or desk-ridden, to the point that we do not get the exercise our bodies need to remain fit. Our declining fitness has become even more evident over the last two decades owing to the pervasiveness of computers, video games, and smart phones, among other conveniences and entertainment devices, among adults and children alike.

Heart disease and cancer (see **Figure 11–1**) make up nearly half of all deaths each year in the United States. Individuals between the ages of 18 and 39 tend to eat a diet higher in fat than those over 60 years of age. Diet is related to a variety of adverse health conditions such as diabetes, cancers, high blood pressure, and high blood cholesterol. Controlling the amount of fat consumed in the daily diet may help control these health conditions.

Wellness programs provide many benefits for employers and employees alike. Both benefit from improved morale, attendance, and reduced healthcare costs. Employees can also experience the benefits of such programs, which can often include employees' family members as well. Employers can measure the effectiveness of programs through a return on investment that is often as much as three to one (three dollars saved for every dollar spent). Reduced absenteeism and sick leave are key effects contributing to the savings achieved.

Figure 11–1. Number of deaths by leading causes of death in the U.S.

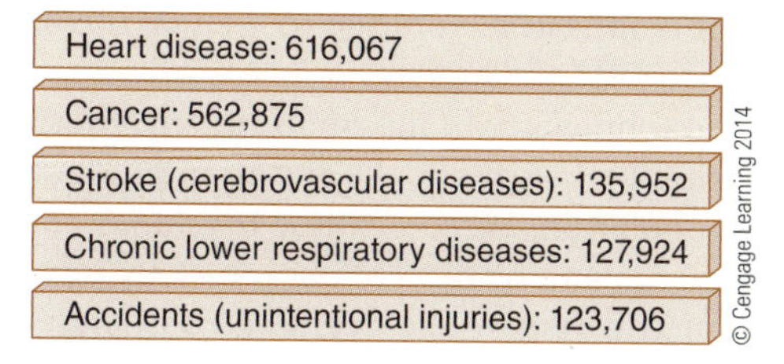

Elements of an Effective Wellness Program

Key elements to implementing and maintaining an effective wellness program include management leadership and employee involvement. Management must set a good example not only by supporting the program, but through active participation. Challenging employees (and their families) to participate and set goals, and providing recognition for achieving those goals, are great motivators for the employee population. For members of management, being a part of that challenge generates additional motivation and success. Another form of management support is through the removal of some on-site temptations such as energy drinks (which are typically diuretics with high amounts of caffeine and potassium) and banning smoking on company property. An organization cannot force an employee to become involved with their wellness program, but communication and support go a long way toward "selling" voluntary participation.

What type of wellness program should an organization use? Because there are varying methods to promote a healthy lifestyle and fitness, an organization has many choices. These may include providing full or partial payment for fitness facilities and/or workshops, access to fitness or nutritional professionals for assistance, internal challenges to lose weight, and pre-work stretching sessions. If multiple choices are not an option for monetary or job-related reasons, a survey could be conducted to determine which activities would best suit the employee group. Providing an opportunity for input and involvement in the decision-making process will increase the likelihood of employee participation. The organization could further benefit from offering programs and information that focuses on problems experienced in the organization (health- or injury-related) that have resulted in claims.

Some low-cost solutions that can be implemented on-site include providing flu shots, safety and health fairs, first aid and CPR training, smoking cessation programs, and blood pressure screening. Many of these can be implemented at little or no cost, with some of these services provided by an occupational clinic used by the organization or sponsored by its insurance provider. Convincing employees that the information and programs provided at work should be taken home to be shared with employees' families is an additional challenge. The key is to provide a continuous message that includes a personal message, statistics, and other information on the benefits for all family members. Implementation of a wellness program for any organization will involve a cultural change that must be driven by the personal values of its leadership and employees.

Ergonomics

Encarta® defines **ergonomics** as the "**study of workplace design**: the study of how a workplace and the equipment used there can best be designed for comfort, efficiency, safety, and productivity." The term itself comes from the Greek word *ergos* (work) and *nomos* (natural laws of). It could also be called human factors engineering, which studies the physical and behavioral interaction between people and their environments (workplace, home, etc.). In the modern era and with more advanced safety organizations, the emphasis has shifted to include the behavioral factor as a key component, rather than design only to "fit" the human.

The objective of ergonomics is to fit the environment (equipment, machine, etc.) to the employee. The design can sometimes be fit specifically for an individual, but most machines and equipment are fit for a standard population. Some of these machines and equipment can be engineered with adjustments to fit a larger group of the population. This process of engineering to fit the population is called **anthropometric design**. The overall goal of ergonomics is to reduce the physical (and sometimes mental) stresses on the body resulting from a given job function and to improve the comfort, health, and safety of the employee. The results should include increased productivity and quality of work as well as a reduction in human error associated with the task.

To better understand the human/machine interaction, it is necessary to understand the demands placed on the worker. These can be classified into three categories: physical, environmental, and mental. Physical demands are those placed on the musculoskeletal system of the body, including such tasks as lifting/lowering, pushing/pulling, and reaching. Walking, standing, or sitting all day are also examples of physical demands. Environmental demands include aspects of the physical environment such as temperature, humidity, vibration, noise, and lighting levels. This category could also include psychosocial factors such as speed of work, shift schedule, length of day, and job satisfaction. Mental demands are those that require, for example, mental calculations, memory, and decision-making skills to perform a task. Combinations of these three demand types are possible as well, adding to the risk factors associated with a task. A ten-hour workday on the airport ramp loading heavy bags in the middle of summer in the deep South can be much more physically demanding than a shorter shift in a different climate, and could lead to fatigue and poor decision making.

An additional risk factor is experienced with new employees, especially in physically demanding positions such as ramp operations. Not only are new employees not familiar with the work methods and safety rules and requirements, but they will have a period during which they will need to be *conditioned* to the job. **Work conditioning** is the period of time it takes an employee to become physically acclimated to the job demands, including the movements and forces placed upon the body to acceptably complete the tasks assigned. While employees will differ in height and reach, which cannot be changed, there will be a period of time in which the employee will gain the strength and stamina required for maximum production. During this time, the employee is at greater risk of injury, and it is important to have greater oversight by a supervisor or mentor to ensure that safe work methods are correctly performed.

One other complicating risk factor in terms of ergonomics concerns the types of equipment used in the field. Airlines for years have employed tugs, belt loaders, and other equipment to minimize handling and lifting by employees. However, many airports and aircraft manufacturers do not consider design for proper ergonomic fit. Bag belts are often installed too low, requiring additional movement and lift force from the employee. Aircraft cargo compartments are often small and short, requiring employees to crouch in awkward positions and postures when handling cargo and bags. These factors are often pervasive among regional airlines as well as many smaller operations and FBOs.

Principles of Preventing Ergonomic-Related Injuries

There are several principles to follow in the prevention of ergonomic-related injuries. First, the principle of working in neutral postures is the foundation for prevention. Working in awkward postures and positions increases physical stresses on the body and reduces its strength, making it more difficult to complete a task. The optimal neutral posture is the one in which the muscles surrounding a joint are equally balanced, avoiding end-range motions or positions. An **end-range motion** or position is one in which the muscle or muscle group is at the end of its range, where it cannot move further, placing it at increased risk of injury if an outside force were introduced that exceeded the limitations of the muscle or muscle group. The **optimum posture** can also be defined as one that provides the most strength and control over movements and the least physical stress on the joint and surrounding tissue. Examples of correct postures include maintaining the natural curve of the back and keeping the neck straight, elbows and shoulders relaxed, and wrists neutral. Poor back posture is typically experienced when the body is bent at the waist rather than at the knees or by slumping over (even slightly) while standing or sitting. Tasks that require the elbows or shoulders to be raised will cause extra force and strength that can quickly fatigue and wear at related muscle groups. The hands and wrists should be parallel to the forearm, which may sometimes require assistance in the form of padding or wrist rests for desk or computer-related job functions.

The second principle is to reduce excessive force. As mentioned, airlines have employed a number of tools such as the tug and bag cart, bag belts, and aircraft container loaders to minimize the amount of work required by the employee. There are several other tools and machines for many other job areas throughout the aviation industry that have served to reduce exertion as well. In aircraft maintenance, for instance, many power tools such as drills and drivers have been designed to reduce force to complete tasks. Stands and body rests have also been specifically designed to put the mechanic into a less taxing position by keeping the work in their "power zone." Excessive force can overload the muscles, creating fatigue and the potential for injury. Furthermore, applying excessive force to perform a task can slow down the effort and interfere with the ability to perform the task well. Consequently, minimizing the exertion required for a task will make it easier and typically faster to perform.

The third principle is to keep objects within easy reach. This can be accomplished by keeping all the parts and tools that one needs frequently within a short reach or movement (either arms, legs, or both). Long reaches often cause a worker to twist, bend, and/or strain, making work more difficult, adding to fatigue. Equipment and work stations must be evaluated to make adjustments for minimizing reaches.

The fourth principle is to reduce excessive motion in terms of the number of movements required to complete a task. Minimizing the number of motions reduces wear and tear on the body and improves efficiency. Repetitive motions waste time and are ultimately inefficient. Obviously, *any* activities that lead to wasting time are ultimately inefficient to the task or process at hand.

The fifth principle is to work at proper heights. Working at improper heights leads to poor postures and/or end-range motions. Although it is hard to avoid this in all situations (such as many aircraft cargo compartments), accommodations can be made in areas such as having adjustable chairs at customer counters and computer work stations, maintenance tail stands, and galley carts designed for the average user population.

The sixth principle is to minimize fatigue and static load. One of the most common sources of fatigue is known as **static load**, holding the same position for a period of time without moving. Static load is especially stressful in combination with high force or weight, and especially if awkward postures are involved. The primary problem with static load is the amount of time that the muscles are worked, even when not in motion. Even when muscles are only lightly tensed over an extended time, pain and fatigue can result.

The seventh principle is to promote physical activity by moving, stretching, and exercising. The human body needs to stretch and exercise. Stretching and exercising, in addition to proper hydration, help to lubricate the muscles and joints. This includes stretching each joint to the full range of motion periodically throughout the day. Doing so will reduce the risk of injury when the employee must take on a load or employ force.

A review of the principles of preventing ergonomic-related injuries will reveal a few common themes; in addition, the principles are interrelated. One of the most obvious is the importance of posture. Most of the principles, in some manner, depend on or relate to proper posture. Another common theme is that violating any of these principles will lead to fatigue, thereby leading to poor decision making, poor postures, poor performance, and greater risk of injury.

Musculoskeletal Disorders

Musculoskeletal disorders are injuries that affect the soft tissue and bones in the body. The parts of the musculoskeletal system are bones, muscles, tendons, ligaments, cartilage, and their associated nerves and blood vessels. Injuries to these tissues are often simply referred to as *ergonomic injuries*. According to the U.S. Department of Labor, Bureau of Labor Statistics, the rate of musculoskeletal disorders (MSDs) per 100 full-time employees in the private industry sector was 33.4 for the 2008 calendar year. The rate jumps to 79.3 for the transportation and warehousing segment within private industry. About two-thirds of all MSD injuries occurred among employees over the age of 35. As alarming as the number of injuries is the fact that MSDs cost about $850 billion annually in the United States.

There are three types of MSD injuries that are most common among employees with physical job requirements such as baggage handlers and aircraft mechanics. They are strain, sprain, and rupture injuries that affect the back, knees, and shoulders. Such injuries are the most debilitating for employees and are typically the most costly for employers.

Back Injuries

A herniated disk, sometimes referred to as a slipped or ruptured disk, most often occurs in the lower back. This is one of the most common causes of lower back pain and sometimes leg pain as well. Between 60 and 80 percent of people will experience lower back pain at some point in their lives. A high percentage of these will have lower back pain caused by a herniated disk.

Although a herniated disk can sometimes be extremely painful, most people will recover within a few months without surgical treatment.

There are many risk factors that contribute to back injury. Among these are age, diet, and fitness level as well as improper lifting. Using the back muscles to lift heavy objects, instead of the legs, can cause a herniated disk. Twisting the trunk of the body while lifting can also add to the back's vulnerability.

The spine (see **Figure 11–2** and **Figure 11–3**) is made up of 24 bones called vertebrae, which are stacked on top of one another. These bones connect to create a canal that protects the spinal cord and together they provide support for the back. Five vertebrae make up the lower back, called the lumbar spine.

Other parts of the spine include the **spinal cord and nerves**, which carry electric signals or messages through the spinal canal between the brain and muscles, and the **intervertebral disks**, which are the flexible disks between the vertebrae. These act as shock absorbers between the vertebrae while walking or running. Intervertebral disks are flat and round and about a half inch thick. They are made up of two components: the **annulus fibrosus**, the tough, flexible outer ring of the disk, and the **nucleus pulposus**, the soft, jelly-like center of the disk.

Prevention of back injuries includes keeping fit. This isn't to say that a person must be a health fanatic, but one should maintain a healthy diet. Being overweight and out of shape increases the risk factors for back injury. Cigarette smoke has also been shown to be a risk factor because of its potential to block the body's ability to deliver nutrients to the disks of the lower back. Smoking can also lead to slower healing, prolonging the pain experienced with back injuries or surgery. It is important to follow basic safety guidelines for lifting and lowering, the primary origin for back injuries. They include:

- Plan ahead to prepare the mind, the body, and the path of travel.
- Position yourself close to the object you want to lift.
- Separate your feet a shoulder-width apart, with one foot slightly ahead of the other to provide a stance that gives a solid base of balance and support.

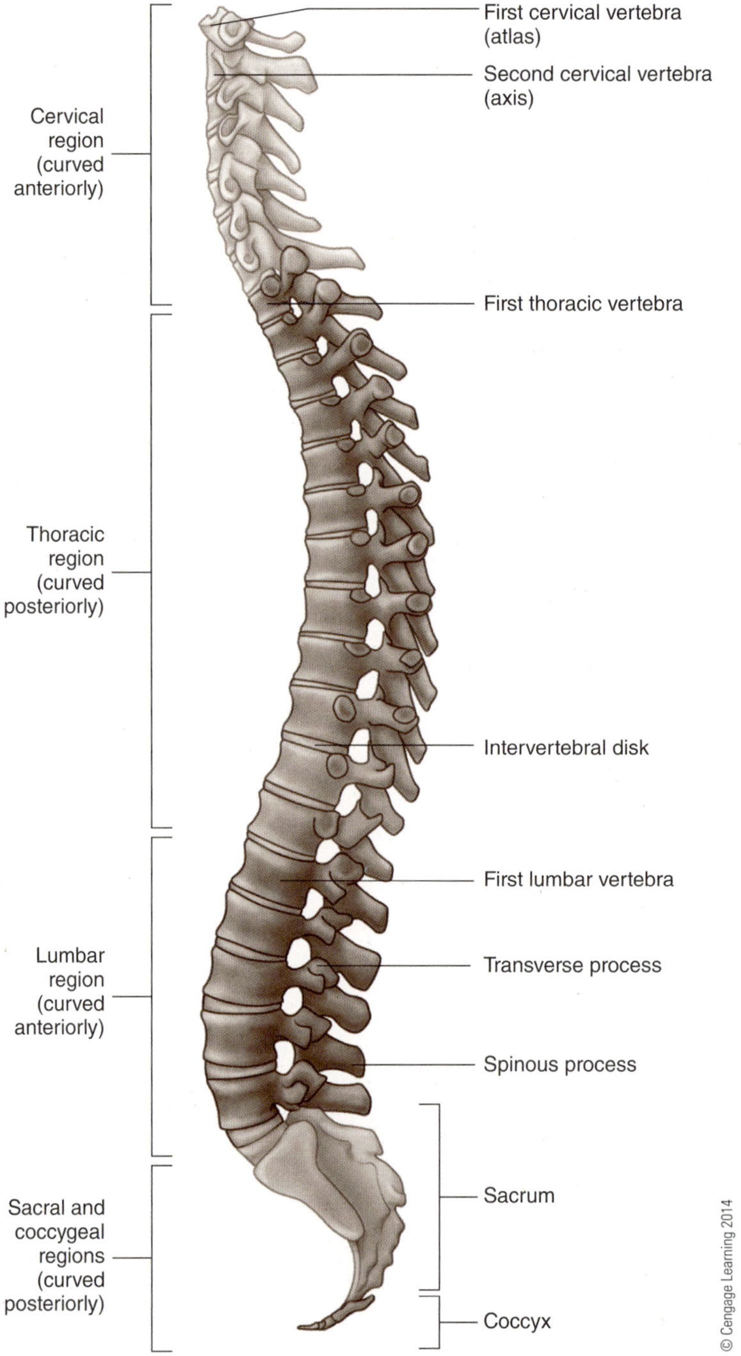

Figure 11–2. Parts of the spine.

Figure 11–3. Posterior and Lateral views of the spine. Included is an illustration of a ruptured disk (middle) and herniated disk (right).

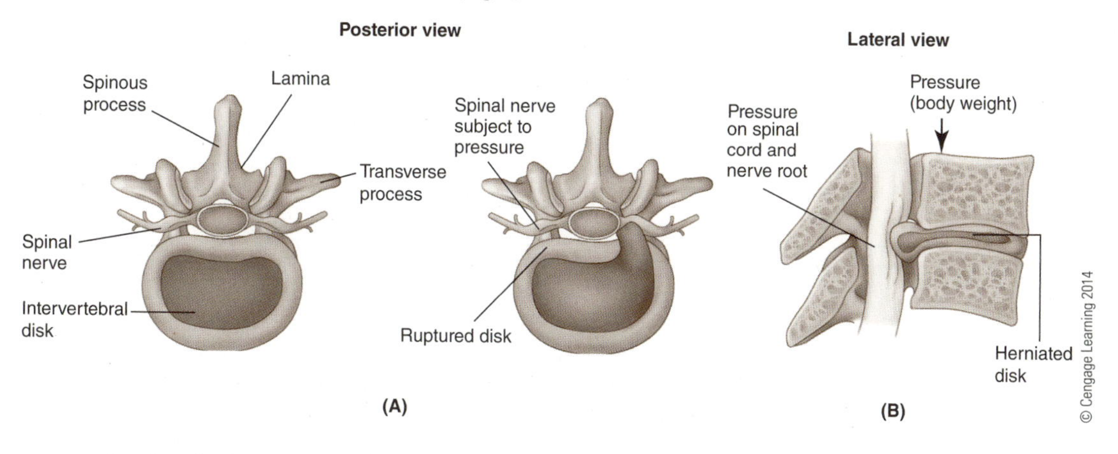

- Bend at the knees, not at the waist, keeping the natural curve of the back.
- Tighten your stomach muscles and keep the head up.
- Begin the lift slowly without jerking to test the weight of the object.
- Lift with the leg muscles as you stand up.
- If you have to move the object to one side, avoid twisting your body. Point your toes in the direction you want to move and pivot in that direction. Keep the object close to you when moving.
- Know your limits and don't try to lift by yourself an object that is too heavy or an awkward shape.
- Use powered or manual equipment available to assist with the lift or lower as necessary.

Shoulder Injuries

Orthopedic surgeons group shoulder problems into the following categories.

- **Instability:** One of the shoulder joints will sometimes move or be forced out of its normal position. This condition is called instability; it can result in a dislocation of one of the joints in the shoulder. Individuals suffering from an instability problem will experience pain when raising their arm. They may also feel as if their shoulder is slipping out of place.
- **Impingement:** Caused by excessive rubbing of the shoulder muscles against the top part of the shoulder blade, called the acromion. Impingement, one of the most common causes of pain in the adult shoulder, can occur during activities that require excessive arm motion above shoulder height.

Rotator cuff injuries (see **Figure 11–4**) account for the majority of shoulder injuries experienced in the workplace. The rotator cuff helps to lift and rotate the arm and to stabilize the ball of the shoulder within the joint. It is made up of four muscles and their tendons, which combine to form a "cuff" over the upper end of the arm. The four muscles of the cuff are attached to the scapula on the back through a single tendon unit. The unit is attached to the side and front of the shoulder on the greater tuberosity (rough projection of a bone where muscles or ligaments are attached) of the humerus.

As with back injuries, there are common methods to prevent shoulder injuries. Maintaining an appropriate fitness level is one key, but following proper safety guidelines is paramount. Most

Figure 11–4. An illustration of the shoulder, including an example of tendinitis (right top) and a ruptured or torn rotator cuff.

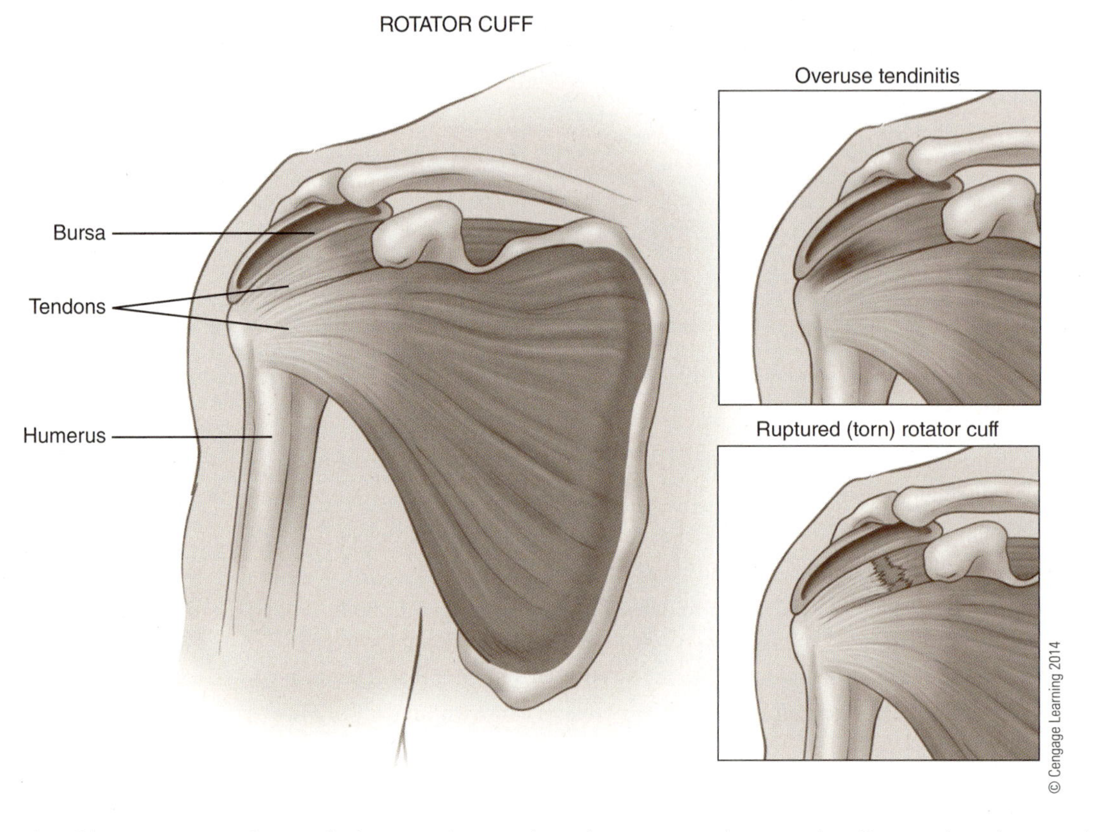

shoulder injuries in the workplace can be attributed to overreaching and pulling rather than pushing. Guidelines for shoulder injury prevention include:

- Pushing carts, dollies, and so on rather than pulling.
- When pushing:
  - Keep the equipment (cart, dolly, etc.) close to the body by bending the elbows. Bending the elbows helps to avoid end-range motion, as does avoiding pulling. An end-range motion involves a group of muscles being stretched to the end of their range or limit of flexibility. Additional movement beyond the muscle group's limit will likely result in injury.
  - Keep the feet a shoulder-width apart and begin the push with one foot ahead of the other to provide added stability and balance.
- Pulling may be necessary in certain situations such as moving a cart or dolly onto a curb, up steps, or through a doorway. Pushing should be resumed as soon as the hazard has been navigated.
- Avoid moving objects to or from a height above shoulder level. Use a ladder, step, or other device to avoid this end-range motion.
- Avoid performing tasks above shoulder height. These tasks will lead to rapid fatigue and greatly increase the risk for injury. Use tools or equipment that place the work in front of the body.

Knee Injuries

The knee (see **Figure 11–5**) is the largest joint in the body and one of the most easily injured. It is made up of the lower end of the thighbone (femur), which rotates on the upper end of the shinbone (tibia), and the kneecap (patella), which slides in a groove on the end of the femur. The knee also contains large ligaments which help control motion by connecting bones and by bracing the joint against abnormal types of motion. The meniscus is a wedge of soft cartilage between the femur and tibia that serves to cushion the knee and helps it absorb shock during motion.

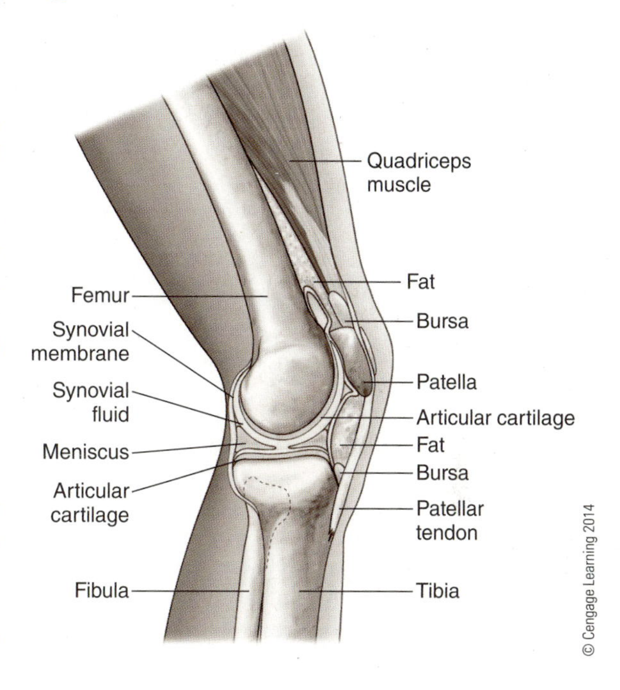

Figure 11–5. Parts of the knee.

Many knee-related injuries in the aviation industry likely stem from poor work habits such as jumping off equipment without using support, assuming postures that are improper, and working long shifts with inadequate footwear. The knee becomes even more vulnerable with age. Degeneration of the meniscus with age, in addition to excessive wear and tear, will lead to irreversible damage which could also lead to arthritis in the knee. Some keys to avoiding injury to the knee include avoiding a twisting motion of the legs and using handrails when ascending or descending equipment, ladders, or stairs. The use of a firm grip on handrails will decrease the amount of force placed on the knee joint during this activity.

General Structure of a Wellness Program

There is no one answer as to how a wellness program could or should be established. The key lies in understanding the potential return versus the investment, developing a strategy that works best for the type and size of the organization, and developing outcome matrices to determine future areas for investment. Organizations have many options, ranging from very little monetary investment to large capital investment. Some of these options are:

- With little to no investment at all, remove vending machine items high in fat and saturated fat and replace them with healthier items that are also appealing. Remove energy drinks that are very high in caffeine and often high in potassium. Caffeine is a diuretic, and high levels increase the risk of dehydration.
- Invest very little by negotiating or using a third-party company to negotiate lower fees on local fitness centers or health clubs for employees. While any effort is positive, this option will likely yield the least results.
- Internal communications using personnel (perhaps from the safety committee) assigned to research and distribute, communicate, and/or demonstrate specific wellness subjects, including stretching

routines. This option not only provides beneficial information to the workgroup and their families; it can also serve as a peer motivator.

- Using a trained nurse or dietary professional to consult with employees on common areas of concern such as blood pressure (hypertension), cholesterol, obesity, smoking cessation, and nutrition. The availability of blood pressure checks and related discussions with a trained professional can be motivating and perhaps life saving.
- Providing partial or full coverage benefits for using a fitness center or health club, either for employees only or including their families. It may be hard to measure return on this investment, but there are options available to encourage and measure its use through positive communications and survey responses.
- Providing an on-site fitness room requires significant capital expense and will involve continuous maintenance costs, depending on the size and complexity. This option is certainly not for most, especially where space is at a premium, but it can be a huge motivator for those employees who can take advantage of such a benefit.

Educating the Workforce

The key factor in changing the behaviors of employees and their families in the direction of healthier lifestyles is effective communication. This can be done on many different levels, including company or linked third-party websites, facility postings, community health and wellness partnerships, and other company communications. One of the most effective means, however, is a joint effort involving a professional resource (either internal or external) and a front-line employee (or health and safety specialist) conducting pre-operational communications and demonstrations. In general, people tend to associate with their peers and learn better through visual response or interaction.

There are many different ideas on effective implementation of a wellness program. As described above, a best practice method for communicating health and wellness information to employees and their families is through a front-line employee (or health and safety specialist). This person (the wellness leader) should be someone of a reasonable fitness level, who does not use tobacco products, and who has a motivating personality. One of the first priorities is to organize the information to distribute to employees. A best practice method for doing this is to divide health and wellness subjects by month, focusing and spending time on just those specific topics. The wellness leader could further drive the process by giving the program a catchy name such as "The Roadmap to Health Starts at the Airport," or by organizing a contest to solicit a name or slogan.

The *roadmap* should be organized well in advance so that planning can be arranged. There are topics that should coincide with national or community-sponsored events such as American Heart Month in February, sponsored by the American Hearth Association and the Centers for Disease Control (CDC), as well as other charities and corporate sponsors. Breast Cancer Awareness Month occurs during the month of October and is sponsored by many charities and corporations as well.

January

Cervical Health Awareness Month: Designated during January by the U.S. Congress to educate the country on issues related to cervical cancer, HPV disease, and the importance of early detection. Approximately 11,000 American women will learn they have cervical cancer each year, and nearly 4,000 will die from an advanced form of the disease. The National Cervical Cancer Coalition (NCCC) is focused on educating women about the importance of the Pap test as a screening tool for cervical cancer/HPV and about vaccines that can further reduce the burden of this devastating disease. For more information, go to http://www.nccc-online.org/.

February

American Heart Month: Heart disease is the leading cause of death in the United States and is a major cause of disability. The most common heart disease in the United States is coronary heart disease, which often appears as a heart attack. In 2009, an estimated 785,000 Americans had a new coronary attack, and about 470,000 will have a recurrent attack. About every 25 seconds, an American will have a coronary event, and about one every minute will die from one.

The chance of developing coronary heart disease can be reduced by taking steps to prevent and control factors that put people at greater risk. Additionally, knowing the signs and symptoms of heart attack are crucial to positive outcomes after having a heart attack. People who have survived a heart attack can also work to reduce their risk of another heart attack or a stroke in the future. For more information on heart disease and stroke, visit the CDC's Division for Heart Disease and Stroke Prevention at http://www.cdc.gov/DHDSP/.

March

National Nutrition Month®: National Nutrition Month® is a nutrition education and information campaign created annually in March by the American Dietetic Association. The campaign focuses attention on the importance of making informed food choices and developing sound eating and physical activity habits. More information can be found at http://www.eatright.org/nnm/.

April

Alcohol Awareness Month: Alcohol abuse can be a problem for persons of almost any age and can affect not only the life of the person using, but their work and the lives of those around them, especially family members. Underage drinking continues to be a huge issue in the United States, with nearly 11 million underage drinkers today. April is designated as Alcohol Awareness Month, an annual public awareness campaign that encourages local communities to focus on alcoholism and alcohol-related issues.

According to the Substance Abuse and Mental Health Services Administration's National Survey on Drug Use and Health, 51.6 percent of Americans age 12 and older had used alcohol at least once in the 30 days prior to being surveyed; 23.3 percent had binged (5+ drinks within 2 hours); and 23.3 percent reported drinking heavily (5+ drinks on 5+ occasions). In the 12–17 age range, 14.6 percent had consumed at least one drink in the 30 days prior to being surveyed, 8.8 percent had binged, and 2 percent drank heavily.

May

National High Blood Pressure Education Month: May is National High Blood Pressure Education Month sponsored primarily by the Centers for Disease Control and Prevention. About 74.5 million people in the United States have high blood pressure, also known as hypertension. Hypertension increases the risk for heart disease and stroke, the first and third leading causes of death in the United States. Researchers estimate that high blood pressure will cost $76.6 billion in direct and indirect costs in 2010. Almost 90 percent of adults aged 45–64 years will develop high blood pressure during the remainder of their lifetime. For more information, visit http://www.cdc.gov/Features/HighBloodPressure/.

June

Sun Safety Week: Sun Safety Week occurs during the month of June and is sponsored by The Sun Safety Alliance, Inc. (SSA). SSA is a communication and education-focused nonprofit organization with the mission to reduce the incidence of skin cancer. SSA is dedicated to creating national awareness of skin cancer as an important health issue. SSA believes that a concerted focus on skin cancer prevention, education, and awareness is the only way to change generations of behavior. For more information on sun safety, go to http://www.sunsafetyalliance.org/index.html.

Men's Health Month: Sponsored by a congressional health education program and its many partners and supporters, Men's Health Month is observed across the country with screenings, health fairs, media appearances, and other health education and outreach activities.

July

Eye Injury Prevention Month: July has been officially recognized as Eye Injury Prevention Month, sponsored by the U.S. Department of Health and Human Services. The focus of this month's activities is based on the protection of the eyes from risks at work and at home through basic awareness and education.

Eye injuries of all types occur at a rate of more than 2,000 per day. About half of these eye injuries occur in American workplaces. The Bureau of Labor Statistics (BLS) found that almost 70 percent of the eye injuries studied occurred as a result of falling or flying objects, or sparks striking the eye. For more information, visit the U.S. Department of Health and Human Services at http://www.foh.dhhs.gov/Public/NYCU/eyeinjury.asp.

August

National Immunization Awareness Month: August is recognized as National Immunization Awareness Month. The goal is to increase awareness about immunizations for all, from infants to the elderly. August is the perfect time to remind family, friends, co-workers, and those in the community to catch up on their vaccinations. Parents are enrolling their children in school, students are entering college, and healthcare workers are preparing for the upcoming flu season. Despite the successes of immunization programs, thousands of people in the U.S. still die from vaccine-preventable diseases.

Immunization is one of the most effective ways to protect children and adults against many common infectious diseases. Keeping individuals healthier through immunizations results in lower associated social and financial costs for families, including time lost from school and work, as well as the expense of medical bills. More information on this topic can be found at the CDC website at www.cdc.gov.

September

National Cholesterol Education Month: September is National Cholesterol Education Month, a good time to get your blood cholesterol checked and take steps to lower it if it is high. National Cholesterol Education Month is also a good time to learn about lipid profiles and about food and lifestyle choices that help you reach personal cholesterol goals. High blood cholesterol affects over 65 million Americans. It is a serious condition that increases your risk for heart disease. Lowering cholesterol levels that are too high lessens your risk for developing heart disease and reduces the chance of having a heart attack or dying of heart disease. The National Heart, Lung, and Blood Institute offers helpful resources to use during National Cholesterol Education Month. For more information, go to http://www.nhlbi.nih.gov/about/ncep/.

October

Breast Cancer Awareness Month: October is National Breast Cancer Awareness Month (NBCAM). Mammography rates have more than doubled for women age 50 and older since the program began in 1985, and breast cancer deaths have declined. Though progress is evident, there are still women who do not take advantage of early detection at all, and others who do not get screening mammograms and clinical breast exams at regular intervals. Women age 65 and older are less likely to get mammograms than younger women, even though breast cancer risk increases with age. Hispanic women have fewer mammograms than Caucasian women and African American women. Women below poverty level are less likely than women at higher incomes to have had a mammogram within the past two years. Breast cancer awareness has received ever-growing attention over the last decade, with high-profile sponsors such as the Susan G. Komen Race for the Cure, the National Football League, and the Mary Kay Foundation.

November

National Diabetes Month: With the growing waistlines and cases of obesity in the United States, the country is facing an epidemic of diabetes, a serious disease that damages the body and shortens lives. In the next four decades, the number of U.S. adults with diabetes is estimated to double or triple, according to CDC scientists. That means anywhere from 20 to 33 percent of adults could have the disease. About 10 percent of American adults have diabetes now. For more information, visit http://www.cdc.gov/Features/LivingWithDiabetes/.

The Great American Smokeout: This event, based around smoking cessation, occurs every year on the third Thursday in November, a week before Thanksgiving. More information can be found at http://www.quitsmoking.com/index.html.

Lung Cancer Awareness Month: A national campaign dedicated to increasing attention to lung cancer issues and primarily sponsored by the Lung Cancer Alliance. By organizing rallies, distributing educational material, holding fund-raising events, contacting Congress, and speaking to the media, those involved can bring much-needed support and attention to a disease that each year kills more people than breast, prostate, colon, and pancreas cancers combined. For more information, visit http://www.lungcanceralliance.org/involved/lcam_month.html.

December

National Drunk and Drugged Driving Prevention Month: December has been designated National Drunk and Drugged Driving Prevention Month, primarily sponsored by Mothers Against Drunk Driving (MADD) and the CDC. This is a time to raise awareness about the consequences of driving under the influence of alcohol and drugs. For more information, visit http://www.madd.org/feature-stories/december/december-is-national-drunk.html or the CDC website at www.cdc.gov.

These are just some of the more common topics that can be covered each month. Other areas that may be added or substituted include:

- Stretching and encouraging physical activity
- Stroke warning signs
- Education about obesity
- Distracted driving
- Home and recreational safety

- Disaster (e.g., fire or weather-related) preparedness
- Heat stress or illness prevention
- Lung and respiratory disease prevention

More important than the number of activities or when they are conducted is making sure that the information is factual and impactful. Information should come from reliable sources such as the Centers for Disease Control and Prevention, the Department of Health and Human Services, or similar trusted organizations. Also important is how the information is distributed. While providing a written message (through handouts, etc.) to employees is helpful, conducting demonstrations where possible is much more impactful. Examples might include leading a demonstration of stretches before a shift, showing slides that include signs of skin cancer, or using a "jar of tar" (quart jar of molasses) to demonstrate the buildup of tar in the lungs from smoking a pack a day of cigarettes over a period of a year.

Case Study: Wellness in Action

Introduction

This case study discusses the effects a wellness program can have on an operation and its employees, as well as the potential benefits for the organization. The program and activities discussed in this study took place in one operation within a large organization that includes a major global commercial cargo and freight air carrier service.

Background Information

The organization began implementing wellness programs over about eight years prior to the events of this study. While many of the elements of the program were driven at the corporate level, much was left up to the local operation level to determine how to use the information, with few instructions to assist.

Over the last four years prior to the events of this study, the corporation improved its methods for communicating to operations and implemented the use of volunteers to serve as "wellness champions." This was a position separate from the safety committee chairperson, with responsibilities to the committee with regard to the wellness program and distribution of wellness information throughout the operation. Each operation within the organization was provided with a wellness champion to accomplish this goal.

In this organization, healthcare costs have decreased slightly since the inception of the wellness program, in an era when costs increase each year for most organizations. However, because of its nature, there is no effective way of knowing or predicting the true cost savings. This study provides an example of one operation's activities and how they not only improved morale within the operation, but also improved the health of a number of its employees and their families.

The Operation's Activity

The wellness champion realized that many of the 100+ members of his operation were overweight. Finding a way to motivate them to lose weight would not only affect their personal health, but improve their attendance and work performance. But what could he do to encourage that kind of motivation and participation?

One night, he walked into the living room where his wife was watching NBC's *The Biggest Loser*. The idea struck him that he could integrate a form of this program into the operation. He developed details of how it might work and approached the safety committee for buy-in and alternative ideas on how to ensure its success. The premise was to weigh in each week on the same day of the week, provide support and encouragement, and recognize successes. The wellness champ would provide stretches, exercises, and nutritional information to assist in their efforts. The committee decided on a grand prize such as a flat screen television or a year's paid gym membership that would be awarded at the end of the program. The prize, they thought, would need to be large enough to keep the interest of the maximum number of participants. The operation would offer the program in four-month cycles that would be offered twice per year.

The Result

The first cycle saw participation of only about 10 percent of the operation. However, the results were very good, with most showing at least some weight loss and a few with significant results. The momentum carried through to the next couple of rounds and, by the third round, participation increased to about 57 percent, with more and more individual successes. In fact, there were some stories of success that carried to the employees' homes, where their family members were challenged to participate with the employee. While there were some who didn't see the program through to the end, the operation saw improved morale, with the majority of the employees looking forward to their weekly weigh-in. Fellow participants helped to encourage others to participate and provided encouragement for those who struggled.

Summary

As part of a successful safety process and an effective safety committee process, the wellness program and other activities have been shown to significantly increase employee involvement in the safety process, improve morale, and improve health while decreasing absenteeism. Successes at the organization varied, with many different activities tried across many different types of operations. Two things have helped to improve the safety process: (1) The organization provides good resources and assistance while encouraging employees to use their imagination and skills to develop new activities, and (2) good activities are packaged into "best practices" with instructions that can be shared throughout the organization.

Chapter Summary

As one can gather, there are many avenues of opportunity for improving the lives of employees and their families. The devotion of a company to a good wellness program will result in a healthier and more motivated workforce. This chapter discussed the ever-increasing cost of healthcare in the United States and the fact that heart disease and cancer make up almost half of the nation's deaths. One of the key contributing factors noted is that the nation has become fatter, perhaps lazier, due to advances in technology such as the use of smart phones, video games, television, and other devices that have become some ingrained in the everyday lives of so many. It is evident that a wellness program is truly valuable, both to an organization's employees and to reducing its healthcare costs. Obviously, not all employees will take advantage of such a program, but the organization will be able to measure success among those that do, and the result will be fewer lost workdays, improved production, and reduction of healthcare costs to both the company and employees.

Methods were discussed in this chapter whereby the company and employees could contribute to the wellness program. The employees' contribution might come by way of employee involvement in the program, volunteering time to develop and/or execute wellness activities for their peers. The key benefits include employee involvement in a venture that will benefit them and improve morale among participants. The employer's contribution comes from providing resources and employees the time for preparation and execution of activities. This chapter also stressed that much of the information learned by the employee and/or provided by the employer can be taken home and shared with families, further contributing to wellness and the reduction of healthcare costs.

Finally, this chapter provided some basic ideas on how to develop a basic plan to educate employees on the subject of wellness and encourage them and their families to participate. Several examples were provided, along with examples of key topics that are typically covered across the country during specific months (e.g., February is American Heart Month).

Chapter Concept Questions

1. For basic ergonomics, the standard countertop height is 36". This is an example of ________________.
2. Explain why a wellness program makes good sense for an organization when it reaches beyond the employee to his/her family.
3. Explain why a newer employee might be at higher risk of developing an injury.
4. Describe some of the common factors that lead to ergonomic injuries.
5. In your own words, explain why posture can be a contributing factor to ergonomic injuries.
6. Effective communication is one of the primary elements of implementing a wellness program in an effort to ________________________.
7. Describe a basic outline of how you would organize a low-cost (less than $10,000 annually) wellness program.

Chapter References

American Academy of Orthopedic Surgeons. 2008. United States Bone and Joint Decade: The Burden of Musculoskeletal Diseases in the United States. Rosemont, IL: AAOS.

American Academy of Orthopedic Surgeons. 2009, February. Meniscal Transplant Surgery. http://orthoinfo.aaos.org/topic.cfm?topic=A00381.

American Academy of Orthopedic Surgeons. 2009, May. Low Back Pain. http://orthoinfo.aaos.org/topic.cfm?topic=A00311.

American Academy of Orthopedic Surgeons. 2009, June. Chronic Shoulder Instability. http://orthoinfo.aaos.org/topic.cfm?topic=A00529.

American Dietetic Association. National Nutrition Month., 2011. http://www.eatright.org/nnm/.

American University Radio. n.d. National Immunization Awareness Month. http://wamu.org/calendar/community_minute/national_immunization_awareness_month.php.

Centers for Disease Control and Prevention. n.d. February Is American Heart Month. http://www.cdc.gov/Features/HeartMonth/.

Centers for Disease Control and Prevention. n.d. National High Blood Pressure Education Month. http://www.cdc.gov/Features/HighBloodPressure/.

DiBerardinis, L. J. 1999. *Handbook of Occupational Safety and Health* (2nd ed.). New York, NY: John Wiley & Sons, Inc.

Gellar, E. S. 2001. *Working Safe: How to Help People Actively Care for Health and Safety* (2nd ed.). Boca Raton, FL: CRC Press.

Lloyd-Jones, D., Adams, R., Carnethon, M., et al. n.d. Heart Disease and Stroke Statistics—2009 Update: A Report from the American Heart Association Statistics Committee and Stroke Statistics Subcommittee. *Circulation.* http://circ.ahajournals.org/cgi/reprint/CIRCULATIONAHA.108.191261v1.

Men's Health Network. n.d. June Is Men's Health Month. http://www.menshealthmonth.org/index.html.

Morrison, K. W. 2010. Living Well. *Safety and Health* 182(4).

Mothers Against Drunk Driving (MADD). n.d. December Is National Drunk and Drugged Driving Prevention Month. http://www.madd.org/feature-stories/december/december-is-national-drunk.html.

National Breast Cancer Awareness Month. http://nbcam.org/.

National Cervical Cancer Coalition. n.d. Global Initiative against HPV and Cervical Cancer. http://www.nccc-online.org/awareness.html.

National Safety Council. 2001. *Accident Prevention Manual for Business & Industry: Administration & Programs* (12th ed.). Itasca, IL: National Safety Council.

U.S. Department of Health and Human Services. n.d. Eye Injury Prevention Month. http://www.foh.dhhs.gov/Public/NYCU/eyeinjury.asp.

U.S. Department of Health and Human Services, National Heart, Lung, and Blood Institute. n.d. National Cholesterol Education Month. http://www.nhlbi.nih.gov/about/ncep/.

U.S. Department of Labor, Bureau of Labor Statistics. 2009. *Injury and Illness Information Report, United States Private Industry for 2008.*

U.S. News and World Report. 2010, November. Reducing Health Care Costs, Improving Care. http://www.usnews.com/science/articles/2010/09/20/reducing-health-care-costs-improving-care.html.

Xu, J., Kochanek, K. D., Murphy, S. L., and Tejada-Vera, B. 2010, May 20. Deaths: Final Data for 2007. *National Vital Statistics Reports* 58(19). http://www.cdc.gov/NCHS/data/nvsr/nvsr58/nvsr58_19.pdf.

GLOSSARY

Accident: An actual unexpected and undesired event that causes injury or death to employees or others, or causes damage to aircraft, equipment, facilities, or other assets.

Accumulation Start Date: The day that a container is full (up to 55 gallons for most hazardous wastes or up to 1 quart for acute hazardous or P-listed wastes).

Accumulation Time: The amount of time starting when a container (up to 55 gallons for most hazardous wastes or 1 quart for acute hazardous or P-listed wastes) is full and ending when the container is removed from the site.

Active Listening: An act that consists of listening carefully to what the person is saying, taking accurate notes, and repeating back to the person what was said in order to ensure clarity and accuracy of information.

Active Risks: Those risks that pose a more obvious hazard and are often direct causes of an accident. An active risk is an action, decision, error, hazard, etc. which results in an immediate undesired event.

Acute Insomnia: Insomnia that may be associated with the occurrence of an immediate life issue (sickness, stress, etc.) that impacts a person's ability to sleep adequately.

Aeronautical Decision Making (ADM): Defined by the FAA as a systematic approach to the mental process used by pilots to consistently determine the best course of action in response to a given set of circumstances.

Aircraft Ground Damage Event: An event in which an aircraft is damaged while on the ground in a non–flight related activity, such as maintenance or ground handling.

Aircraft Ground Handling: Usually described as the reception and dispatch of aircraft upon arrival and preparation for departure.

Airport Familiarization: The practice of learning the layout of the airfield and where things are located within the airport operations area.

Airport Operations Area (AOA): Consists of the runways, taxiways, ramps, and roadways located within the operations portion of an airport.

Annulus Fibrosus: A tough, flexible outer ring of a disk in the spine.

Anthropometric Design: The process of engineering to fit the population.

Aqueous Solution: A solution in which water is the solvent.

Attractants: Anything that attracts wildlife to an airfield; typically food, water, and shelter.

Auditing: Generally synonymous with conducting a comprehensive inspection of the various areas within an organization.

Autonomy: The principle of recognizing and regarding each person as an individual being, possessing their own unique characteristics, identity, strengths, weaknesses, personality, etc.

Auxiliary Power Unit (APU): A small turbine engine that provides power for some of the systems of an aircraft, usually used during ground handling.

Aviation Safety Action Program (ASAP): A voluntary safety program employed by participating organizations as a means of reporting and recording information related to possible and actual safety issues and hazards.

Behavioral Observation Process: A process established to conduct observations for review and to provide feedback on safe and at-risk behaviors in an effort to reinforce safe behaviors and change at-risk behaviors.

Beneficence: Acting in ways that are good for other people.

Board of Inquiry: A group consisting of senior NTSB staff members and chaired by the board member presiding over the proceedings of an aircraft incident.

Business Ethics: The systematic examination of decisions (before and after they are made) to ensure that the decisions are aligned with the moral standards of a given society and its culture.

Capital Improvement Program (CIP): An economic tool used to plan for the acquisition of items that are expensive to obtain.

Cargo Handling: Similar to the handling of baggage, except that the packages and materials handled are often varied and large in size, weight, and shape.

Causal Factors: The individual issues or events that contributed directly or indirectly to the occurrence of an event under investigation.

Certificate Actions: Enforcement actions that may be taken against the certificate(s) of a person or organization. This includes type certificates, manufacturing production certificates, aircraft airworthiness certificates, airman certificates, air carrier operating certificates, air navigation facility certificates, and air agency certificates.

Chronic Insomnia: The most serious form of insomnia, in which a person is in a continual state of sleep deprivation that may last days, weeks, months, or even years.

Cockpit Voice Recorder (CVR): Records flight crew conversation, radio transmissions, and other flight deck sounds.

Company Materials (CoMat): Items that are owned or controlled by a company and are shipped within the organization in support of various parts of their overall operations.

Complaint Resolution Officials (CRO): A person who assists passengers with disabilities in a variety of ways, including addressing any concerns or complaints from a passenger with a disability.

Compressed Gas: Any material or mixture having in the container an absolute pressure exceeding 40 pounds per square inch (p.s.i.) at 70°F or, regardless of the pressure at 70°F, having an absolute pressure exceeding 104 p.s.i. at 130°F; or any liquid flammable material having a vapor pressure exceeding 40 p.s.i. absolute at 100°F as determined by ASTM Test D-323. Refer to §261.21 for information describing ignitability of a compressed gas.

Conditionally Exempt Small Quantity Generators (CESQG): Facilities that produce less than 100 kg of hazardous waste, or less than 1 kg of acutely hazardous waste, per calendar month. A CESQG may only accumulate less than 1,000 kg of hazardous waste, 1 kg of acutely hazardous waste, or 100 kg of spill residue from acutely hazardous waste at any one time.

Consequence Management: A process of explaining the consequences of a given behavior. This may be done by discussing with the employee what a safe behavior might do for the employee or what an at-risk behavior might do to the employee (e.g., placing a bag on the belt without twisting will result in a lesser chance of back muscle strain).

Contained Spill: A spill that is captured either by secondary containment (an outside container or pallet capable of holding any spilled product) or by a diking system (a curbed or ramped area capable of containing the spill, usually seen in aviation around a fueling island).

Control: A form of mitigating a risk.

Conventional Decision Making (CDM): A decision that can be likened to a norm; that is, the typical way in which a person makes decisions. CDM involves making a decision sparked by the recognition that something has changed or that an expected change did not occur.

Cost-Benefit Analysis (CBA): The practice of calculating and comparing the actual and projected costs of a decision against the actual and potential benefits of the decision.

Crew Member Self-Defense Training (CMSDT): A program that provides basic self-defense training to airline crew members who participate voluntarily.

Crew Resource Management (CRM): The proper use of all available resources by a person or crew, including people, information, and available equipment.

Customer Service Agent: Employee who works the ticket counter, gate, or service counter of an aviation organization.

Dangerous Goods: Products meeting hazardous characteristics as defined by the Environmental Protection Agency (EPA) or found in the DOT or International Air Transport Association (IATA) Dangerous Goods Regulations.

Deadman Switch: A device consisting of a cable and switch, sometimes on a truck and sometimes in the user's control.

Direct Costs: Costs directly connected to the development and production of a specific product. This term may also be used to describe the direct cost related to an injury or damage, such as medical or repair costs.

Direct Discharges (or "point source" discharges): Wastewater discharges from sources such as pipes and sewers.

Duty (or deontological ethics): The belief or perception of what a person or organization should do or is obligated to do.

Economics: The study of the allocation of scarce resources within a society or organization.

Egoism: What is good for oneself or acting in accordance with self-interest.

Employee Involvement: The concept of employees being involved or empowered with various attributes of the safety processes of the organization.

End-Range Motion: The position at which an extremity (elbow, wrist, shoulder, etc.) is straightened or bent to or near its maximum potential. May also be described as the position where the muscle or muscle group is at its end range, where it cannot move further, placing it at increased risk of injury if an outside force is introduced exceeding the limitations of the muscle or muscle group.

Ergonomics: The study of workplace design; the study of how a workplace and the equipment used there can best be designed for comfort, efficiency, safety, and productivity.

Error Chain: One issue linked to the next and so on, which causes an undesired event to occur.

Ethical Climate: The organization's overall climate with respect to ethical practices and beliefs.

Ethics: Rules for behavior, based on beliefs about how things should be; involves assumptions about humans and their capacities, logical rules extending from these assumptions, and notions of what is good and desirable.

External Decisions: Those decisions made within an organization that are likely to create one or more impacts outside the organization.

Fatigue: Being tired, sleepy, or weary. From an operational perspective applicable to aviation, fatigue has been defined by the FAA as a condition characterized by increased physical and/or mental discomfort, with reduced capacity for work, reduced efficiency in task accomplishment, loss of motivation or capacity to respond to stimulation, and typically accompanied by weariness and tiredness.

Federal Air Marshals: The primary law enforcement group within the TSA, deployed on flights around the world and in the United States.

Federal Aviation Administration (FAA): The agency of the DOT that serves as the regulatory body governing civil aviation interests and activities in the United States.

Field Condition Report: Identifies existing airport conditions that are other than clean and dry.

Flammable Liquid: Defined by the DOT, EPA, and OSHA with slight differences. For purposes of this text, the DOT definition is: A liquid having a flashpoint of not more than 60°C (140°F).

Flashpoint: The minimum temperature at which a liquid gives off vapor in sufficient concentration to form an ignitable mixture in air near the surface of the liquid.

Flight Data Recorder (FDR): A device that monitors aircraft parameters such as altitude, airspeed, flight control settings and movements, engine rpm, pitch, heading, and other measurements.

Flight Operational Quality Assurance (FOQA): A program that uses recorded data from aircraft to improve safety and efficiency.

Foreign Object Debris (FOD): Any type of foreign object(s) on the ground in ramps, aircraft parking areas, taxiways, runways, and other areas frequented by aircraft.

Frangible Mounts: "Breakaway" types of mounts for lighting on or at an airport runway or taxiway.

Frequency (or exposure opportunity): How often a work activity occurs that produces a hazard. Frequency is influenced by the scope of exposure and how often the exposure exists.

Full: As it relates to a hazardous waste satellite accumulation drum, the drum's contents have reached the most extruded portion of the drum's upper ring (approximately four inches from the top of the drum).

Gap Analysis: The process of assessing an organization's current state based on a given set of criteria. The "gap" is the deficiency that exists between the desired outcome and the current state.

Go-Team: Consists of a group of individuals dispatched to an accident location to conduct the on-site investigation.

Grooving: Channels water through grooves below the grade of a runway to prevent hydroplaning.

Group Chairperson: These persons are experts in areas such as witness interviewing, aircraft systems and structures, aircraft maintenance, operations, meteorology, human factors, air traffic control, and other specific areas pertinent to an accident.

Hazard: The source of a problem or danger that poses a risk of injury or damage to property or the environment.

Hazard Analysis (also referred to as a job hazard analysis or JHA): A detailed breakdown of each of the jobs or tasks expected within each operation to determine where hazards exist, using historical data, experts, regulations, and logical thinking to determine and prioritize risk.

Hazardous Waste: A solid waste which exhibits the characteristics of ignitability, corrosivity, reactivity, or toxicity, or is on the EPA's F-List, K-List, P-List, or Q-List.

Hazardous Waste Manifest: The shipping document, EPA form 8700-22 and, if necessary, EPA form 8700-22A, originated and signed by the generator in accordance with the instruction included in the appendix to 40 CFR Part 262.

Human Factors (HF): A desciption of people in their working and living environments and their relationships with technology, equipment, machines, policies, procedures, regulations, other people, and the environment in which all are located.

Ignitable: A characteristic of hazardous waste defined as a waste with a flashpoint less than 140°F.

Impervious Surface: A surface that will not absorb any moisture, such as a nonporous texture that is impenetrable.

Incident: A safety-related occurrence (other than an accident) associated with flight or ground operations which affects or may affect the safety of operations. (Modified from the ICAO definition to include ground operations.)

Incident Commander: The person within the organization in charge of the command center and its activities during the emergency.

Incident Rate: The rate of employee injuries based on OSHA's calculation compared to 100 full-time employees.

Indirect Costs: Costs that are present but not associated directly with the development and production of a specific product.

Insomnia: A disruption of a person's normal sleep cycle and/or a person experiencing inadequate sleep.

Internal Decisions: Those decisions made within the organization that are likely to create one or more impacts within the actual organization.

Internal Evaluation Program (IEP): A proactive approach to safety by providing the means to continually assess the current state of the various departments of an aviation organization with respect to compliance or noncompliance with any corresponding regulatory agencies and company policies, safety, areas of risk and hazards, areas of deficiency, and other related issues.

Job Safety Analysis (JSA): A process that breaks down job tasks, identifies each risk element (lifting, chemical, etc.), and details measures that should be taken (specific proper tasks or methods, proper personal protective equipment that should be used, etc.) to complete the task in a safe manner.

Judgment: The assessment of a situation; a cognitive process that enables us to make conclusions.

Justice: To give others what is owed to them and what they deserve; treatment with rewards and discipline with fairness and consistency.

Large Quantity Generators: Facilities that generate more than 1,000 kg of hazardous waste per calendar month, or more than 1 kg of acutely hazardous waste per calendar month.

Latent Risks: Areas within the system that pose a hazard (contributory factors) but are not always obvious and tend to lurk beneath the surface. Latent risks are the result of an action, policy, or decision made long before an accident occurs and its accompanying areas of hazard; consequences may be dormant within the system for a long time before the occurrence.

Law of Diminishing Returns: States that as equal quantities of a variable factor are increased and other factors remain constant, a point is reached beyond which the addition of one more unit of the variable factor will result in a diminishing rate of return.

Likelihood (or chance of occurrence): The most subjective element of a process in that it is often misunderstood.

Loss Type Potential: What type of loss exposures exist. For aviation, what is considered the "worst-case scenario" is usually an aircraft crash.

Lost Workday Rate: Rate of employee injuries that result in lost days of work (other than the day of injury) based on OSHA's calculation of lost work day (LWD) injuries per 100 full-time employees or 2,000 work hours.

Marshaller (or aircraft guide): A person who has been trained to visually and (sometimes) audibly direct the movement of an aircraft on the ground.

Moral Standards: A societal measuring stick for determining moral norms and principles (e.g., cheating, theft, lying, bribery, coercion, consequences, honesty, integrity, telling the truth, etc.) and helping to establish the baseline for determining whether an action or decision is right or wrong in a society.

Movement Area: Area of the airport consisting of runways and taxiways at towered airports, which a vehicle operator must receive permission to enter.

Musculoskeletal Disorders: Injuries that affect the soft tissue and bones in the body.

Mutual Aid Agreement: An agreement between two or more organizations to render aid in the event of specified emergencies.

National Transportation Safety Board (NTSB): An independent agency of the federal government responsible for investigating all civil aviation accidents in the United States and significant railroad, highway, marine, and pipeline accidents.

Near-Miss: An actual undesired "close-call" type of event that does not cause injury or damage, but could have.

Non-Contained Spill: A spill without secondary containment, usually at a location other than where the product is stored or while in transit. For example, uncontrolled spills can occur during the fueling process, because of improper loading of a fuel truck, or as a result of the dropping of chemical drums or containers during transit.

Non-Malfeasance: To be free from intentional harm.

Non-Sole Source: An ASAP report where there is existing evidence of the event outside of ASAP and the FAA would have been aware of the event without the ASAP program.

Nonspecific Threat: May be related to one or more targets, but there is doubt as to their credibility because of inaccurate locale descriptions, no known affiliations, etc.

Notices to Airmen (NOTAM) System: An avenue for providing timely information to airport users describing conditions around the airport that may affect aircraft operations.

Nucleus Pulposus: A soft, jelly-like center of the disk of the spinal cord.

Opportunity Cost: The real cost of something is what you give up in order obtain it. This concept includes not just how much money was spent to purchase something, but also the potential benefits that were sacrificed by not obtaining something else.

Optimum Posture: Posture that provides the most strength and control over movements and the least physical stress on joints and surrounding tissue.

Organic Matter: Of, relating to, or derived from living organisms such as wood or manure.

Organizational Culture: How a company or other organization actually is, functionally and environmentally.

Ownership: Assigning a specific person or group the full responsibility of implementing corrective action in the organization.

Oxidizer: A substance such as a chlorate, permanganate, inorganic peroxide, or a nitrate that yields oxygen readily to stimulate the combustion of organic matter.

Party System: The process by which the NTSB includes representatives from certain organizations that may have a vested interest or expertise in segments of an accident under investigation.

Pavement Condition Index (PCI): Evaluates the condition of an airport's pavements so the airport operator can determine what pavement needs attention.

Porous Friction Course (PFC): An enhancement to the pavement of a runway to help eliminate the possibility of hydroplaning during a rain event. PFC allows water to drain through the pavement.

Predictive Method: A process used to identify and anticipate problems and issues well before the fact based on how an organization functions.

Preventive Maintenance: The practice of replacing critical parts or components of a piece of equipment or machinery before failure occurs.

Proactive Method: Identifying issues that are a concern or are potentially problematic.

Probability (or opportunity for occurrence of an unwanted event): A combination of frequency and likelihood.

Probable Cause: Considers all factors which are believed to be contributory to the occurrence of an accident under investigation.

Public Hearing: A scheduled public meeting in which the particulars and known factual information pertinent to an accident investigation are discussed in detail.

Qualitative Assessment: The process by which a risk assessment is conducted based on recognized hazards.

Reactive: Changes are made only when and after something bad has occurred such as an aircraft accident.

Residual Risk: The product left after risk reduction or system safety efforts have been employed.

Resource Conservation and Recovery Act (RCRA) of 1976: Governs the regulation of solid and hazardous waste.

Risk: The probability and severity of harm.

Risk Analysis: A proactive approach to identifying the most probable threats of loss to an organization and analyzing the potential effects of loss to the organization (system or procedure) from these threats.

Risk Assessment: The process of evaluating existing controls and assessing their adequacy to the potential threats of loss to the organization.

Risk Reduction: The process by which an organization can mitigate risk through a change in process, procedures, equipment, behavior, etc.

Risk Statement: A clear and concise description of the risk.

Runway Incursion: Any unauthorized intrusion onto a runway, regardless of whether or not an aircraft presents a potential conflict.

Runway Safety Area: The areas off the ends and sides of runways that should be inspected daily to ensure no ruts or erosion is present due to a vehicle or aircraft exiting the runway.

Safe: Being free from harm or risk.

Safety Culture: Internal attitude, level of motivation, and approach that characterizes the organization with respect to how seriously safety is regarded.

Safety Management System (SMS): A system composed of policies, processes, procedures, and practices adopted and implemented to enhance the overall safety of an organization.

Satellite Accumulation Area: A designated area at the point where hazardous waste is generated, which is under the control of an operator. Up to 55 gallons of hazardous waste (or 1 quart of acute hazardous waste) may be accumulated in a satellite accumulation area without starting the accumulation time clock.

Scarce: Refers to the fact that resources are finite and limited; there is only so much of any resource that is available.

Scope of Exposure: A variable that may be customized to meet the needs of the assessment.

Severity: The degree of harm for which an unwanted event has potential.

Situational Awareness: Continuous awareness of the immediate environment and accounting for any changes taking place that may affect an individual's level of alertness. May incorporate technology as an enhancement to awareness.

Small Quantity Generators: Facilities that generate between 100 kg and 1,000 kg of hazardous waste per calendar month.

Sole Source Report: A report where the reporter is the single point of the information contained within an ASAP report, and had the reporter not filed an ASAP report the event would have been unknown to the FAA and likely to the company as well.

Solid Waste: Any material that has fulfilled its intended purpose and is being discarded, has been abandoned, or is inherently waste-like. Solid waste includes both hazardous and non-hazardous waste and may take the form of a solid, semi-solid, liquid, or gas.

Spalling: Refers to the physical deterioration of paved areas where the surface of the material is in a state of "crumbling" or is appreciably fractured.

Specific Threat: Allows for positive target identification or is made by a person claiming to be a member of an organization or group that is judged credible (known terrorist groups or organizations).

Spill: The accidental release of liquid or solid materials from their proper container, whether from container failure, upset, or unintentional drainage.

Squib: A small explosive device used to activate the engine fire-extinguishing system of an aircraft.

Stakeholders: An aviation stakeholder is any individual or group that may have a vested interest in, or may be directly or indirectly impacted by, the decisions and operations of an aviation organization.

Standard Threshold Shift (STS): A change in hearing threshold, relative to the baseline audiogram for that employee, of an average of 10 decibels (dB) or more at 2,000, 3,000, and 4,000 hertz (Hz) in one or both ears.

Static Load: Holding the same position for a period of time or not moving.

Stationary Hydrant Fueling System: Siphons fuel directly from the storage tanks to the aircraft.

Step Analysis: This type of analysis identifies multiple levels of performance such as achieving the basic standards, a higher standard such as OSHA VPP (for safety), and "world-class" performance.

Storm Water Pollution Prevention Plans (SWPPPs): Should identify potential sources of pollution that may reasonably be expected to affect the quality of storm water discharges associated with industrial activity at a facility.

Stress Tolerance Level: A person's individual threshold for handling stress. If the number or intensity of stressors becomes too great, the person becomes susceptible to overload.

Strikes: TSA surprise security inspections of air cargo at selected air carrier facilities and airports.

System Safety Process: A proactive approach of identifying, assessing, and eliminating or controlling safety-related hazards to an acceptable level to prevent injuries and accidents.

Systematic: Safety management programs and initiatives are applied throughout the aviation organization as a part of their overall operational plan.

Systems Approach: The study of how to understand the connections and interactions between the various components of a given system (e.g., social, economic, and environmental).

Threat: An expression of intent to harm another human being or property.

Toxic: Having the quality of being hazardous to human health. For waste purposes, a waste that will leach chemicals at a concentration greater than the regulatory level provided in the Maximum Concentration of Contaminants for the Toxicity Characteristics Table found in 40 CFR Part 261 and is considered a toxic hazardous waste.

Transient Insomnia: The most common form or insomnia, usually rather short-term in nature, perhaps lasting only a few days.

Transportation Security Administration (TSA): The branch of the federal government charged with providing oversight and regulation of security in the various primary transportation modes.

Transportation Security Officers: The TSA agents who work as aviation passenger and baggage security screeners in commercial service airports.

Treatment/Storage/Disposal (TSD): The processes involved in converting a hazardous substance to a less dangerous one that is then later moved to a designated EPA site for hazardous waste.

Trend Analysis: The practice of analyzing data submitted over a period of time and identifying common issues and problems or "trends" that have occurred over a specified time frame.

Triage: Persons with the more serious injuries are attended to first until adequate resources for all arrive.

Underwater Locator Beacon (ULB): An internally installed radio beacon automatically activates when the recorders are immersed, transmitting from depths up to 14,000 feet, to make recovery from water more likely.

Unidentified Risk: Risk that has not been determined, perhaps overlooked, or unrealized.

Utilitarianism: The concept of doing what is believed or perceived to be best for the greatest number of people.

Voluntary Protection Program (VPP): Establishes a cooperative relationship between OSHA, management, and labor at worksites that have implemented comprehensive health and safety management systems that exceed minimum federal and state standards.

Watch List: A large database containing the names of known terrorists, persons suspected of terrorist activity, and persons having or suspected of having connections to terrorist groups.

Wildlife Hazard Management Plan: A document that will incorporate best practices on how to control the wildlife on the airfield.

Will Carry Airline: An airline that has chosen to carry select hazardous materials, otherwise known as dangerous goods.

Wing Walkers: Employees that have been trained to assist the marshaller with aircraft movement by appropriately signaling to the marshaller if the way is clear for the travel path of the aircraft wings.

Work Conditioning: The period of time it takes an employee to become physically acclimated to the job demands, including the movements and forces placed upon the body, to acceptably complete the tasks as assigned.

Worksite Analysis: Analyzing the history of events (injuries, damage, etc.) that have occurred within an organization over a given period of time.

INDEX

G

H

S